AF622453

INTRODUCTION TO SYNTHETIC APERTURE RADAR USING PYTHON AND MATLAB®

INTRODUCTION TO SYNTHETIC APERTURE RADAR USING PYTHON AND MATLAB®

Andy Harrison

ARTECH HOUSE
BOSTON | LONDON
artechhouse.com

Library of Congress Cataloging-in-Publication Data
A catalog record for this book is available from the U.S. Library of Congress.

British Library Cataloguing in Publication Data
A catalogue record for this book is available from the British Library.

Cover design by Robert Oakley

ISBN 13: 978-1-63081-864-7

Accompanying software at:
https://sarbook.github.io
access code: $PythonSARBook!

© 2022
Artech House
685 Canton Street
Norwood, MA 02062

All rights reserved. No part of this book may be reproduced or utilized in any form or by any means, elec-tronic or mechanical, including photocopying, recording, or by any information storage and retrieval system, without permission in writing from the publisher.

All terms mentioned in this book that are known to be trademarks or service marks have been appropriately capitalized. Artech House cannot attest to the accuracy of this information. Use of a term in this book should not be regarded as affecting the validity of any trademark or service mark.

10 9 8 7 6 5 4 3 2 1

Contents

Preface

Synthetic aperture radar (SAR) can prove to be a very broad and complex subject area, especially for those new to the field. Even so, I firmly believe even the most difficult topics can and should be presented in a clear and concise manner such that the student can fully comprehend the material. Education is not the aggregation of facts; it is the accumulation of understanding and knowledge about particular subjects. That said, there are several SAR handbooks, briefings, and online references containing useful material with little or no explanation or analysis. These are valuable resources and have their place as references for students and engineering professionals. However, there is a tendency to indiscriminately use equations, generalities, and approximations in situations where not applicable. The intent of this book is to provide the reader with a straightforward and succinct introduction to, and explanation of, critical elements that go into the generation of SAR imagery. To reinforce those SAR concepts, a tool suite consisting of Python® notebooks and corresponding MATLAB® live scripts is included with this book. This gives the user a mechanism to analyze and predict SAR performance for various scenarios and applications. It also provides the user with tools to create, manipulate, and work with SAR imagery. The user has full access to the Python notebooks and MATLAB live scripts to modify for their particular application. Several examples using the tool suite are given at the end of each chapter. These examples, combined with the end of chapter problem sets, make this an excellent choice for a senior or first-year graduate level course in SAR, as well as professionals seeking to enter the field.

While there are numerous topics in the field of SAR, this book focuses on the more crucial concepts. The book begins with a brief history of the invention and development of SAR. This is followed by a discussion of sensors and applications.

Included in the introductory chapter are detailed instructions for obtaining and executing the accompanying software. Chapter 2 begins with basic SAR imaging principles and then moves into a detailed coverage of stripmap and spotlight imaging. Next, modes designed to cover exceptionally large areas are discussed, including ScanSAR, TOPSAR, and SweepSAR. The chapter proceeds with interferometric SAR, which is used to obtain very precise measurements of surface topology, and polarimetric SAR, which is beneficial in determining the nature of the target scattering. Bistatic and multistatic data collection geometries are then presented, followed by inverse SAR imaging. The chapter concludes with a discussion about digital beamforming and its usage in various imaging modes. Chapter 3 covers the details of image formation algorithms. First, one-dimensional range profiles are created with matched filtering and stretch processing. The chapter then moves into two-dimensional imaging algorithms, including range Doppler, polar format, and filtered backprojection methods. The simultaneous iterative reconstruction technique (SIRT), simultaneous algebraic reconstruction technique (SART), and multiplicative algebraic reconstruction technique (MART) are then presented. The chapter concludes with a brief discussion of some nontraditional image formation methods. Chapter 4 is dedicated to three-dimensional SAR imaging. To begin, a review of the development of three-dimensional imaging is given. This review includes airborne, spaceborne, through-wall, and inverse SAR imaging. The point spread function is developed and spatial resolution is examined in three-dimensions. Next, the polar format and filtered backprojection algorithms are extended to three-dimensions and the chapter concludes with three-dimensional imaging examples. Chapter 5 deals with autofocus techniques and begins with background and motivation for SAR autofocus, followed by the signal error model used in this book. As an introduction to nonparametric autofocus algorithms, inverse filtering is presented as it is intuitive and relatively simple to implement. Map drift, phase gradient, and minimum entropy autofocus algorithms are then covered. The chapter concludes with a discussion of several advanced autofocus methods. Chapter 6 is concerned with image registration and begins with an overview of the different algorithm categories. Next, phase correlation is used to register images that are translated, rotated, and scaled with respect to one another. Feature extraction and matching algorithms including Harris corner, scale invariant feature transform (SIFT), and speeded-up robust features (SURF) are then covered. Oriented FAST and rotated BRIEF (ORB) is presented as a viable alternative to SIFT and SURF. The chapter concludes with methods for extracting the geometrical warp function between two images. Chapter 7 investigates various limiting factors on SAR system performance, including image quality, revisit times, mapping rates, onboard processing, and downlink performance. The chapter begins with an analysis of spatial resolution and signal-to-noise ratio, which influence image contrast and speckle. The radar general image quality equation (RGIQE) and the Radar National Imagery Interpretability Rating Scale (RNIIRS) assessments of image quality are also presented. The chapter concludes with an overview of data processing, storage, and downlink performance considerations. Chapter 8 provides an overview of current trends and future directions for SAR systems and imagery. There has recently been a huge surge of interest in commercial SAR applications and systems and as data becomes more affordable and accessible, advancements in data analysis, image processing, and ease of use will follow. This will invariably lead to new and different use cases and will make for exciting times in the field of SAR.

I would like to express my gratitude to ICEYE, Umbra Lab, and Lincoln Laboratory for generously providing the imagery and datasets used throughout this book. Colby Gibson deserves special acknowledgment and thanks for the generation of signature data of a Learjet and backhoe. This data was used in two- and three-dimensional image formation techniques covered in Chapters 3 and 4. My sincere appreciation to Will Halcomb for very generously giving his time and effort to reviewing material in this book. His comments, suggestions, and corrections notably improved the overall quality of this book. I would like to also thank my good friend Bassem Mahafza who urged me to author technical books, gave me invaluable opportunities and advice, and without whom this book would not have been possible.

Most importantly, I would like to thank my wife, Lacon, my son, Conner, and my daughter, Vada, for their love, support, patience, understanding, and sacrifice during the countless hours committed to this endeavor.

Chapter 1

Introduction

A picture is worth a thousand words.

–Fred R. Barnard

Synthetic aperture radar (SAR) is an active sensing system capable of producing high-resolution imagery of landscapes, planets, space debris, drones, ships, and even humans. The question is often raised, "why SAR?" While there are a number of benefits of SAR, the biggest advantage it has over other forms of imaging is the ability to provide high-resolution, 24-hour, all-weather imagery [1]. Passive imaging sensors such as electro-optical (EO), infrared (IR), and hyperspectral are severely degraded and often completely obscured by clouds, smoke, canopies, darkness, rain, and snow. Figure 1.1 shows SAR imagery of the Mississippi River Delta. In contrast, Figure 1.2 shows an example of SAR imagery of the Muscat International Airport in Oman. While these are two very different types of imaging scenes, landscape and urban, the ICEYE constellation is capable of producing very high-quality imagery of various types of scenes using different imaging modes [2].

SAR makes use of the flight path of the platform to synthesize a much larger aperture than the physical antenna. Radar signals are transmitted and received at successive locations along the synthetic aperture and signal processing coherently combines these signals to create the imagery. As shown in later chapters, SAR systems can produce two-, three-, and four-dimensional (space and time) images. This

Figure 1.1 SAR imagery of the Mississippi River Delta, 2020. Courtesy of ICEYE.

chapter begins with the history and development of SAR, followed by basic scattering phenomenology. A discussion of various applications and current systems, and an overview of the software tool suite accompanying this book is then given. The chapter concludes with detailed instructions for obtaining and running the Python® software in Google Colaboratory and the MATLAB® software in Live Editor.

Figure 1.2 SAR imagery of Muscat International Airport, Oman, 2021. Courtesy of ICEYE.

1.1 HISTORY

The term radar was first used in the 1940s by the United States Signal Corps working on those systems for the United States Navy. Immediately following World War II, advancement in radar technology slowed significantly. However, the 1950s saw the emergence of several new technologies including highly accurate tracking radars, klystron amplifiers for high-power, long-range systems, and the emergence of SAR systems [3].

Carl Wiley came up with the original principles of synthetic aperture radar at Goodyear Aircraft Company in June 1951 while working on a correlation guidance system for the Atlas intercontinental ballistic missile program [4]. Using the Doppler shift associated with a moving antenna, and analyzing the frequency content of the return signals, Wiley was able to create higher resolution radar images using a smaller antenna than required for traditional systems. In early 1952, Wiley constructed a concept system named Doppler Unbeamed Search Radar which was deployed on a C-47 aircraft and was the first system to use SAR to create imagery. The original system had a resolution of 500 feet which was improved to 50 feet by the late 1950s [5]. Wiley applied for a patent in 1954 for what was named Simultaneous Buildup Doppler [6]. While this is widely considered the first SAR patent, researchers at the University of Michigan and the University of Illinois developed a similar approach of processing signals over a synthetic aperture [5]. The University of Illinois' Control Systems Laboratory laid the groundwork for radar systems with improved angular resolution and simultaneous range focusing [7]. This work was included in a 1952 Department of Defense study named "The Eyes of the Army" [8] and in 1953 produced stripmap imagery as part of Project Wolverine. In the summer of 1953, an Army, Navy, and Air Force sponsored program at the University of Michigan's Willow Run Research Center resulted in the plans for the development of a more practical SAR system and by 1957 airborne SAR began providing remarkable imagery.

The vast amount of processing required for SAR was beyond the limits of analog processors, therefore optical signal processing techniques were employed in the first SAR systems. This approach produces high-quality images very quickly. The main disadvantage was the output had to be photographically recorded, which created a time delay and made the equipment heavy. The optical components were susceptible to vibration and had to be aligned with great care and precision. Many of the early SAR systems transferred the data to a ground station for image creation and processing [9]. The development of solid-state circuit components in the 1970s made it possible to process the radar return signals digitally in real time. These components were low-cost and lightweight which led to SAR systems being incorporated in small airborne platforms capable of advanced image processing techniques for a variety of applications.

The public acknowledgment of SAR came in 1960 with a press release about an experimental Army system [10]. Civilian applications of SAR first became practical in the 1970s. The National Oceanic and Atmospheric Administration and the Jet Propulsion Laboratories collaborated to develop SAR technologies to be used on a satellite to make oceanic observations. In June 1978, Seasat was launched, making it the first civilian application of SAR as well as one of the earliest Earth-observing satellites [11]. Even though Seasat only operated for 105 days, it measured wave heights, atmospheric moisture, features of sea ice and ocean topography, and more importantly demonstrated the feasibility of a SAR sensor on a spaceborne platform [11].

Following the success of the Seasat project, the European Space Agency (ESA) made the decision to include a SAR sensor in the final design of the European remote sensing (ERS) Earth-observing satellites ERS-1 and ERS-2. These two systems provided global measurements of the Earth's environment [12, 13]. These were significantly important SAR systems that far exceeded expected life spans, and provided well over one million images of different scenes [13]. The ERS systems had a standard image size of 100 × 100 km with a spatial resolution of 25 × 25 m [12, 13]. For some imaging modes, the data rate was too high to record the data onboard and could only be used when in range of ground stations used by ERS systems. In 2002, Envisat launched as a successor to ERS [14] Envisat was a very large Earth observation satellite with 10 sensors including a more advanced SAR sensor, the advanced synthetic aperture radar (ASAR). ASAR was an extension of the SAR radar from the ERS program and was built by Matra Marconi Space [14]. The Envisat mission concluded in 2012 with the loss of communication with the satellite.

Spaceborne SAR designs in the USSR were originally developed during 1963–1965 [15]. Plans for a 1978 launch of a manned orbit station with SAR were cancelled [16]. This was during the same time frame the United States launched Seasat [11]. The Ekor-1 SAR system was used on the Kosmos-1870 spacecraft during 1987–1989, with a spatial resolution of 20 m [17]. The Ekor-1 system also employed optical aperture synthesis. Improvements to this system were made and the Ekor-A1 was deployed on the Almaz-1 spacecraft with a spatial resolution of 10–12 m [16, 17]. This upgraded system took advantage of spacecraft roll to significantly reduce revisit times and took part in joint experiments with AIRSAR, SIR-C/X-SAR, ERS-1, and ERS-2 [16].

Over the past decades there have been numerous advances in radar hardware, computing, image processing techniques, and imaging algorithms. These advancements have created a wide range of SAR systems and missions covered in later chapters.

1.2 FUNDAMENTALS

Figure 1.3 shows a basic SAR imaging geometry where the platform moves in the along-track direction while transmitting and receiving electromagnetic energy [18, 19]. Referring to Figure 1.3, the resolution of the SAR system in the ground range dimension is causally related to the bandwidth of the radar waveform, while the resolution in the cross-range dimension is related to the wavelength, physical antenna size, and synthetic aperture length. A larger antenna gives finer cross-range resolution for a fixed wavelength. For example, to achieve a cross-range resolution of 10 m from a spaceborne satellite operating at X-band would require an antenna of approximately 2,550 m. An antenna of this size is impractical, and this is where the synthetic aperture concept comes into play. A smaller, practical size antenna is used to collect a series of radar measurements that are combined in such a way as to synthesize a much larger effective antenna. This concept of synthesizing an aperture is the basis for all modern SAR systems. Modifications to the basic geometry given in Figure 1.3 are used to optimize coverage, resolution, processing, and other imaging parameters. Recent technology advances in antenna design and processing have made it possible to produce high-resolution images across exceptionally large image coverage areas [19, 20].

1.2.1 Operating Frequency

SAR systems transmit electromagnetic energy in different frequency bands depending on the intended application. Table 1.1 gives commonly used frequency bands and applications for SAR systems [1]. The penetration of the signal transmitted by the SAR system into the scene is fundamentally related to the wavelength of the electromagnetic energy. Signals transmitted at longer wavelengths penetrate deeper into soils and vegetation. For example, X-band signals scatter from the top layer of a forest canopy, while C-band signals penetrate the top layer and experience a mix

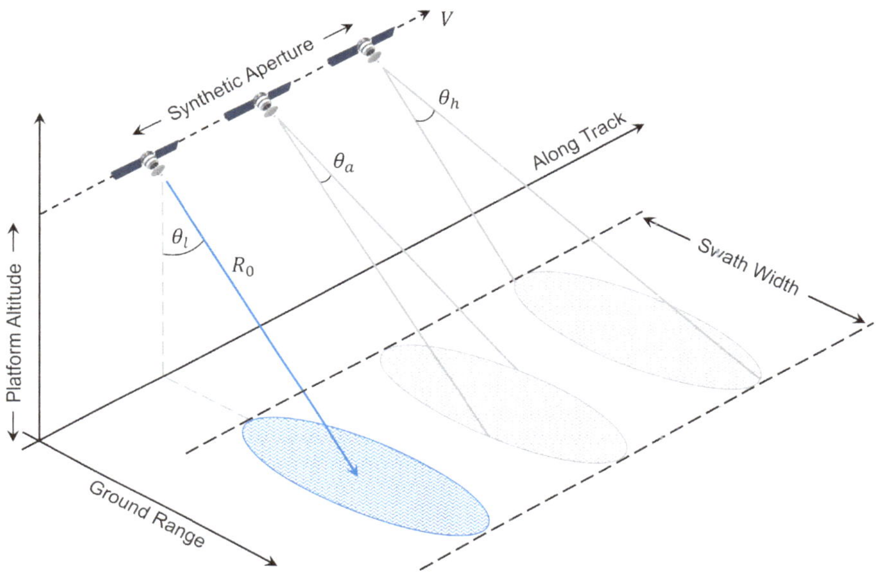

Figure 1.3 Basic stripmap SAR imaging geometry.

of surface and volume scattering. On the other hand, L-band and P-band signals have much deeper penetration and experience more volume scattering as well as double-bounce scattering as illustrated in Figure 1.4 [19]. The choice of operating frequency also depends on factors such as atmospheric attenuation, RF hardware limitations, sampling requirements, digital processing, and data downlinks.

1.2.2 Sensor Polarization

In addition to frequency diversity, SAR systems can transmit and receive electromagnetic energy with different polarization. Polarization refers to the orientation of the electric field of an electromagnetic wave. In general, the electric field traces out an elliptical path as a function of time. For radar applications there are two important special cases of elliptical polarization, namely linear and circular polarization. Most SAR systems are linearly polarized with newer sensors having full polarization capability.

Table 1.1
Frequency Bands Commonly Used by SAR Systems

Frequency Band	*Frequency Range (GHz)*	*Applications*
UHF (ultrahigh frequency)	0.3–1	Foliage penetration, subsurface imaging, and biomass estimation
L (long)	1–2	Foliage penetration, subsidence monitoring, biomass and vegetation mapping, interferometry
S (short)	2–4	Agriculture, ocean, ice, and subsidence monitoring
C (compromise)	4–8	Large area mapping, change detection, ice monitoring, ocean navigation
X (secret during WWII)	8–12	High resolution, snow and ice monitoring, urban imaging, low penetration
Ku (under K)	12–18	Snow monitoring and very high resolution
K (*kurz*)	18–24	Through-wall, automotive, and small UAV systems
Ka (above K)	24–40	Scan on receive, interferometry, very high resolution

For linearly-polarized systems, the electric field vector is always in the same plane and may be either vertical (V) or horizontal (H). Vertically-polarized waves tend to be less affected by reflections over the transmission path, while horizontal polarization tends to be less affected by man-made objects [3]. Note, care must be taken with SAR nomenclature. When discussing SAR imagery, HV is commonly described as a signal transmitted in horizontal polarization and received in vertical polarization. However, when discussing the scattering matrix, the matrix element S_{HV} represents a signal transmitted in vertical polarization and received in

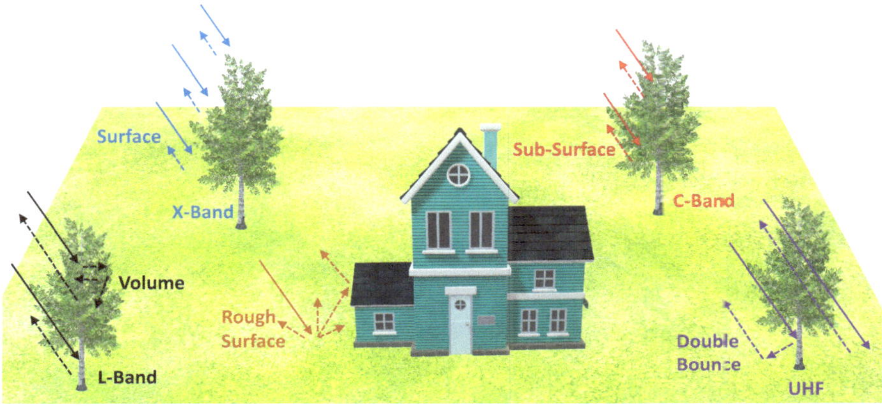

Figure 1.4 Scattering mechanisms and penetration by frequency band.

horizontal polarization. This difference can lead to confusion for those new to the field, especially when presented within the same educational materials[1] [21, 22].

The polarimetric data contains information about the structure of the area being imaged. Referring to Figure 1.4, rough surface scattering, including scattering by bare soil and water, is strongest in the VV component. Double-bounce scattering from buildings, trees, or dense vegetation results in dominant HH scattering. Strong cross-polarization (i.e., VH or HV) is seen in volume scattering, including scattering from leaves and branches as well as some marsh grasses in wetland areas [19, 23]. Applications that are made possible from observables using polarimetric SAR include ocean wave and wind mapping, feature tracking, glacier velocities, separation of dry and wet snow, soil moisture and surface roughness estimations, identification of crop types, forest classification, man-made target identification, permittivity estimation, and surface slope estimation [1]. Figure 1.5 illustrates the advantages of SAR systems capable of full-polarization data collection with a three-polarization color overlay of Iceland's Vatnajökull Glacier.

1 A literature search has not revealed the origin of the deviation from scattering matrix theory. See the Appendix for more details.

Figure 1.5 Polarimetric SAR imagery of Iceland's Vatnajokull Glacier on June 3, 2012 (HH, HV, VV). Courtesy NASA/JPL.

1.3 APPLICATIONS

There are numerous applications of SAR and advances in relevant technologies are pushing SAR systems into previously unthought-of areas. This is particularly true with the recent push in commercial SAR constellations. SAR has been employed in the remote sensing of Earth and other planets, three-dimensional imaging of distant space objects, imaging of orbiting space debris, military applications, and as a navigational aid. Each of these areas have diverse goals, unique problems and limitations, and specialized SAR imaging systems. Chapter 8 covers future trends in more detail.

1.3.1 Remote Sensing

Since SAR can produce high-quality imagery from various altitudes, in all weather conditions, and operate both day and night, it has become one of the most beneficial tools for the remote sensing of Earth and its environment. This includes surface mapping, glacier variations, sea ice observation, wind patterns, rainfall, erosion, storm surges, and drought prediction. SAR has been used to aid disaster relief efforts by providing information concerning flooding, water flow, damage to infrastructure, and prudent evacuation routes. Three years prior to Hurricane Katrina, spaceborne SAR was used to detect subsidence with high spatial resolution under parts of New Orleans, LA [24, 25]. Along with mapping Earth's features, SAR has been used to map distant planets. In 1989 the Magellan spacecraft, shown in Figure 1.6, was deployed from the cargo bay of the Space Shuttle Atlantis [26]. The Magellan mission resulted in high-quality SAR imagery of 98% of the surface of Venus with a resolution of 120 $\times$ 300 m. A snapshot of SAR imagery of the surface of Venus is given in Figure 1.7. The data showed that 85% of Venus' surface is covered in volcanic flows and that surface features can exist for millions of years [26, 27]. Figure 1.8 shows SAR imagery of impact craters and volcanic layering on the surface of Venus.

A technique known as interferometric SAR (InSAR) can detect surface changes on the order of centimeters. InSAR combines two or more SAR images of the same scene. The images created from different platform positions reveal the topography of the scene, while images taken at different times show deformation in the scene. This technique can be used to monitor tectonic deformation due to earthquakes. This method was first used for studying earthquakes in 1992 and has been employed extensively since then [28]. InSAR has also seen use in the monitoring of and prediction of volcanic eruptions, including associated surface deformations, magma distribution variations, volcanic edifices spreading, and tectonic dislocations [29, 30]. InSAR is also used to derive digital elevation models. The National Aeronautics and Space Administration (NASA) Shuttle Radar Topography Mission used this approach aboard Space Shuttle Endeavour in 2000 to obtain elevation measurements of the Crater Highlands along the East African Rift in Tanzania [31]. The imagery of this region is shown in Figure 1.9.

The health of forests and agriculture is related to soil and watershed quality and is crucially important in climate interactions. Dense forests and agriculture

Figure 1.6 NASA's Magellan spacecraft deployed from Space Shuttle Atlantis. Courtesy NASA/JPL.

covering enormous areas of land are often difficult if not impossible to access. This makes airborne and satellite SAR advantageous for mapping deforestation and degradation, prediction of forest fires, estimating standard forest height, and surveying forest biomass [28]. Table 1.2 gives a summary of systems used in remote sensing applications along with their key characteristics [3].

Figure 1.7 SAR imagery of Venus created by the Magellan spacecraft. Courtesy NASA/JPL.

Table 1.2
Examples of SAR Systems Used in Remote Sensing Applications

Sensor	*Operational Dates*	*Band*	*Polarization*	*Notes*
Seasat (NASA/JPL)	1978	L	HH	First civilian SAR satellite
Radarsat-1 (CSA)	1995–2013	C	HH	First Canadian SAR satellite
Radarsat-2 (MDA)	2007–Present	C	Full	Improvements over Radarsat-1 include finer spatial resolution and added imaging modes
TerraSAR-X & TanDEM-X (DLR)	2007–Present 2010–Present	X	Full	First spaceborne bistatic radar
Radarsat Constellation Mission (CSA)	2019–Present	C	Compact	Three-satellite Earth observation constellation
ERS-1	1991–2000	C	VV	First ESA Earth observation system
ERS-2	1995–2011	C	VV	Flew in tandem with ERS-1 for nine months
ICEYE-X1, X2	2018–Present	X	VV	First microsatellite SAR, largest SAR constellation
Umbra-SAR	2021–Present	X	HH or VV	Finest resolution commercially available
Capella-Denali	2018–Present	X	HH	First U.S. commercial SAR satellite

Figure 1.8 Three impact craters and volcanic layering on the surface of Venus created by the Magellan spacecraft. Courtesy NASA/JPL.

1.3.2 Space Debris Monitoring

Orbital debris presents a serious hazard for satellites and human spaceflight. Since November 2000, the International Space Station (ISS) has been continuously inhabited by a crew of up to six astronauts. Due to orbital debris, ISS makes debris avoidance maneuvers approximately once per year [32]. Also, satellite operations including SAR, internet, communications, navigation, reconnaissance, and weather forecasting are nearly universal today. These factors make it imperative to not only understand the current debris environment provided by measurement systems, but to also predict the future debris environment using modeling and simulation tools.

Figure 1.9 Topography of Crater Highlands in Tanzania. Courtesy NASA/JPL.

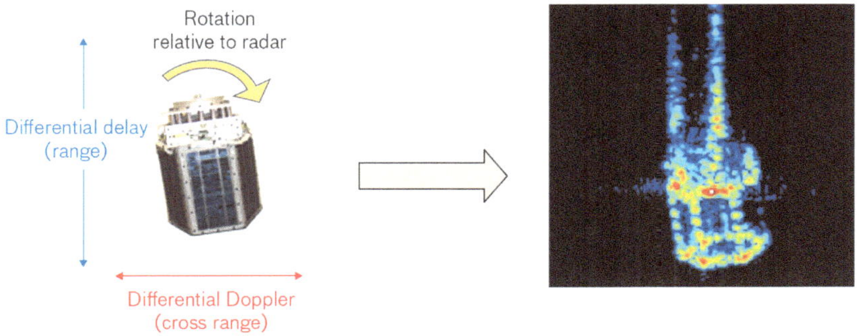

Figure 1.10 ISAR image created by the relative rotation of the target to the radar Reprinted with permission courtesy of MIT Lincoln Laboratory, Lexington, Massachusetts.

Therefore, NASA's Orbital Debris Program Office was founded in 1979 to tackle these challenging problems.

For altitudes of 1,000 km or less, NASA depends on measurements from ground-based radars to characterize the distribution of small debris [32]. In addition, measurements from ground-based radars may also be used to generate high-resolution imagery through a technique known as inverse synthetic aperture radar (ISAR). ISAR is analogous to SAR, the difference being the synthetic aperture is created by the movement of the target rather than the radar platform, as illustrated in Figure 1.10. This topic is covered in detail in Chapter 2. One of the first radars of this type was the Advanced Research Projects Agency (ARPA) Lincoln C-band Observables Radar (ALCOR), which was capable of 50 cm resolution [33, 34]. Shortly thereafter, ARPA sponsored the X-band Haystack long-range imaging radar (LRIR) capable of 25 cm resolution, followed by the Ka-band millimeter-wave (MMW) radar with a resolution of 6 cm, and most recently the W-band Haystack Ultra-wideband Satellite Imaging Radar (HUSIR) achieving a resolution of 3 cm. The compact range imagery of a NASA satellite model, shown in Figure 1.11, depicts the difference in image resolution between 1 GHz and 8 GHz [33].

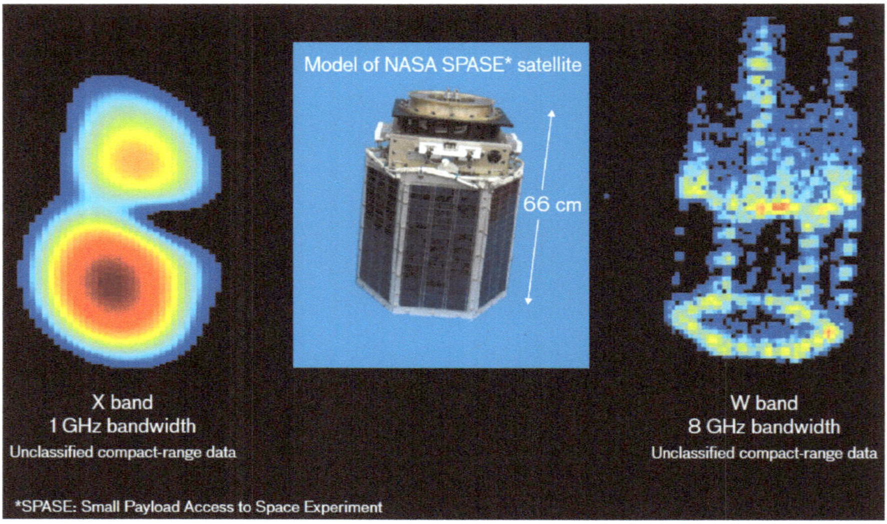

Figure 1.11 Compact-range imagery data of a NASA Small Payload Access to Space Experiment satellite model to depict the difference in resolution between 1-GHz and 8-GHz radar bandwidths. Reprinted with permission courtesy of MIT Lincoln Laboratory, Lexington, Massachusetts.

1.3.3 Through-Wall Sensing

Through-wall radar imaging systems allow for gathering of information about the location, classification, and movement of targets from behind walls, barriers, and other obstructions. This information can be critical to military and law enforcement conducting operations in urban areas, as well as search and rescue efforts. These systems typically employ ultrawideband (UWB) waveforms, also referred to as impulse or carrier-free waveforms and current systems have bandwidths as large as 8 GHz [35]. Note, the Federal Communications Commission (FCC) defines ultrawideband to be a waveform with a bandwidth greater than 25% of the center frequency. Research and development of UWB systems increased rapidly with the FCC's Report and Order in 2002 which allows unlicensed UWB operation of imaging, vehicle radar, and communication systems [36, 37]. UWB systems have very fine range resolution arising from short pulse duration. Also, the low spectral density of these waveforms results in low health hazard to humans.

The earliest UWB through-wall imaging system was publicly announced in 2003 by Time Domain Corporation [35]. Since then, there has been much research and development performed by groups such as the Defense Advanced Research Projects Agency (DARPA), the Canadian Department of National Defense, Eureka Aerospace, CyTerra Corporation, and Camero-Tech. Advancements in hardware, computing, and imaging algorithms have led to sophisticated systems such as the Xaver™ 800, which can display real-time, three-dimensional imagery for intelligence, surveillance, and reconnaissance [38]. The aperture for through-wall systems is synthesized by using an antenna array with individual elements placed in a grid layout. Each antenna element transmits and receives the radar signal thus creating the aperture. This application of three-dimensional radar imaging continues to see much attention from commercial, government, and academic communities, and the reader is referred to several excellent sources on the subject [35, 39–42].

1.3.4 Military Applications

Due to the performance, flexibility, and varying techniques, SAR is a valuable tool in military applications. One such application is the detection of buried or hidden explosives such as landmines and improvised explosive devices (IEDs). Much of the work in this area has focused on material identification and the detection of command wires [43, 44]. The Spectrally Agile Frequency-Incrementing Reconfigurable (SAFIRE) radar was developed in 2016 by the U.S. Army Research Laboratory (ARL) [45, 46]. SAFIRE is an ultrawideband forward-looking ground-penetrating radar that is mounted on a vehicle. It is a stepped-frequency system that is highly reconfigurable and more immune to radio-frequency interference than traditional short-pulse ultrawideband systems. The SAFIRE system has been tested against various target emplacements and imaged broadside landmines buried in soil [47].

Ground clutter and foliage were a problem for early airborne and battlefield radars as many of these systems operated at wavelengths not suitable for penetrating foliage. Foliage played a significant role in concealing troops and equipment in Vietnam, which was problematic for tactical engagements. During this time, the first foliage penetration (FOPEN) radar systems were developed. These included the Camp Sentinel Radar-II and the multipurpose foliage penetration radar (M-FOPEN) [48, 49]. While these systems were successful in detecting enemy movement in foliage, to detect stationary tactical targets required the development of low-frequency foliage-penetrating SAR systems. Early systems such as synthetic aperture dual frequency radar (SADFRAD) proved the ability to detect man-made objects from natural clutter with a SAR system [50]. Since that time, SAR and FOPEN technologies have been continuously refined, including advanced automatic target detection and recognition algorithms, automatic cueing algorithms, and ultra-wideband systems. In 2004, development began on the tactical reconnaissance and counter concealment enabled radar (TRACER) to be deployed on both manned and unmanned platforms. The system was developed to detect vehicles, buildings, and large metallic objects in broad areas of dense foliage, forested areas, and wooded terrain and was successfully tested on an unmanned aerial vehicle in 2011 [51].

Another military application of SAR technology is reconnaissance. In 1960, Lockheed Martin began work on the SR-71 long-range, high-altitude, strategic reconnaissance aircraft. In 1964, the SR-71 was operating with the KP1 SAR system developed by Goodyear. It could produce imagery with a resolution of 30 feet at a range of 100 miles from a height of more than 80,000 feet and a speed over three times the speed of sound [5, 52]. Throughout the 29-year operation of the SR-71, the SAR systems were continuously improved. In the early 1970s, the AN/UPD-3 SAR developed by Goodyear achieved a resolution of five feet and flew on a KC-125 aircraft [5]. The 1970s also saw a series of prototype SAR systems built by Goodyear for the U-2 aircraft, including the first long range SAR with a resolution of one foot [5, 53]. The CARABAS-II VHF SAR system was designed and operated by the Swedish Defence Research Agency and is capable of efficient detection of concealed man-made objects hidden in dense foliage suing change detection techniques [54]. The CARABAS-II operational frequency band is 20–90 MHz, resulting in a spatial resolution of approximately 2.5 m. Design of the system began in 1994 and the first flight tests were conducted in 1996. Since then, the system has flown numerous missions in Sweden, France, Finland, and the United States [54].

1.3.5 Alternative Navigation

As civilian and government applications and operations tend to be overreliant on the Global Positioning System (GPS) for position location and navigation, this can lead to very serious vulnerabilities. GPS signals are particularly susceptible to several jamming and spoofing techniques [55]. One possible solution is a SAR-based navigation aid. Since SAR systems actively transmit and receive signals, there is less sensitivity to various attacks [55, 56]. One approach that is gaining a lot of interest is based on the correlation of the SAR imagery with digital elevation models and often existing maps (e.g., Google Earth) created from satellite and aerial imagery, topography, ocean bathymetry, and other geographic data [57]. The SAR estimates are then used in conjunction with inertial navigation systems to provide accurate navigation when GPS is unavailable [55, 56]. Other approaches use neural networks to model the errors in flight path, due to loss of GPS measurements, relative to the start of the synthetic aperture or some known global reference [55].

1.4 SOFTWARE

To aid reader comprehension of the concepts covered in the subsequent chapters, a SAR tool suite is included with this book. The tool suite is written in the Python and MATLAB programming languages [58, 59] and gives the reader a mechanism to generate SAR images, predict the performance of SAR systems for various applications, and analyze images resulting from different scenarios.

The tool suite makes use of Google's Colaboratory, or Colab for short, and MATLAB's Live Editor to provide the user with interactive notebooks and live scripts that help the reader gain valuable insight about the topic of interest. Full access to both the Python notebooks and MATLAB live scripts gives the user the ability to customize and extend each tool for unique applications or include as part of a report as both Colaboratory and Live Editor allow for the notebooks and live scripts to be saved in portable document format (PDF). Several examples using the tool suite are given at the end of each chapter.

The Python and MATLAB sets of code are obtained from the GitHub repository at `https://sarbook.github.io`. The user enters the access code as shown in Figure 1.12. The user may clone the repository or download as a zip file, as illustrated in Figure 1.13. Once the software has been obtained, the structure of the subfolders is identical, as the MATLAB Live Editor scripts correspond to each of the Google Colab notebooks. More information about Google Colab and MATLAB Live Editor is given in the following sections.

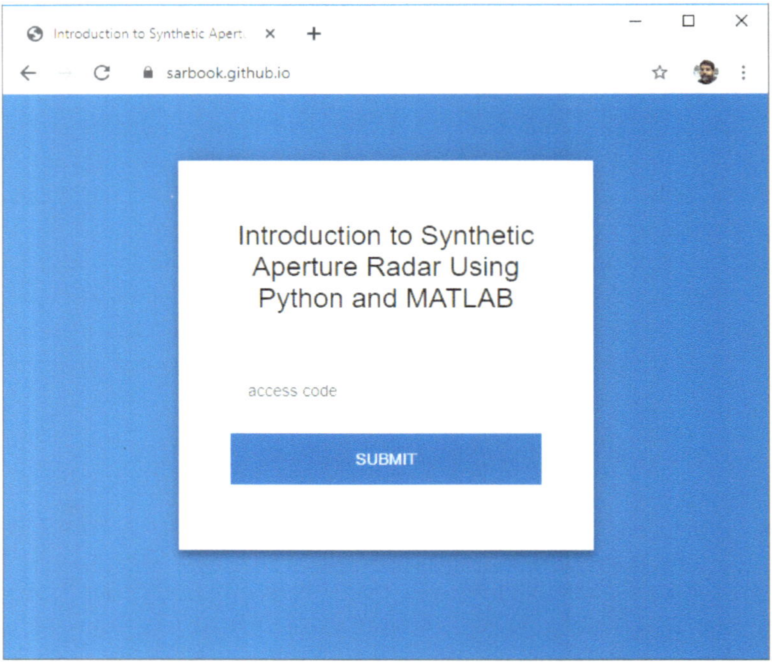

Figure 1.12 Login screen to access the software accompanying this book.

Watch this step-by-step guide for obtaining the software.

1.4.1 Python

Python is a popular programming language created by Guido van Rossum and released in 1991 [59]. Python is developed under an Open Source Initiative (OSI) approved license, making it free to use and distribute, even for commercial use. Python's license is administered by the Python Software Foundation [59]. The SAR tool suite also makes use of additional packages, including SciPy, which is a Python-based system of open-source software for mathematics, science, and engineering [60]. It also makes use of the core package, NumPy, which is a fundamental package for scientific computing, and includes a powerful N-dimensional array object, useful

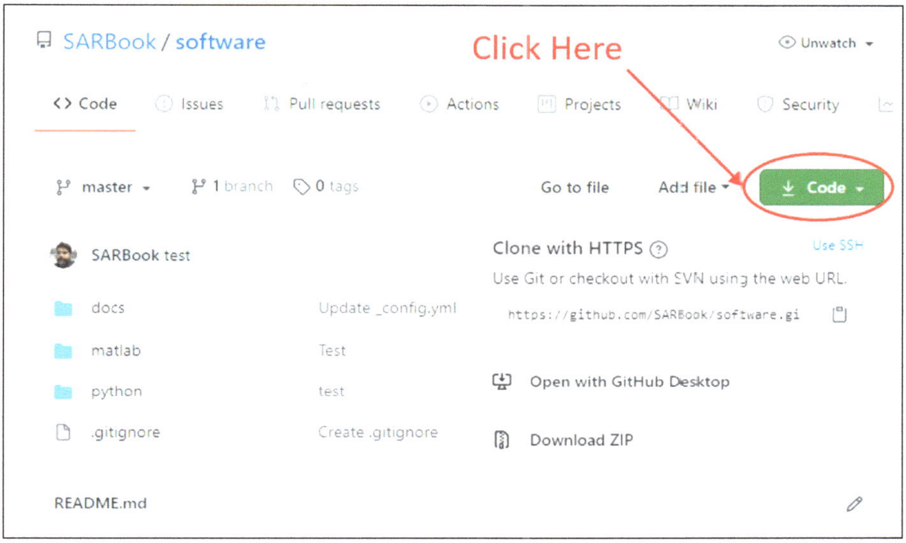

Figure 1.13 GitHub repository for the software accompanying this book.

linear algebra, Fourier transform, and random number capabilities [61]. For plotting, the tool suite takes advantage of Matplotlib, a Python plotting library that produces publication-quality figures in a variety of formats and interactive environments across platforms [62]. For more information about Python and its features visit `https://www.python.org`.

1.4.1.1 Google Colaboratory

Google Colaboratory is an easy-to-use Python environment, allowing the user to write and execute Python in a browser [63]. There is no setup required on the user's machine, projects are saved and easily shared through Google Drive, and free access to graphics processing units (GPUs) is provided. A Google Colab notebook is an executable document composed of cells, each of which can contain code, formatted text, HTML, LaTeX, and images, as shown in Figure 1.14. The packages mentioned in Section 1.4.1 are available in Google Colab without any required setup and are imported into the user's project as shown in Figure 1.15.

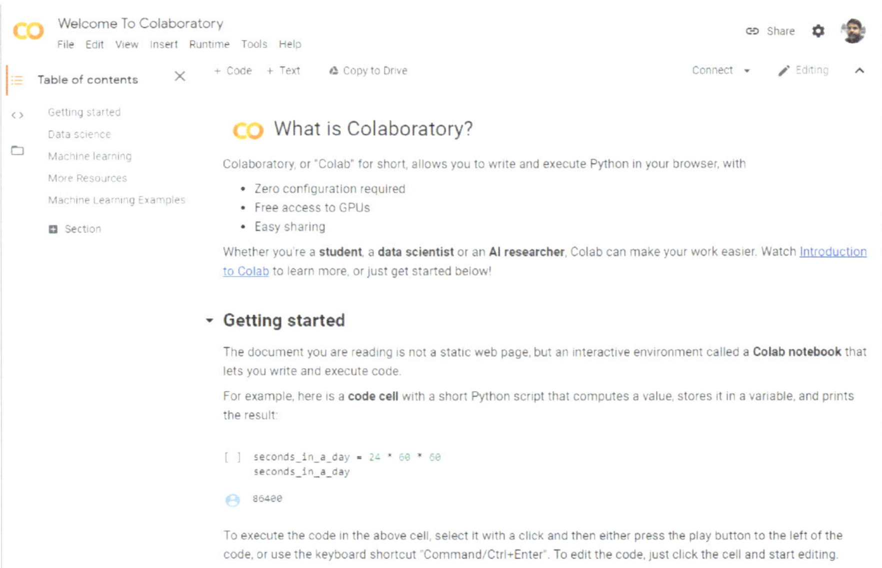

Figure 1.14 Example Google Colaboratory notebook.

Watch this introduction to Google Colaboratory.

1.4.2 MATLAB

MATLAB is a desktop programming environment and has become a ubiquitous tool for engineers and scientists performing design and analysis [58]. MATLAB has been used for a wide range of applications in industry and academia, including deep learning and machine learning, signal processing and communications, image and video processing, control systems, test and measurement, computational finance, and computational biology. Being a matrix-based system allows MATLAB to work well with SAR imaging datasets. While often used as a desktop tool, MATLAB scales to run on clusters, GPUs, and clouds, which can prove beneficial when working with very large SAR imagery. Finally, MATLAB offers several toolboxes

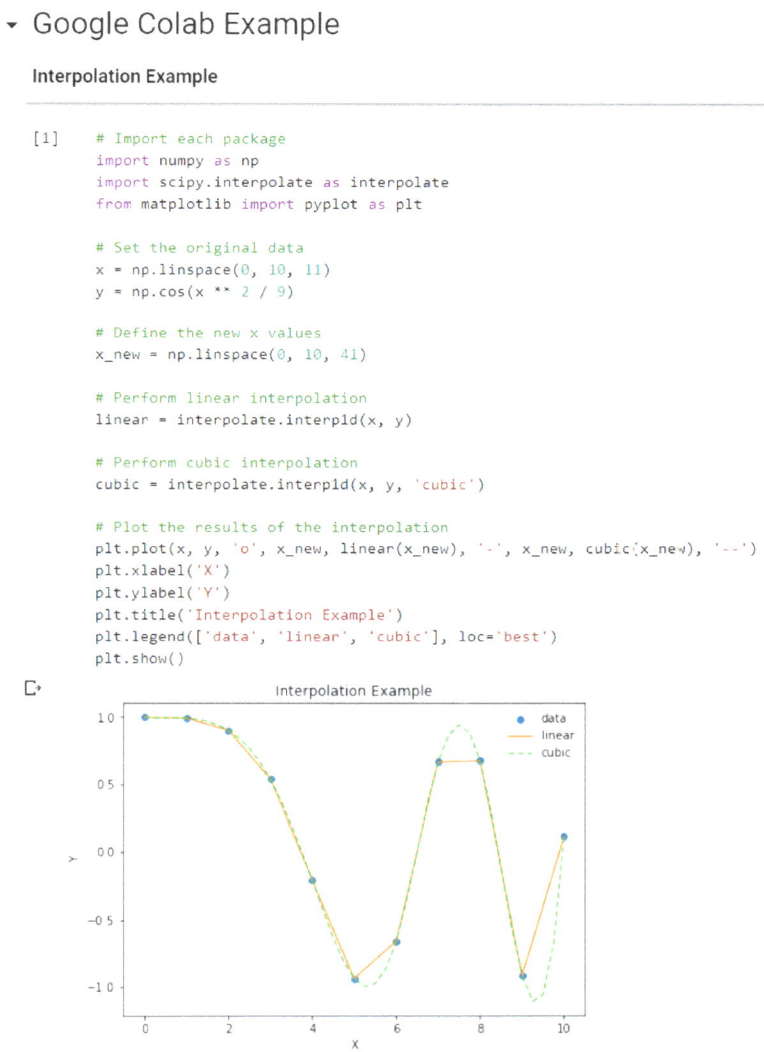

Figure 1.15 Simple interpolation example in Google Colaboratory.

including those for image analysis. For more information about MATLAB visit https://www.mathworks.com.

1.4.2.1 Live Editor

Live Editor is included with MATLAB and allows the user to create an executable notebook that combines code, output, formatted text, images, equations, and hyperlinks [64]. This proves to be an excellent tool for developing the examples given in this book. The notebook can even be divided into sections that can be run independently, allowing the reader to change certain imaging parameters to see the effect on the imagery, and the output and visualizations are viewed next to the code. Live Editor can be used in MATLAB and in MATLAB Online™. A simple example of Live Editor corresponding to the example in Figure 1.15 is given in Figure 1.16.

Watch this introduction to Live Editor.

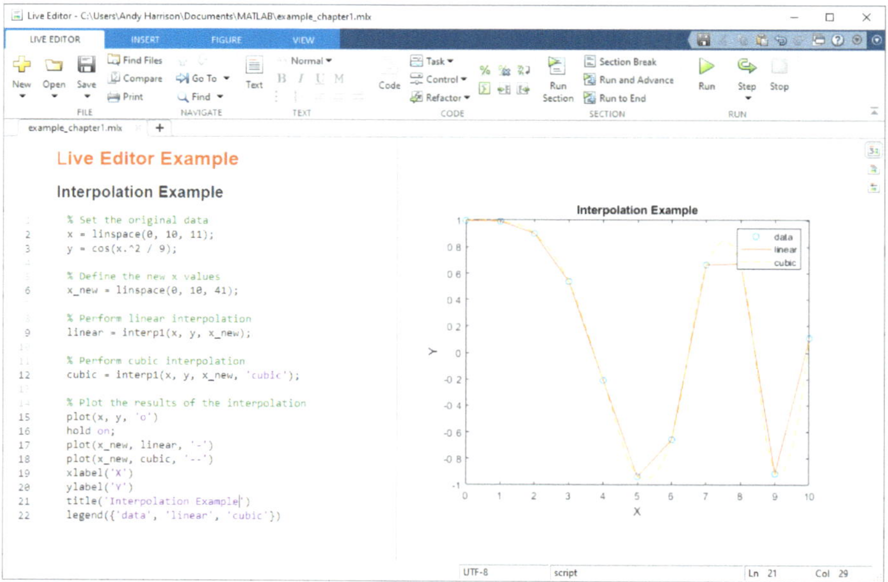

Figure 1.16 Simple interpolation example using MATLAB Live Editor.

PROBLEMS

1.1 Describe the differences between Carl Wiley's original technique, Simultaneous Doppler Buildup, and SAR imaging methods used in modern systems.

1.2 What were the major limitations of early SAR systems and what steps were taken to overcome these limitations?

1.3 For an airborne SAR operating at C-band and a range of 30 km, what would be the dimension of a physical antenna required to realize 20 m resolution? Hint: $D = \lambda R/\delta_a$ where, D is the antenna dimension, λ is the wavelength, R is the range, and δ_a is the azimuth resolution. All values have units of meters.

1.4 Describe the differences in X, C, L, and P-band scattering from wet soil.

1.5 Referring to Table 1.1, what would be an appropriate frequency band for monitoring the health of corn crops during dry summer months?

1.6 Considering the scenario in Problem 1.5, what polarization, VV, HH, or VH, would be a good choice and why?

1.7 Expound on the major technological advances in radar hardware, software, computing, and algorithmic techniques that lead to today's advanced SAR imaging systems.

1.8 Obtain the SAR tool suite accompanying this book, as outlined in Section 1.4. Execute the Google Colab notebook *software/python/Chapter1/example.ipynb* and the Live Editor notebook *software/matlab/Chapter1/example.mlx*. Verify that the output of the Google Colab notebook matches that of the Live Editor notebook.

References

[1] A. Moreira, P. Prats-Iraola, M. Younis, G. Krieger, I. Hajnsek, and K. Papathanassiou. "A tutorial on synthetic aperture radar." *IEEE Geoscience and Remote Sensing Magazine*, 1(1):6–43, 2013.

[2] ICEYE. Example SAR data from ICEYE. https://www.iceye.com/downloads/datasets. Accessed: 2021-09-21.

[3] A. Harrison. *Introduction to Radar Using Python and MATLAB*. Artech House, Norwood, 2020.

[4] C. Wiley. "Synthetic aperture radars." *IEEE Transactions on Aerospace and Electronic Systems*, AES-21(3):440–443, 1985.

[5] S. Lasswell. History of sar at lockheed martin (previously Goodyear Aerospace). In R. Trebits and J. Kurtz, editors, *Radar Sensor Technology IX*, volume 5788, pages 1–12. International Society for Optics and Photonics, SPIE, 2005.

[6] C. Wiley. Pulsed doppler radar methods and apparatus. US Patent US3196436A, August 1954.

[7] C. Sherwin, J. Ruina, and R. Rawcliffe. Some early developments in synthetic aperture radar systems. *IRE Transactions on Military Electronics*, pages 111–115.

[8] Office of the Chief Signal Officer. Problems of battlefield surveillance. Technical report, Defense Technical Information Center, 1953.

[9] G. Stimson. *Introduction to airborne radar*. Hughes Aircraft Company, El Segundo, 1983.

[10] J. Cutrona, E. Vivian, E. Leith, and G. Hall. A high-resolution radar combat intelligence system. *IRE Transactions on Military Electroncis*, pages 127–131, April 1961.

[11] NASA - Jet Propulsion Laboratory. Seasat. https://www.jpl.nasa.gov/missions/seasat/, June 2018. Accessed: 2020-9-9.

[12] E. Attema, Y. Desnos, and G. Duchossois. Synthetic aperture radar in Europe: ERS, Envisat, and beyond. Technical Report 1, Johns Hopkins APL, 2000.

[13] European Space Agency. ERS Instruments. https://earth.esa.int/eogateway/missions/ers, 2020. Accessed: 2022-5-8.

[14] European Space Agency. Envisat. https://earth.esa.int/eogateway/missions/envisat, 2020. Accessed: 2022-5-8.

[15] J. Greenberg. A system look at satellite-borne high resolution radar. *RCA Review*, 28(4):679–709, 1967.

[16] V. Verba, L. Neronskiy, and I. Osipov. Russian spaceborne imaging radars: Scientific and technicalachievements and priority perspectives of development. Technical report, VEGA Corporation.

[17] P. Salganik, G. Efremov, L. Neronskiy, M. Kulikovsky, and A. Kurmaev. Radar remote sensing of Earth from Kosmos-1870 satellite. *Space Research of the Earth*, (2):70–79, 1990.

[18] NASA. What is synthetic aperture radar. https://earthdata.nasa.gov/learn/what-is-sar, 2020. Accessed: 2020-10-1.

[19] A. Flores-Anderson, K. Herndon, R. Thapa, and E. Cherrington. The SAR handbook: Comprehensive methodologies for forest monitoring and biomass estimation. Technical report, NASA Space Science and Technology Center, Huntsville, 2019.

[20] ICEYE. SAR Product Guide. Technical Report Version 4.2, ICEYE, Espoo, 2021.

[21] Government of Canada. Polarization in radar systems. https://www.nrcan.gc.ca/maps-tools-and-publications/satellite-imagery-and-air-photos/satellite-imagery-products/educational-resources/tutorial-radar-polarimetry/polarization-radar-systems/9567, 2014. Accessed: 2022-5-5.

[22] Government of Canada. The scattering matrix. https://www.nrcan.gc.ca/maps-tools-and-publications/satellite-imagery-and-air-photos/satellite-imagery-products/educational-resources/tutorial-radar-polarimetry/the-scattering-matrix/9537, 2016. Accessed: 2022-5-5.

[23] X. Duan and C. Jones. Coherent microwave scattering model of marsh grass. *Radio Science*, 52:1578–1595, 2017.

[24] T. Dixon. Space geodesy: Subsidence and flooding in New Orleans. *Nature*, (441):587–588, 2006.

[25] S. Kim, S. Wdowinski, F. Amelung, T. Dixon, and S. Hong. X-band InSAR observations in New Orleans, Louisiana. In *IGARSS 2008 - 2008 IEEE International Geoscience and Remote Sensing Symposium*, volume 4, pages IV – 514–IV – 517, 2008.

[26] NASA. NASA Science Solar System Exploration. https://solarsystem.nasa.gov/missions/magellan/in-depth/, July 2019. Accessed: 2020-9-20.

[27] A. Siddiqi. Beyond Earth: A chronicle of deep space exporation, 1958–2016. Technical report, NASA, Washington, 2018.

[28] D. Massonnet, M. Rossi, C. Carmona, F. Adragna, G. Peltzer, K. Feigl, and T. Rabaute. The displacement field of the Landers earthquake mapped by radar interferometry. *Nature*, 364(6433):138–142, 1993.

[29] D. Massonnet, D. Briole, and A. Amaud. Deflation of Mount Etna monitored by spaceborne radar interferometry. *Nature*, 375(6532):567–570, 1995.

[30] P. Rosen, S. Hensley, H. Zebker, F. Webb, and E. Fielding. Surface deformation and coherence measurements of Kilauea Volcano, Hawaii, from SIR-C radar interferometry. *Journal of Geophysical Research*, 101:23109–23125, 1996.

[31] NASA. NASA-ISRO SAR Mission. http://nisar.jpl.nasa.gov/mission/get-to-know-sar/. Accessed: 2020-09-05.

[32] J. Murray, C. Blackwell, J. Gaynor, and T. Kennedy. Haystack Ultra-Wideband Satellite Imaging Radar Measurements of the Orbital Debris Environment: 2014-2017. Technical report, National Aeronautics and Space Administration, 2019.

[33] M. Czerwinski and J. Usoff. Development of the Haystack ultrawideband satellite imaging radar. *Lincoln Laboratory Journal*, 21(1):24–44, 2014.

[34] W. Camp, J. Mayhan, and R. O'Donnell. Wideband radar for ballistic missile defense and range-doppler imaging of satellites. *Lincoln Laboratory Journal*, 12(2):267–269, 2000.

[35] Y. Yang. *Development of a Real-time Ultra-wideband See Through Wall Imaging Radary System*. Ph.D. Thesis, University of Tennessee, 2008.

[36] A. Ameti, R. Fontana, E. Knight, and E. Richley. Ultra wideband technology for aircraft wireless intercommunications systems (AWICS) design. Reston, 2003.

[37] R. Fontana. Recent system applications of short-pulse ultra-wideband (UWB) technology. *IEEE Transactions on Microwave Theory and Techniques*, 52(9):2087–2104, 2004.

[38] Camero-Tech. Camero Xaver 800. http://www.camero-tech.com/xaver-products/xaver-800/. Accessed: 2021-02-12.

[39] Office of the OSD/DARPA. Assessment of ultra-wideband (UWB) technology. Technical report, Ultra-Wideband Radar Review Panel R-6280, Washington, 1990.

[40] R. Fontana. *A Brief History of UWB Communications*. Multispectral Solutions, Inc.

[41] G. Ross. *Early Motivations and History of Ultra Wideband*. Anro Engineering, Inc., Lexington.

[42] T. Barrett. *History of Ultra Wideband (UWB) Radar*. UCI, Vienna.

[43] B. Gonzalez-Valdes, J. Martinez-Lorenzo, and C. Rappaport. Dual band SAR processing for low dielectric contrast buried IED detection. In *2013 IEEE Antennas and Propagation Society International Symposium (APSURSI)*, pages 1080–1081, 2013.

[44] M. Otten, M. Cuenca, and M. Fernandez. IED command wire detection with multi-channel drone radar. In *EUSAR 2021; 13th European Conference on Synthetic Aperture Radar*, pages 1–4, 2021.

[45] B. Phelan, K. Ranney, K. Gallagher, J. Clark, K. Sherbondy, and R. Narayanan. Design of ultrawideband stepped-frequency radar for imaging of obscured targets. *IEEE Sensors Journal*, 17(14):4435–4446, 2017.

[46] B. Phelan et al. System upgrades and performance evaluation of the spectrally agile, frequency incrementing reconfigurable (SAFIRE) radar system. In K. Ranney and A. Doerry, editors, *Radar Sensor Technology XXI*, volume 10188, pages 397–406. International Society for Optics and Photonics, SPIE, 2017.

[47] K. Ranney et al. Initial processing and analysis of forward- and side-looking data from the Spectrally Agile Frequency-Incrementing Reconfigurable (SAFIRE) radar. In K. Ranney and A. Doerry, editors, *Radar Sensor Technology XXI*, volume 10188, pages 557–564. International Society for Optics and Photonics, SPIE, 2017.

[48] L. Novak T. Bryant, G. Morse and J. Henry. Tactical radars for ground surveillance. *Lincoln Laboratory Journal*, 12(2):341–354, 2000.

[49] L. Surgent. Evaluation of the Multipurpose Foliage Penetration Radar (M-FOPEN) in Hawaii. Technical report, U.S. Army Land Warfare Laboratory, Aberdeen Proving Ground, 1974.

[50] Air Force Cambridge Research Laboratories. Report on Research at AFCRL. Technical report, Air Force Systems Command, Bedford, 1975.

[51] Lockheed Martin. Tracer. https://www.lockheedmartin.com/en-us/products/tracer.html, April 2018. Accessed: 2020-9-4.

[52] J. Nelson. Side-looking radar imagery. *Electronics World*, 74(2):42–44, 1965.

[53] Wikimedia Foundation. Lockheed U-2. https://en.wikipedia.org/wiki/Lockheed_U-2, 2020. Accessed: 2020-9-26.

[54] L.M.H. Ulander, P.-O. Frolind, A. Gustavsson, H. Hellsten, T. Jonsson, B. Larsson, and G. Stenstrom. Performance of the carabas-ii vhf-band synthetic aperture radar. In *IGARSS 2001. Scanning the Present and Resolving the Future. Proceedings. IEEE 2001 International Geoscience and Remote Sensing Symposium (Cat. No.01CH37217)*, volume 1, pages 129–131, 2001.

[55] T. White. GPS-Denied Navigation Using Synthetic Aperture Radar Images and Neural Networks. Master's thesis, Utah State University, 2021.

[56] Z. Reid. Leveraging 3D Models for SAR-based navigation in GPS-denied Environments. Master's thesis, Wright State University, 2018.

[57] Google. Google Earth. https://earth.google.com, 2022. Accessed: 2022-4-30.

[58] The Mathworks, Inc., Natick, Massachusetts. *MATLAB version 9.8 (R2020a)*, 2021.

[59] P.S. F. Python software foundation. https://www.python.org/psf. Accessed: 2020-5-5.

[60] E. Jones et al. SciPy: Open source scientific tools for Python. http://www.scipy.org/. Accessed: 2021-06-23.

[61] T. Oliphant. A guide to NumPy. http://www.numpy.org, 2006. Accessed: 2020-8-9.

[62] J. Hunter. Matplotlib: A 2d graphics environment. *Computing in Science and Engineering*, 9(3):90–95.

[63] Google. Google Colaboratory. https://colab.research.google.com/notebooks/intro.ipynb, September 2020. Accessed: 2020-9-5.

[64] The Mathworks, Inc., Natick, Massachusetts. *MATLAB Live Editor*, 2021.

Chapter 2

Imaging Modes

Over the past decades there have been numerous technological advancements that have paved the way for the cutting-edge SAR systems in use today. These advancements have affected all aspects of SAR imaging including antenna design, signal and data processing, image creation techniques, and sensor platforms. Also, with new technology comes novel designs for imaging modes, which have evolved from basic stripmap modes to advanced modes capable of target discrimination, navigation, volcanic activity monitoring, earthquake prediction, and imaging of deep space objects, to name a few. This chapter begins with some of the basic principles of SAR imaging and then moves on to two of the basic imaging modes, stripmap and spotlight. The subsequent sections cover newer and more advanced imaging modes. Finally, the chapter concludes with several Python and MATLAB examples to reinforce some of the basic concepts associated with various imaging modes.

2.1 BASIC PRINCIPLES

SAR distinguishes itself from other remote sensing techniques, such as optical and infrared, in that it provides high-resolution imaging, day or night operation, and can perform in all weather conditions [1]. In general, SAR imaging modes are based on a side-looking geometry and take advantage of the motion of the platform to synthesize an aperture. As the platform moves, radar pulses are transmitted sequentially.

The received energy is then coherently combined to produce SAR imagery. The transmitted pulse is commonly a linearly frequency modulated (LFM) waveform, also known as a *chirp* waveform which provides an excellent compromise between range resolution and energy on the target [2]. The time related to the transmission and reception of electromagnetic energy in the range dimension is often referred to as *fast time*, as it is related to the speed of light. The time related to the motion of the platform in the cross-range direction is referred to as *slow time*. As a SAR platform moves, the range to the target varies. This range variation is causally related to the azimuth phase of the received signal. SAR signal processing takes full advantage of the phase information in the received energy when forming imagery.

In contrast to other sensing systems, raw SAR data is not useful for visualization or interpretation. The raw data must go through various signal and data processing to be appropriate for viewing and further exploitation. Referring to Figure 2.1, each row in the raw data is the complex-valued (amplitude and phase) signal reflected from targets in the scene and received by the radar for a given platform position in the along-track direction. The raw data depends on the sensor's design and mission. For example, the raw data may consist of single channel data, polarimetric data, interferometric data, or some combination of these. The methods used to process the raw data therefore change with the type of data collected. Referring again to Figure 2.1, the objective of signal processing is to focus the energy dispersed from targets in the scene.

Depending on the SAR sensor design and mission, numerous data products are available. These include single look complex (SLC), multilook intensity (MLI), precision image (PI), geocoded ellipsoid corrected (GEC), geocoded terrain corrected (GTC), as well as the metadata accompanying SAR data products. The reader is referred to these excellent sources for further information on more mission specific data products [3–6].

2.2 STRIPMAP

Stripmap SAR imaging is the original imaging mode used in the first SAR systems in the 1950s and is today used by many SAR systems on satellites, aircraft, and UAVs [1, 7, 8]. As a SAR platform moves along its flight path, the sensor illuminates and images a specific portion of Earth's surface. This area is often referred to as the

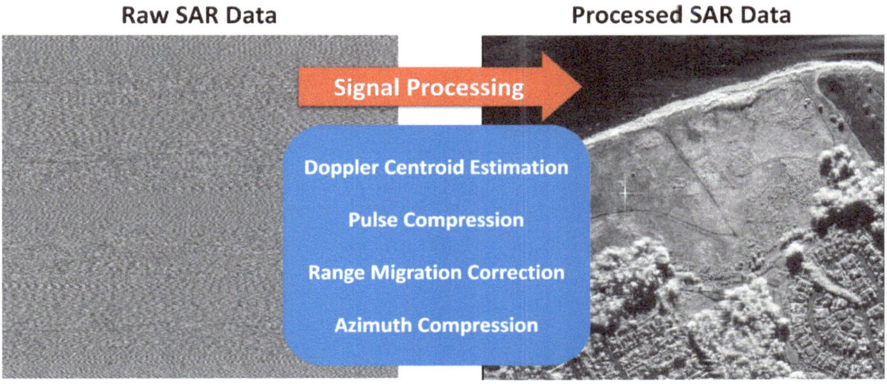

Figure 2.1 Illustration of raw SAR data along with processed image data. Courtesy of Umbra Lab, Inc.

ground swath. The ground swath for spaceborne SAR can vary between tens and hundreds of kilometers. For traditional stripmap SAR imaging, the antenna beam is pointed at a fixed look angle, θ_l, as the ground swath is illuminated by a continuous sequence of pulses. The basic geometry for stripmap SAR imaging is illustrated in Figure 1.3. The area on the ground illuminated by a single pulse is, in general, an ellipse. For a sequence of pulses, the ellipse moves along the ground following the platform's flight path.

Watch this animation of stripmap SAR imaging.

The width of the ground swath is related to the footprint of the SAR system's antenna, and is expressed as

$$W_g \approx \frac{\lambda R_0}{D_v} \qquad \text{(m)}, \tag{2.1}$$

where

λ = wavelength (m),
R_0 = slant range to the swath midpoint (m),
D_v = antenna dimension in the vertical direction (m),
W_g = width of the ground swath (m).

The expression in (2.1) is the classical Rayleigh resolution for a finite aperture [9, 10]. The swath width is shown in detail in Figures 1.3 and 2.2.

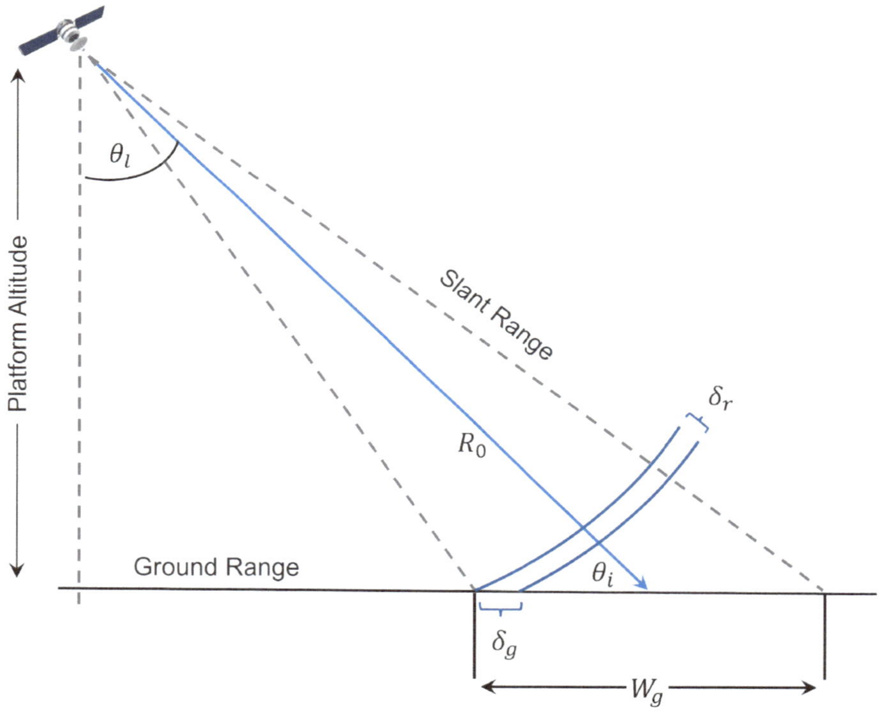

Figure 2.2 Ground swath, ground-range resolution, and slant-range resolution for stripmap SAR.

2.2.1 Resolution

The resolution along the slant range direction is causally related to the waveform bandwidth as

$$\delta_r = \frac{c}{2B} \qquad \text{(m)}, \tag{2.2}$$

where

c = speed of light (m/s),
B = waveform bandwidth (Hz),
δ_r = slant-plane range resolution (m).

The expression in (2.2) is the theoretical limit for range resolution. Sampling rates, quantization, and windowing reduce the resolution achieved by the radar. For example, a Hanning window reduces the resolution by a factor of approximately 1.6 [11]. Referring to Figure 2.2, the resolution in the ground plane is related to the resolution in the slant plane and the incident angle, θ_i, as

$$\delta_g = \frac{c}{2B\cos(\theta_i)} \qquad \text{(m)}. \tag{2.3}$$

The cross-range direction is often referred to as the azimuth direction, and the cross-range resolution is based on the physical size of the antenna, D_a, in that direction. This is simply expressed as

$$\delta_a = \frac{D_a}{2} \qquad \text{(m)}. \tag{2.4}$$

From (2.4), cross-range resolution is independent of range and smaller antennas give finer cross-range resolution. While this may seem counterintuitive at first, a smaller physical antenna produces a larger beamwidth. This leads to longer illumination times and a larger effective synthetic length, and therefore a finer cross-range resolution. The expression in (2.4) assumes the synthetic aperture covers the support band of the targets being imaged, as shown in Figure 2.3. The support band is given by

$$b \in \left[-\frac{L_{span}}{2} - \frac{R_0\,\lambda}{2D_a}, \frac{L_{span}}{2} + \frac{R_0\,\lambda}{2D_a} \right] \qquad \text{(m)}, \tag{2.5}$$

where

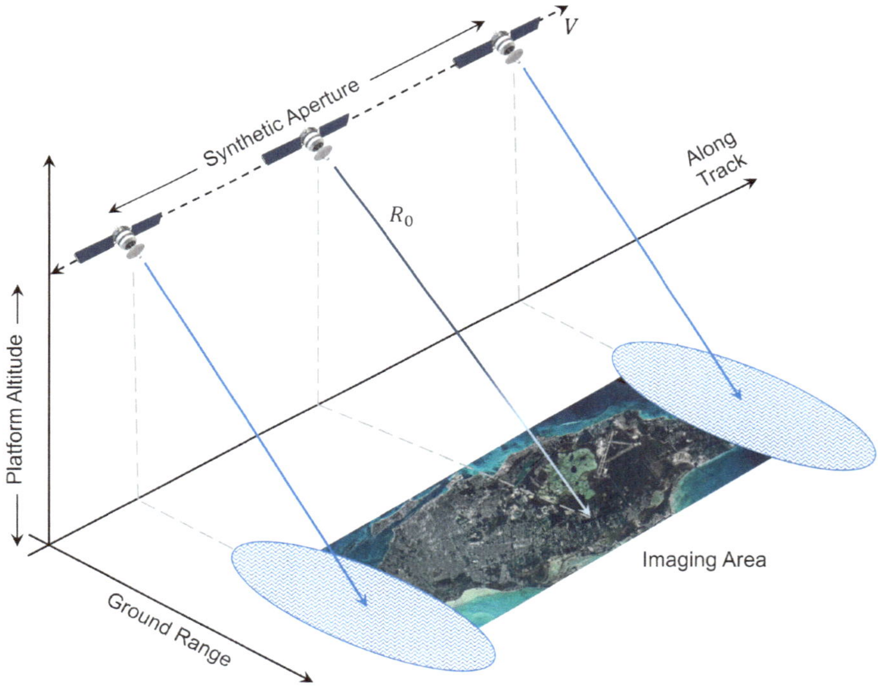

Figure 2.3 Stripmap imaging geometry and support band.

L_{span} = span of the imaged area in cross-range (m),
λ = wavelength (m),
b = support band of the area to be imaged (m).

The support band is a subset of the synthetic aperture interval $\left[-L_{sa}/2, L_{sa}/2\right]$, where L_{sa} is the length of the synthetic aperture. For example, consider the scenario depicted in Figure 1.3, consisting of a single point scatterer. Since the span of a point target is 0, the support band is

$$b \in \left[-\frac{R_0\,\lambda}{2D_a}, \frac{R_0\,\lambda}{2D_a}\right] \qquad \text{(m)}, \tag{2.6}$$

and the synthetic aperture must encompass this support band as

$$L_{sa} > \frac{R_0\,\lambda}{D_a} \qquad \text{(m)}. \tag{2.7}$$

The expressions in (2.2) and (2.4) are based on approximating the target spectral support as a rectangular region as illustrated in Figure 2.4. The region is described by the span

$$\Delta k_x = 2\big(k_{max} - k_{min}\big) \qquad \text{(rad/m)}, \tag{2.8}$$

$$\Delta k_y = \frac{4\pi}{D_a} \qquad \text{(rad/m)}. \tag{2.9}$$

The support region represents the bandwidth of the target function and does not provide information about the variations in the underlying target function. This approximation results in a point spread function that is a two-dimensional sinc function and is expressed as

$$psf = \text{sinc}\left(\frac{\Delta k_x\, x}{2\pi}\right)\text{sinc}\left(\frac{\Delta k_y\, y}{2\pi}\right) \tag{2.10}$$

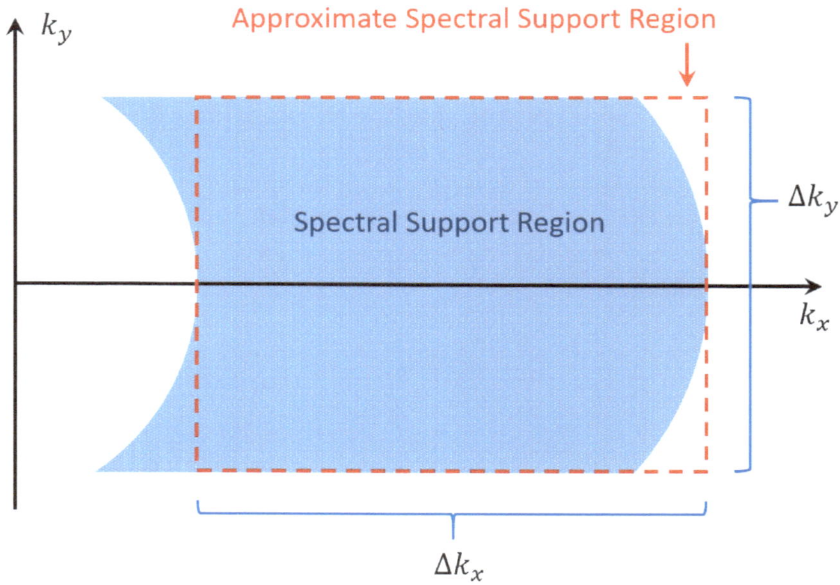

Figure 2.4 Stripmap imaging spectral support for a point target.

where the sinc function used in this book is defined as

$$\text{sinc}(x) = \begin{cases} 1 & \text{for } x = 0 \\ \dfrac{\sin(\pi x)}{\pi x} & \text{otherwise.} \end{cases} \tag{2.11}$$

The resulting range and cross-range resolutions are given by

$$\delta_x = \frac{2\pi}{\Delta k_x} \qquad \text{(m)}, \tag{2.12}$$

$$\delta_y = \frac{2\pi}{\Delta k_y} \qquad \text{(m)}. \tag{2.13}$$

Substituting (2.8) and (2.9) into (2.12) and (2.13) leads to

$$\delta_x = \frac{c}{2B} \qquad \text{(m)}, \tag{2.14}$$

$$\delta_y = \frac{D_a}{2} \qquad \text{(m)}, \tag{2.15}$$

which is consistent with (2.2) and (2.4). This is a good approximation when the bandwidth is much smaller than the radar center frequency. However, the approximation does not hold for wide bandwidth systems where the bandwidth is comparable to the radar center frequency. For those systems, the point spread function is shift-varying and its shape resembles that of a funnel. Figure 2.5 illustrates the point spread function and associated resolution for stripmap SAR.

2.2.2 Pulse Repetition Frequency

As the SAR platform moves along the flight path with a velocity, v, the radar transmits and receives pulses at a rate known as the pulse repetition frequency (PRF). To prevent aliasing of higher frequencies back into lower frequencies, the PRF must be large enough to cover the signal Doppler bandwidth. Referring to Figure 2.6, the Doppler frequency shift for a stationary target is given by

$$f_D = \frac{2v}{\lambda}\sin(\theta_t) \qquad \text{(Hz)}. \tag{2.16}$$

The maximum Doppler shift occurs at the antenna half beamwidth, $\theta_a/2$. Substituting into (2.16) gives the Doppler bandwidth, which is written as

$$B_D = 2f_D = \frac{4v}{\lambda}\sin\left(\frac{\theta_a}{2}\right) \qquad \text{(Hz)}. \tag{2.17}$$

Substituting $\theta_a \approx \lambda/D_a$ and using the small angle approximation, $\sin\theta_a \approx \theta_a$, gives the required PRF as

$$PRF \geq B_D \approx \frac{2v}{D_a} \qquad \text{(Hz)}. \tag{2.18}$$

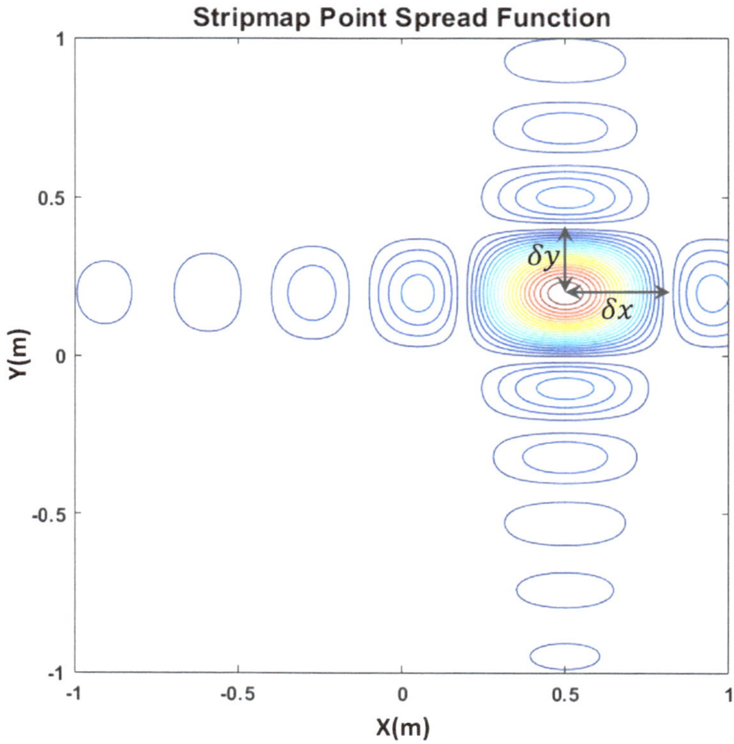

Figure 2.5 Point spread function for stripmap SAR with a target location of (0.5, 0.25) meters.

Rearranging (2.18) in terms of the pulse repetition interval (PRI) gives the minimum spatial sampling requirement of

$$v \times PRI \leq \frac{D_a}{2} \qquad \text{(m)}. \tag{2.19}$$

This means at least one pulse transmission and reception is required when the platform has moved by half of the physical antenna azimuth dimension. While decreasing the size of the antenna increases cross-range resolution, it increases

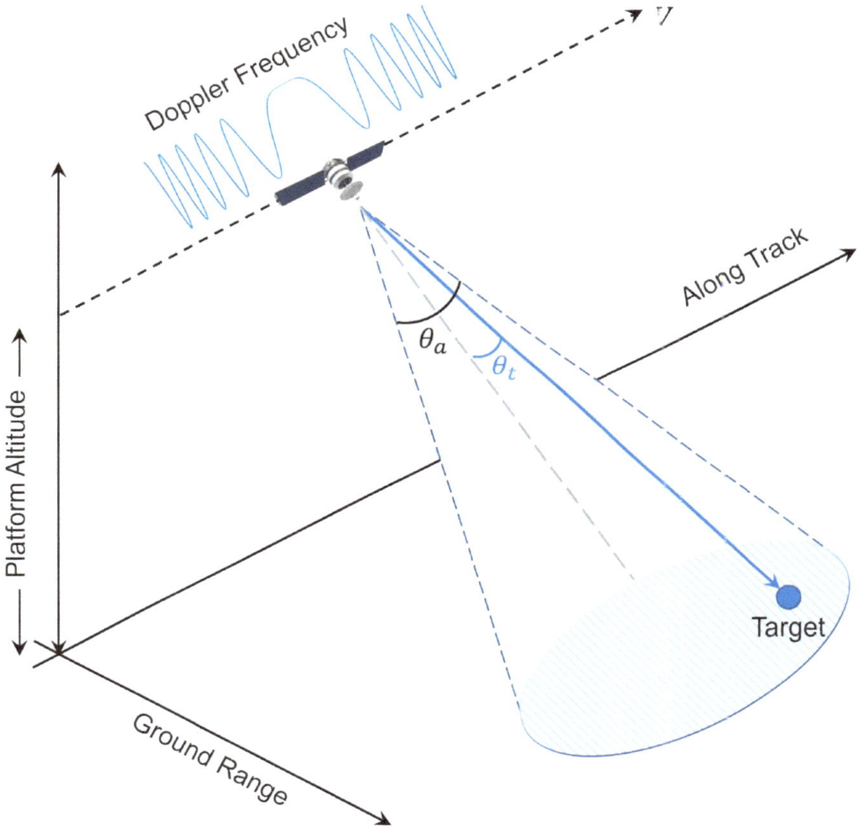

Figure 2.6 Doppler frequency and bandwidth for a stationary target.

the required PRF which in turn increases system complexity and cost. Another constraint on the PRF is the unambiguous range, illustrated in Figure 2.7, and is expressed as

$$R_u \leq \frac{c(PRI - \tau)}{2} \qquad \text{(m)}, \tag{2.20}$$

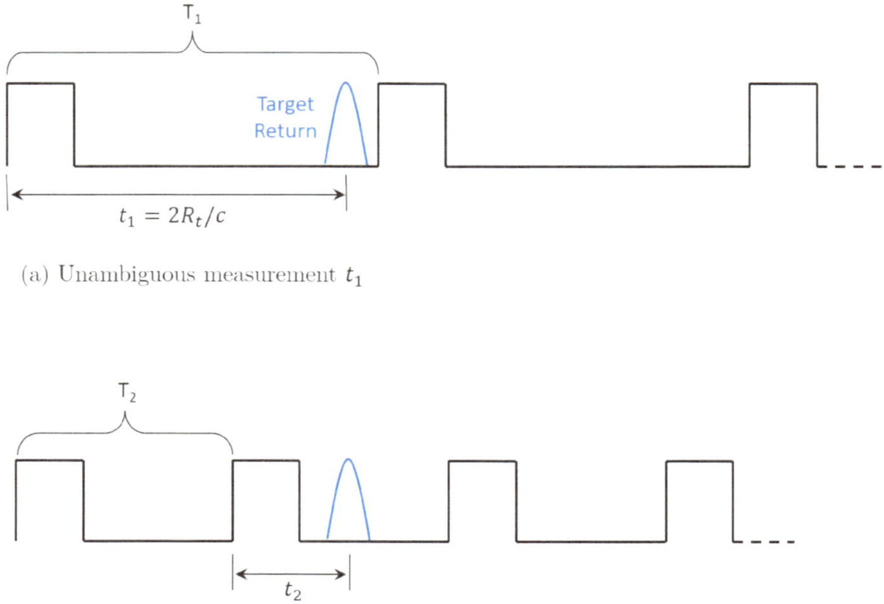

(a) Unambiguous measurement t_1

(b) Ambiguous measurement t_2

Figure 2.7 Ambiguous vs unambiguous radar range measurements.

where τ is the length of the transmitted pulse in seconds. Referring to Figure 2.7, suppose the SAR system transmits a pulse with a PRI of T_1 seconds. If the round-trip time to the target is less than the PRI, the target return signal arrives before the next pulse has been transmitted. Therefore, the measured time, t_1, is associated with the first pulse and the range measurement is unambiguous, as illustrated in Figure 2.7(a). However, if the round-trip time to the target is greater than the PRI, the target return signal arrives after the next pulse has been transmitted. This creates an ambiguity as to whether the target return signal was due to the first or second pulse, as shown in Figure 2.7(b).

Watch this animation of unambiguous range measurements.

If the pulsewidth is small compared to the PRI, it may be ignored, and the upper and lower limits for the PRF are written as

$$\frac{2v}{D_a} \le PRF \le \frac{c}{2R_u} \qquad \text{(Hz)}. \tag{2.21}$$

This may be written in terms of the ground swath as

$$\frac{2v}{D_a} \le PRF \le \frac{c}{2W_g \cos(\theta_i)} \qquad \text{(Hz)}. \tag{2.22}$$

The left-hand side of the inequality in (2.22) is proportional to the cross-range resolution. Therefore, increasing the width of the ground swath, W_g, may require a decreased cross-range resolution. While this is rarely an issue for airborne SAR, this could become problematic for spaceborne systems. Rewriting (2.22) for the maximum mapping rate, vW_g, gives

$$vW_g \le \frac{D_a}{2}\left(\frac{c}{2\cos(\theta_i)}\right) \qquad (\text{m}^2/\text{s}). \tag{2.23}$$

This illustrates the conflict between mapping rate and cross-range resolution as a higher mapping rate requires a wider antenna, which results in a degradation in resolution.

2.2.3 Signal-to-Noise Ratio

The signal-to-noise ratio (SNR) for a single point scatterer in a SAR scene may be written in terms of the cross-range resolution as [2, 12, 13]

$$SNR = \frac{P_{avg}\, A_e^2\, \sigma}{8\pi\, \lambda\, R^3\, \delta_a\, v\, k\, T_0\, F\, L}, \tag{2.24}$$

where

P_{avg} = average transmitted power (W),
A_e = antenna effective aperture (m^2),
σ = point target radar cross section (m^2),
λ = wavelength at center frequency (m),
R = target range (m),
δ_a = cross-range resolution (m),
v = platform velocity (m/s),
k = Boltzmann constant (J/K),
T_0 = reference temperature (K),
F = noise figure,
L = system losses.

For distributed targets, such as clutter, the radar cross section may be written in terms of the range and cross-range resolutions as

$$\sigma = \sigma_0 \, \delta_a \, \delta_r \qquad (\text{m}^2), \tag{2.25}$$

where σ_0 is the backscattering coefficient for the distributed target. Substituting (2.25) into (2.24) gives

$$SNR = \frac{P_{avg} \, A_e^2 \, \sigma_0 \, \delta_r}{8\pi \, \lambda \, R^3 \, v \, k \, T_0 \, F \, L}. \tag{2.26}$$

The expression in (2.26) is often referred to as the clutter-to-noise ratio (CNR) when the distributed target is clutter. The SNR for point targets increases with finer cross-range resolution and is independent of range resolution. The SNR for distributed targets decreases with finer range resolution and is independent of cross-range resolution. In general, an SNR greater than 3 dB is required for acceptable imagery products.

2.2.4 Noise Equivalent Sigma Zero

The sensitivity of SAR systems is often given in terms of the noise-equivalent sigma zero (NESZ), which is a measure of the backscattering coefficient that produces an SNR of one. NESZ is written as

$$NESZ = \frac{8\pi \, \lambda \, R^3 \, v \, k \, T_0 \, F \, L}{P_{avg} \, A_e^2 \, \delta_r}, \tag{2.27}$$

and captures the effect of system noise on image quality. NESZ may be predicted analytically from system parameters and is commonly used to drive system design. By taking measurements over dark areas of a SAR image, such as calm waters, NESZ may be characterized. Lower NESZ values are required for imaging scenes where targets with low-backscattering intensity need to be detected. NESZ is also a function of range resolution. Wider bandwidths lead to better range resolution, but results in more noise in the imagery. NESZ is covered in more detail in Section 7.2.

2.2.5 Squint Angle

A flexible imaging technique used with stripmap systems is to point the antenna beam ahead of or behind the platform, as illustrated in Figure 2.8, where θ_s is the *squint angle*. This allows for timely imaging without the need for making a platform course change. While the Doppler centroid is ideally equal to zero for the broadside case of Figure 2.6, squinting the antenna beam creates a shift in the spectrum of the return signal, as shown in Figure 2.9. A similar Doppler shift is also present in the case of a spaceborne SAR over a rotating Earth [14]. Errors in the estimate of the Doppler centroid degrade SNR, sidelobe levels, and ambiguity levels. While the Doppler centroid could be estimated from the flight path and platform attitude data, measurement uncertainties limit the practical use of this approach. Alternatively, time-domain and frequency-domain techniques have been developed to estimate the Doppler centroid from the received signals [14–17].

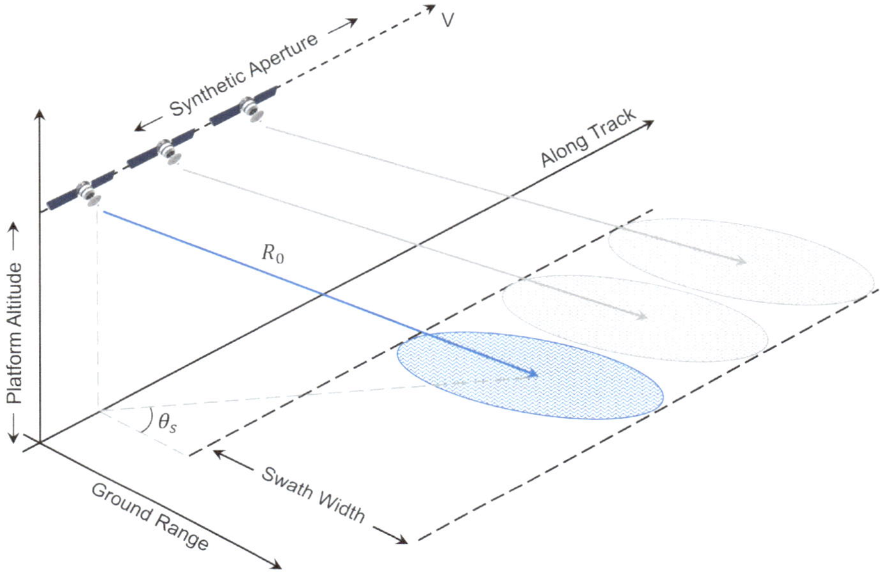

Figure 2.8 Stripmap SAR squint angle geometry.

2.3 SPOTLIGHT

Unlike stripmap SAR, where the area illuminated by the radar beam moves with the flight path, spotlight SAR uses mechanical or electronic beam steering to keep the target area illuminated for a longer period, as illustrated in Figure 2.10 [18, 19]. While this reduces the scene size that may be imaged, it leads to higher resolution images than stripmap imaging. The scene size is often restricted to a defined size (e.g., 5 km $\times$ 5 km). While spotlight-mode SAR is more contemporary than stripmap, that does not mean it is a better choice in all situations. When imaging large areas for reconnaissance or mapping purposes, stripmap SAR is often a better choice. Stripmap and spotlight SAR may also be used in conjunction with one another. For example, stripmap SAR may be used to image a large area and identify areas of interest which are then imaged with spotlight SAR to provide higher quality images [8]. As spotlight and stripmap SAR are different imaging modes, their

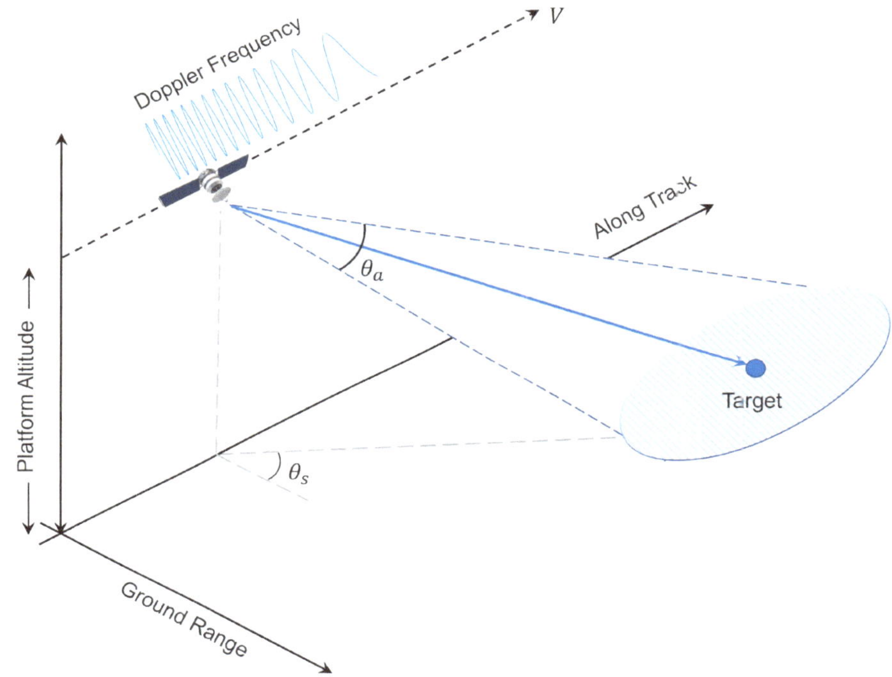

Figure 2.9 Doppler frequency and bandwidth for squint angle operation.

data collection and image reconstruction methods may also differ depending on the application.

Watch this animation of spotlight SAR imaging.

2.3.1 Resolution

For spotlight SAR imaging, the resolution along the slant range and ground range direction is the same as stripmap mode, which was given in (2.2) and (2.3).

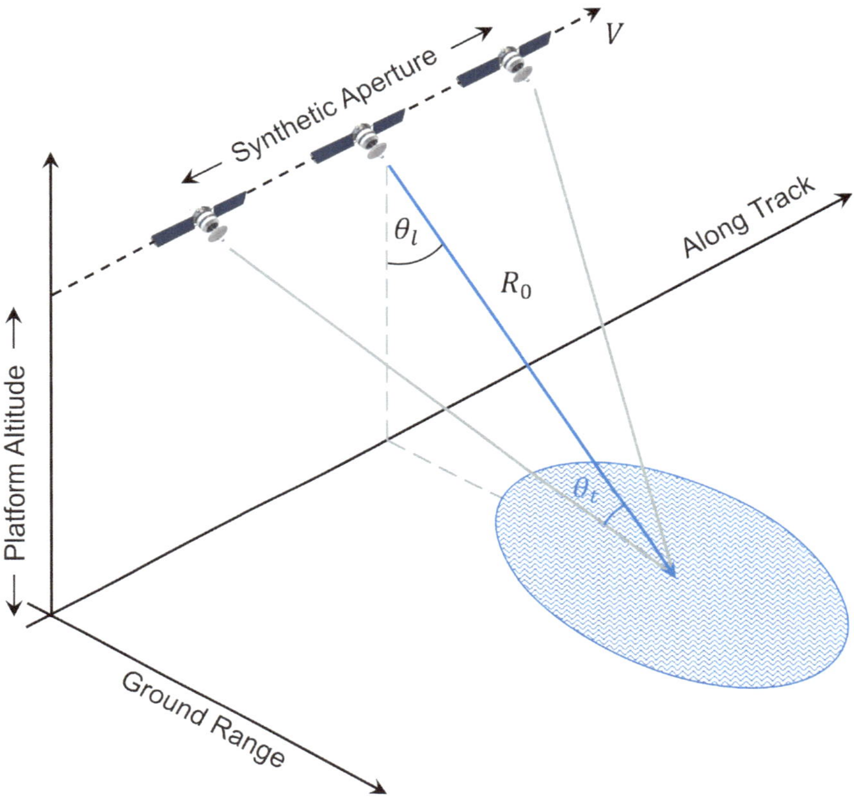

Figure 2.10 Basic spotlight SAR imaging geometry.

The cross-range resolution is related to the angle subtended by the synthetic aperture, and is expressed as

$$\delta_a = \frac{\lambda}{4\sin\left(\dfrac{\theta_{sa}}{2}\right)} \qquad \text{(m)}, \tag{2.28}$$

where θ_{sa} is the angle subtended by the synthetic aperture. Substituting

$$\sin\left(\frac{\theta_{sa}}{2}\right) = \frac{L_s}{2R_t} \tag{2.29}$$

into (2.28) gives

$$\delta_a = \frac{\lambda R_t}{2L_s} \qquad \text{(m)}, \tag{2.30}$$

where R_t is the range to the target. Often, the small angle approximation of $\sin\theta_{sa} \approx \theta_{sa}$ is used in (2.29) to write the cross-range resolution as

$$\delta_a \approx \frac{\lambda}{2\theta_{sa}} \qquad \text{(m)}. \tag{2.31}$$

Similar to the stripmap case, the expressions in (2.30) and (2.31) are based on approximating the target spectral support as a rectangular region, as shown in Figure 2.11. For the spotlight case

$$\Delta k_x' = 2\big(k_{max} - k_{min}\big) \qquad \text{(rad/m)}, \tag{2.32}$$

$$\Delta k_y' = 2k_c\big[\sin(\theta_t^+) - \sin(\theta_t^-)\big] \qquad \text{(rad/m)}, \tag{2.33}$$

where $\theta_t^{\pm}$ is the angle to the point target from the endpoints of the synthetic aperture, and k_c is the wavenumber at the center frequency. The point spread function is written as

$$psf = \text{sinc}\left(\frac{\Delta k_x'\, x'}{2\pi}\right)\text{sinc}\left(\frac{\Delta k_y'\, y'}{2\pi}\right). \tag{2.34}$$

The resulting range and cross-range resolutions are calculated by substituting (2.32) and (2.33) into (2.12) and (2.13) which gives

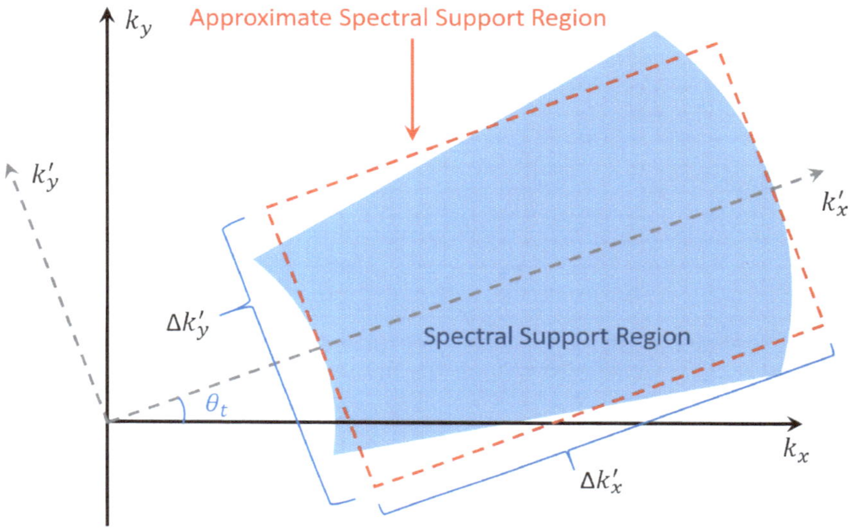

Figure 2.11 Spotlight imaging spectral support for a point target located at θ_t.

$$\delta x' = \frac{c}{2B} \qquad \text{(m)}, \tag{2.35}$$

$$\delta y' = \frac{\lambda\, R_t}{2L_s \cos(\theta_t)} \qquad \text{(m)}. \tag{2.36}$$

From (2.34) and (2.36), the point spread function and resolutions are dependent on the position of the target (i.e., spatially varying). If the targets are located such that the angular variation between their positions is small, then (2.35) and (2.36) are consistent with (2.2) and (2.30).

2.3.2 Pulse Repetition Frequency

Referring to Figure 2.10, the Doppler shift for a stationary target is given by

$$f_D = \frac{2v}{\lambda}\sin(\theta_t) \qquad \text{(Hz)}. \tag{2.37}$$

The maximum Doppler shift occurs at the largest aspect angle, θ_{max}. Substituting into (2.37) gives the Doppler bandwidth, which is written as

$$B_D = 2f_D = \frac{4v}{\lambda}\sin(\theta_{max}) \qquad \text{(Hz)}. \tag{2.38}$$

Substituting $\theta_{max} \approx L_{sa}/2R_0$ and using the small angle approximation, $\sin\theta_{max} \approx \theta_{max}$, gives the required PRF as

$$PRF \geq B_D = \frac{2vL_{sa}}{\lambda\, R_0} \qquad \text{(Hz)}. \tag{2.39}$$

Rearranging (2.39) in terms of the PRI gives the minimum spatial sampling requirement of

$$v \times PRI \leq \frac{\lambda\, R_0}{2L_{sa}} \qquad \text{(m)}. \tag{2.40}$$

The PRF for spotlight SAR is also bound by the unambiguous range as given in (2.20). The PRF may then be expressed in terms of upper and lower limits as

$$\frac{2vL_{sa}}{\lambda\, R_0} \leq PRF \leq \frac{c}{2R_u} \qquad \text{(Hz)}. \tag{2.41}$$

2.3.3 SNR

To calculate the SNR for spotlight SAR imaging, begin with the radar range equation, which is given by [2]

$$SNR = \frac{P_{avg}\, A_e^2\, \sigma}{4\pi\, \lambda^2\, R^4\, k\, T_0\, B\, F\, L}, \tag{2.42}$$

where

P_{avg}	=	average transmitted power (W),
A_e	=	antenna effective aperture (m^2),
σ	=	point target radar cross section (m^2),
λ	=	wavelength (m),
R	=	target range (m),
k	=	Boltzmann constant (J/K),
T_0	=	reference temperature (K),
B	=	receiver bandwidth (Hz),
F	=	noise figure,
L	=	system losses.

The SNR given in (2.42) is at the output of the receiver and does not include processing gains. For spotlight SAR imaging, range and azimuth processing gains are factored in to give an image SNR expressed as

$$SNR_{image} = SNR \times G_{range} \times G_{azimuth}, \tag{2.43}$$

where

SNR	=	SNR at receiver output,
G_{range}	=	gain due to range processing,
$G_{azimuth}$	=	gain due to azimuth processing,
SNR_{image}	=	SNR for the image.

The range processing gain is due to the pulse compression process, whether matched filtering, stretch processing, or other techniques are employed [2]. The compression gain for matched filtering is equal to the time-bandwidth product which is a figure of merit representing the ideal increase in the SNR at the output of the matched filter compared to the input. The time-bandwidth product is given by $B\,\tau$, where B is the bandwidth in Hz, and τ is the pulsewidth in seconds. However, perfect matched filtering is difficult to achieve. Also, windowing is used to reduce range sidelobes, which in turn widens the main lobe and reduces the range processing gain. Including a loss term, the range processing gain is written as

$$G_r = \frac{B\,\tau}{L_r}. \tag{2.44}$$

The azimuth processing gain is a result of coherently integrating the sequence of pulses transmitted and received by the SAR system. Coherent integration would ideally result in an increase in the SNR by a factor of N, the number of pulses. However, integration loss always occurs in practical systems and this ideal value is not achieved. Integration loss may be due to component instability, environmental changes, target and platform motion, and fluctuations in target scattering [2]. Therefore, a loss term is included in the azimuth processing gain, which gives

$$G_a = \frac{N}{L_a}. \tag{2.45}$$

The number of pulses transmitted and received is a function of the synthetic aperture length, the platform velocity, and the PRF. The number of pulses is found from

$$N = T_{sa} \times PRF = \frac{L_{sa}}{v} \times PRF. \tag{2.46}$$

Substituting (2.44), (2.45), and (2.46) into (2.43) allows the SNR for the image to be written as

$$SNR_{image} = \frac{P_{avg}\,A_e^2\,\sigma\,\tau\,L_{sa}\,PRF}{4\pi\,\lambda^2\,r^4\,k\,T_0\,F\,L_{total}\,v}, \tag{2.47}$$

where the loss terms have been combined into a single term, L_{total}.

2.3.4 Squint Angle

Similar to stripmap, spotlight SAR may also point the radar beam ahead of or behind the platform, as illustrated in Figure 2.12. The squint angle also changes the Doppler frequency and therefore the Doppler centroid must be estimated [14–17]. Based on the image formation algorithm, high-squint spotlight SAR may become more complicated due to the range of a scatterer varying by more than the range resolution over the length of the aperture [20, 21]. This is known as range migration and will be discussed in more detail in Chapter 3.

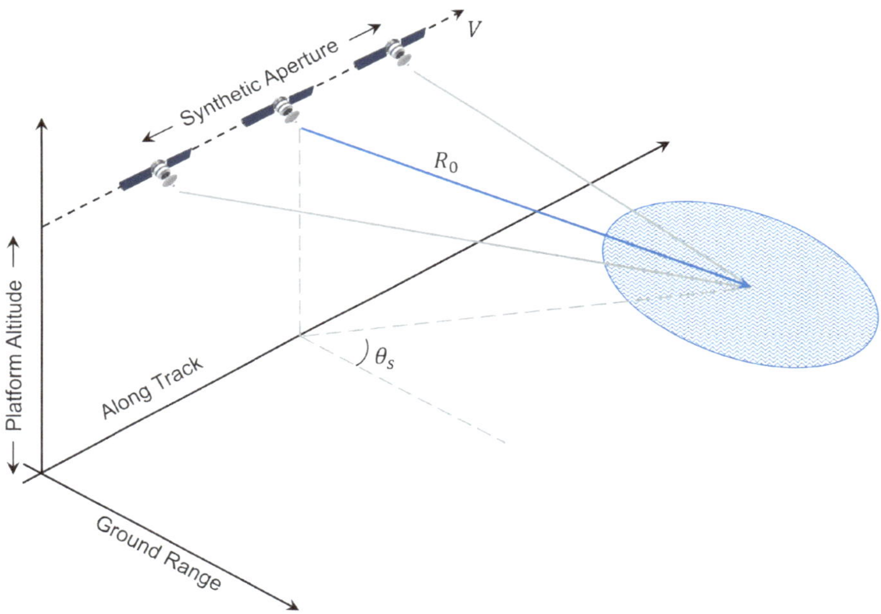

Figure 2.12 Spotlight SAR squint angle geometry.

2.4 SCANNING SYNTHETIC APERTURE RADAR

Scanning synthetic aperture radar (ScanSAR) is an imaging mode designed to image exceptionally large areas by increasing the swath width. This increase in swath width is accomplished by scanning the radar beam in elevation to illuminate multiple subswaths, as illustrated in Figure 2.13. The scanning repeats continuously and image processing techniques are employed to combine the subswaths into a single wide-swath SAR image. The subswaths are illuminated for a shorter period compared to traditional stripmap imaging, which limits the Doppler extent for each target, thus the azimuth resolution is degraded compared to stripmap mode [1]. For example, TerraSAR-X is capable of either four or six adjacent ground subswaths resulting in swath widths as large as 270 km in standard mode and up to 1,500 km in extended mode, while spatial resolution is reduced to 40 m [5]. Also, the shorter illumination times will cause target returns to be weighted differently in the azimuth direction by distinct portions of the antenna pattern. This leads to a repetitive modulation of the amplitude called scalloping. The shorter illumination time also results in varying azimuth resolution and SNR. The contribution of point targets with different azimuth locations is shown in Figure 2.14. Note the difference in both Doppler frequency and amplitude. Depending on the application, the effects of these artifacts may be reduced with various techniques such as multilook processing [22, 23].

Watch this animation of ScanSAR imaging.

2.5 TERRAIN OBSERVATION BY PROGRESSIVE SCAN

Terrain observation by progressive scan (TOPSAR) is an imaging mode designed to replace ScanSAR. TOPSAR is similar to ScanSAR in that the radar beam is scanned in elevation. However, the radar beam is also scanned in the azimuth direction for each subswath, as shown in Figure 2.15. TOPSAR achieves the same coverage size and image resolution as ScanSAR, but with more consistent image quality and nearly uniform SNR. The scanning of the radar beam in the azimuth direction results in variations in the Doppler centroid. These variations result in errors in registration

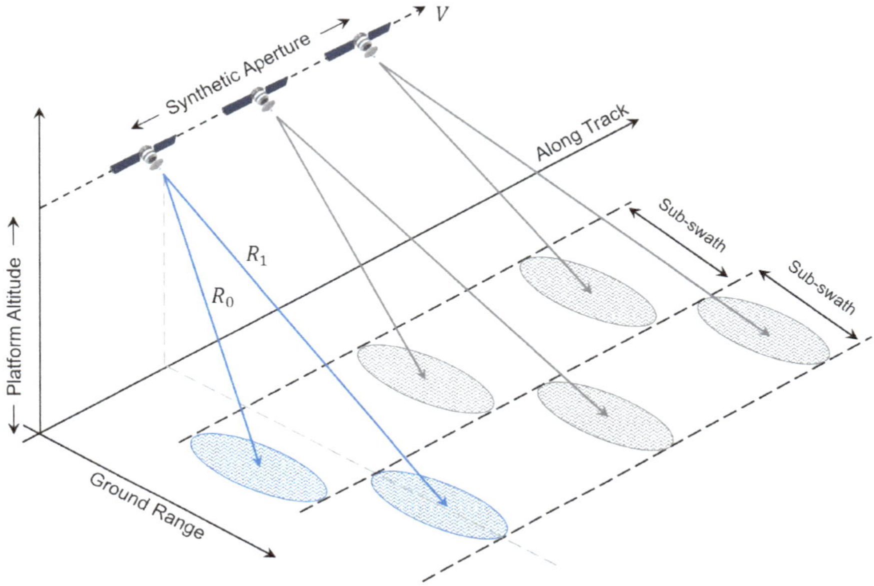

Figure 2.13 ScanSAR imaging geometry with two subswaths.

of the subswaths and must be accounted for in data processing [24].

Watch this animation of TOPSAR imaging.

2.6 SWEEPSAR

SweepSAR is an imaging mode designed to simultaneously provide very large swath widths and fine resolution while operating at longer wavelengths than ScanSAR or TOPSAR. This is accomplished by transmitting energy in a single radar beam and then receiving energy through a multibeam feed network. The basic imaging geometry for SweepSAR is illustrated in Figure 2.16. Using an analog feed network on receive is simpler to implement and requires lower data rates than digital techniques. The feed network is designed to steer the receiving beams in elevation.

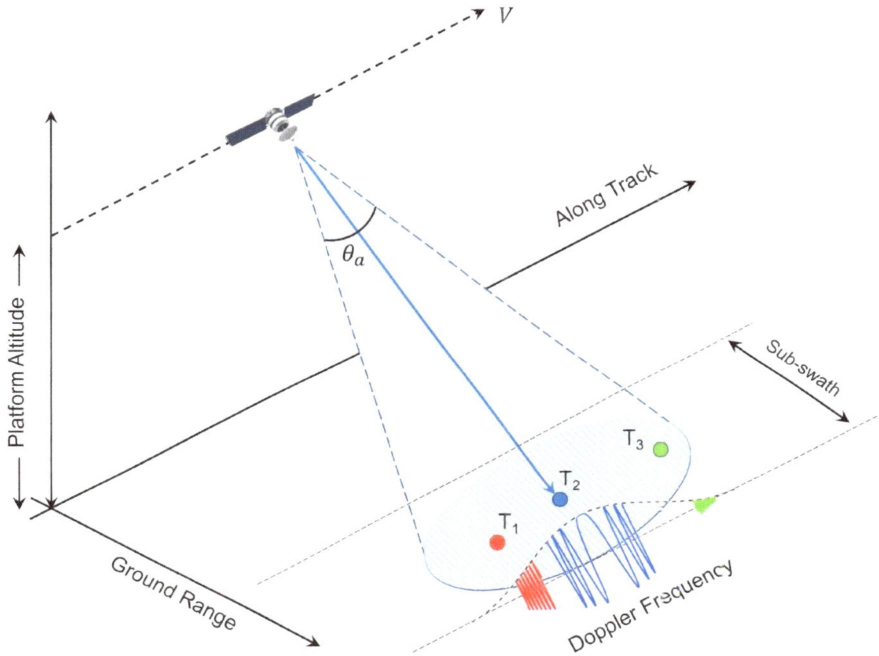

Figure 2.14 ScanSAR Doppler frequencies and amplitudes.

While the SweepSAR mode offers advantages in swath width and resolution over ScanSAR, a disadvantage of this method is a higher data rate and an increased number of receivers. Another potential advantage is lower power consumption when using a reflector style antenna [25]. This limits flexibility over phased array antenna systems. The NASA-ISRO Synthetic Aperture Radar (NISAR) L-band SAR instrument was the first spaceborne sensor to employ SweepSAR [26, 27].

Watch this animation of SweepSAR imaging.

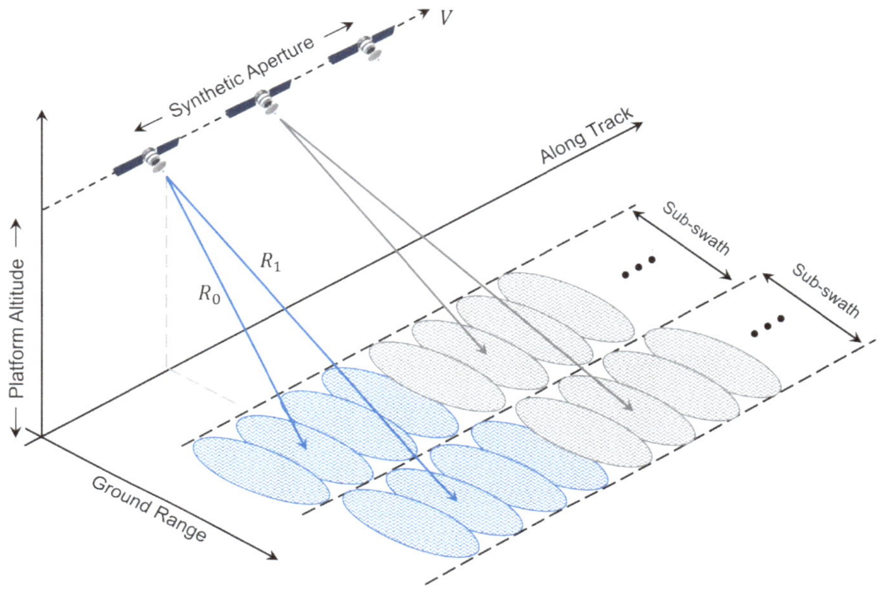

Figure 2.15 TOPSAR imaging geometry with two subswaths.

2.7 INSAR

InSAR is an imaging mode which uses two SAR images created at slightly different viewing angles to obtain very precise measurements of surface topology. This is accomplished by creating images on multiple passes along slightly different trajectories or in a single pass with displaced phase center antennas (DCPA), as shown in Figure 2.17. The phase of a radar return signal depends on the range to the scene. By taking the phase difference of the SAR images, InSAR can detect surface changes on the order of centimeters. This is extremely useful for creating digital elevation models, monitoring surface deformations due to earthquakes, and predicting volcanic eruptions [28–30]. Figure 2.18 shows a comparison of the amplitude only SAR image versus InSAR image of Kilauea volcano in Hawaii [31]. The interference fringes clearly highlight the elevation changes in the crater as well as the summit.

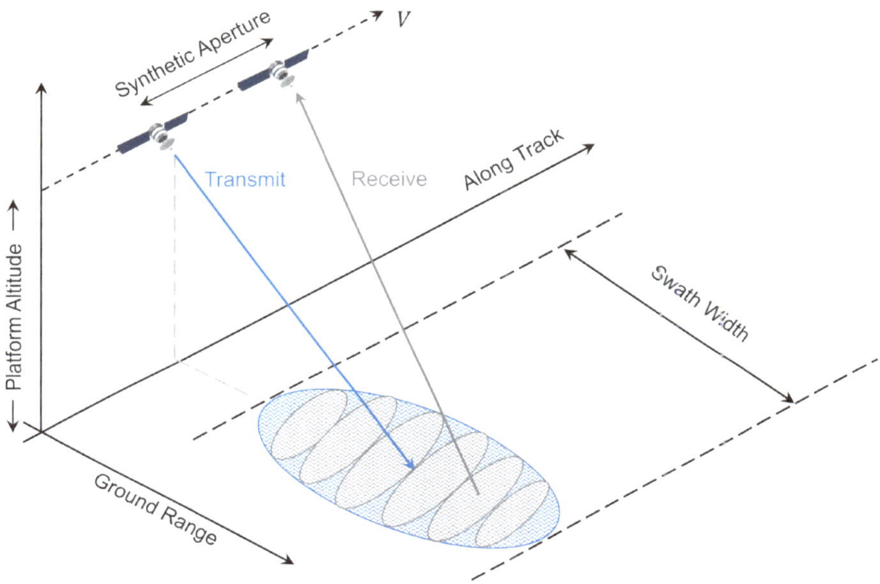

Figure 2.16 SweepSAR imaging geometry.

2.8 POLARIMETRIC SAR

Polarimetric SAR enhances the capability of SAR imaging by transmitting and receiving energy in multiple polarizations. Polarization refers to the orientation of the electric field of an electromagnetic wave. Since the antenna is the mechanism by which the radar radiates energy, it is the antenna design that determines the polarization of the transmitted electromagnetic wave [2]. Most spaceborne SAR systems operate in linear polarization, which may be horizontal or vertical. Circular polarization may be achieved by transmitting horizontal and vertical polarized signals phase shifted by an odd multiple of $\pi/2$ (see the Appendix). The energy received in different polarizations is a function of the electrical and geometrical properties of the target area. For example, rough surfaces tend to have a depolarizing effect on the transmitting energy, while smooth surfaces do not. For soils, the ratio of the horizontal to vertical polarization is an indicator of moisture. Figure 2.19 shows a

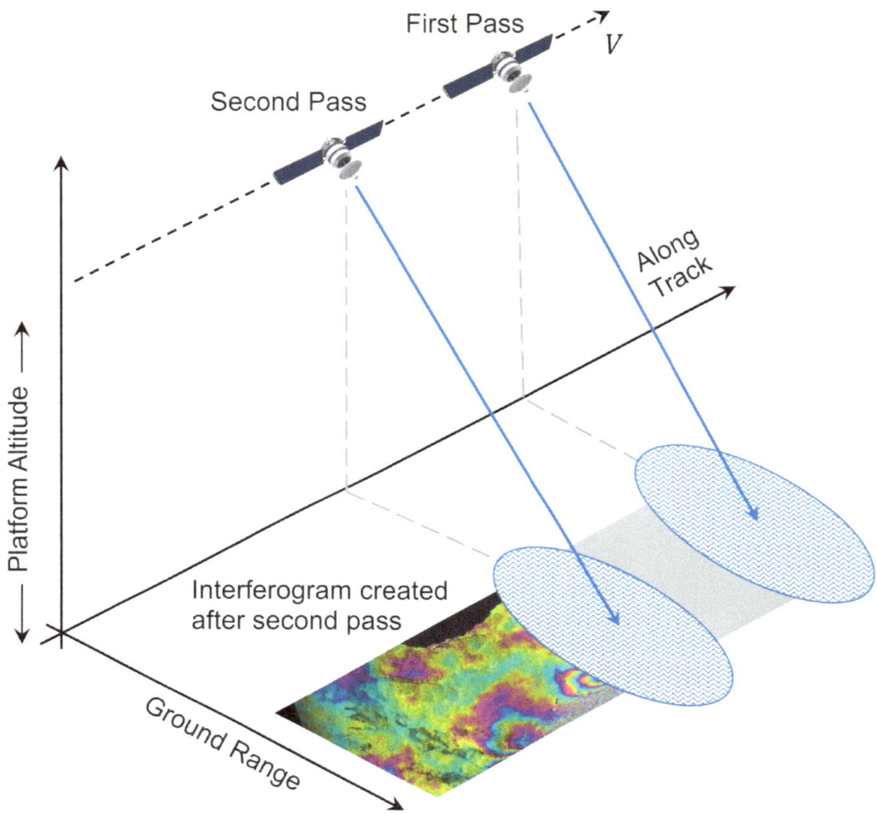

Figure 2.17 Interferometric SAR imaging geometry.

comparison between optical and polarimetric SAR images of the Kliuchevskoi volcano eruption in Kamchatka, Russia [32]. In this figure, red represents L-band HH, green is L-band HV, and blue is C-band HV. Another useful technique is to transmit circular polarization and receive both horizontal and vertical polarization. This is often referred to as compact-pol. Compact-pol offers better power balance between receive channels while keeping the discrimination ability of dual polarization [33].

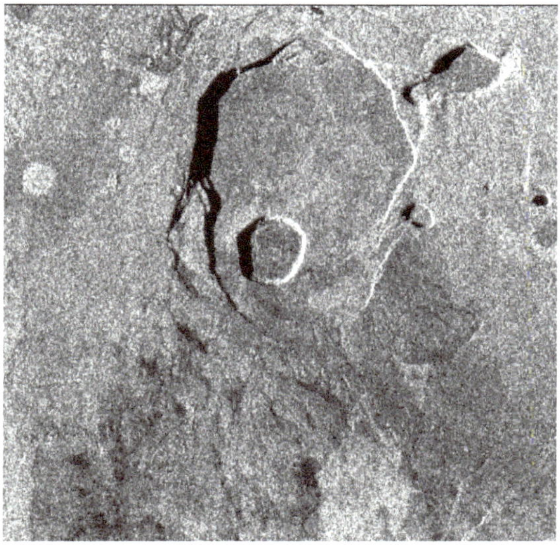

(a)

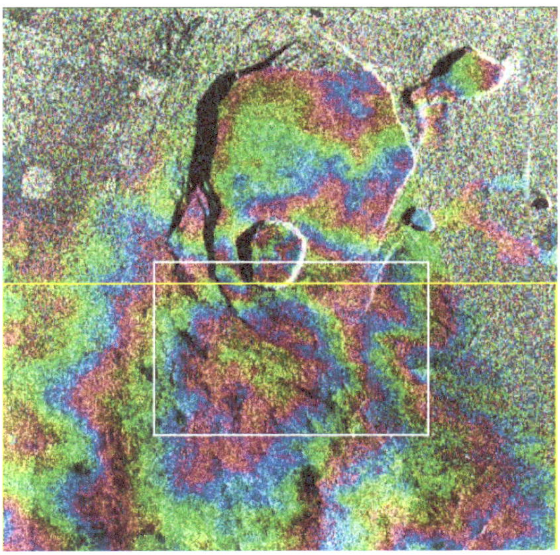

(b)

Figure 2.18 Comparison of (a) amplitude only and (b) interferometric SAR images. Courtesy NASA/JPL.

(a)

(b)

Figure 2.19 Comparison of (a) optical and (b) polarimetric SAR images. Courtesy NASA/JPL.

2.9 BISTATIC AND MULTISTATIC SAR

Bistatic and multistatic SAR systems have transmitting and receiving hardware that is spatially separated. Figure 2.20 presents a simplified diagram for a bistatic SAR system. This configuration has the potential to be less susceptible to jamming and physical attacks as the receiver can be completely passive and more difficult to detect and may use illuminators of opportunity. The diversity in the receiving angle of bistatic systems is advantageous when imaging target areas which have low monostatic backscatter. While beyond the scope of this book, the point spread function and corresponding spatial resolution for bistatic SAR imaging may be found and the reader is referred to [34].

Figure 2.21 illustrates a generic configuration for multistatic SAR imaging. There may be a single transmitting sensor and multiple receiving sensors, or each sensor may be capable of transmitting and receiving. By using orthogonal waveforms, all sensors may simultaneously transmit and then separate the returns on receive, as shown in Figure 2.22. By providing simultaneous SAR images from different viewing geometries, detection, classification, and discrimination of target areas is improved [35–37]. Multistatic SAR systems enable precise target tracking, cross-track and along-track interferometry, increased resolution, and three-dimensional imaging [38–40]. Multistatic experiments with TerraSAR-X and TanDEM-X show the need for more research and development of simulations and predictions for this imaging mode [41]. While these systems have unique challenges to overcome, there is great potential for innovative imaging techniques and products.

Watch this animation of bistatic SAR imaging.

2.10 ISAR

ISAR is an imaging mode that takes advantage of the motion of the target rather than the platform motion. ISAR is analogous to spotlight SAR imaging as shown in Figure 2.23. The difference being spotlight SAR uses the platform motion to collect data over the integration angle θ_I, while ISAR uses the target's motion to create the synthetic aperture. ISAR methods have seen extensive use in the classification

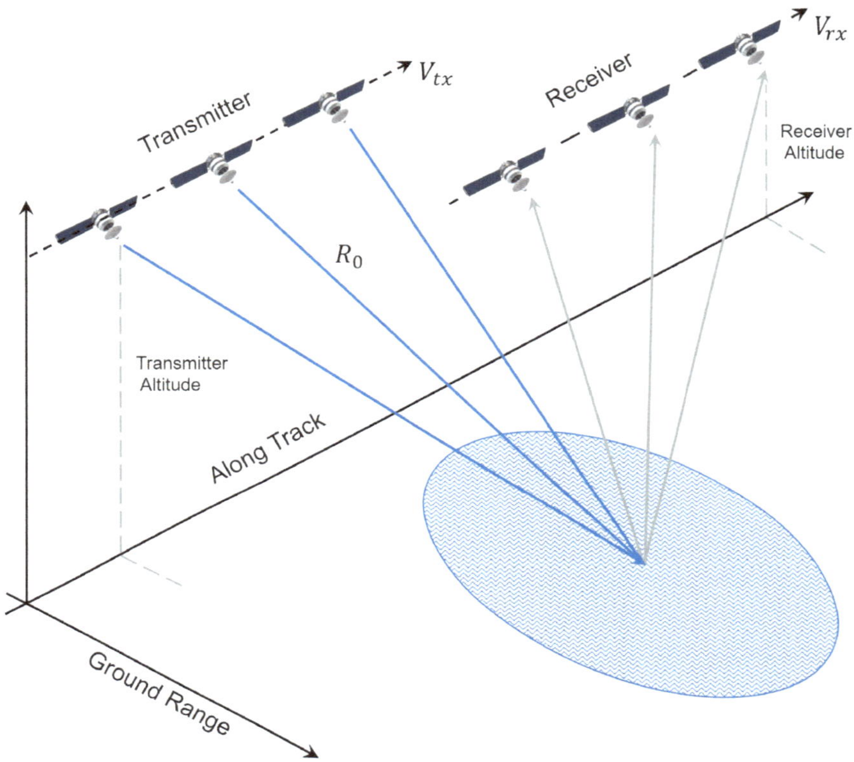

Figure 2.20 Bistatic imaging geometry with a spatially separated transmitter and receiver.

of ships and other maritime crafts [42, 43]. Inverse imaging techniques have also been used in the imaging of deep space objects such as asteroids [44]. To generate the synthetic aperture shown in Figure 2.23 requires the target motion to have a rotational component perpendicular to the radar line of sight. In general, targets to be imaged will have both translational and rotational motion. Translational motion is the main source of motion error that must be accounted for as it results in unaligned data, time-varying Doppler shift, and scatterers that traverse resolution bins. The rotational motion also produces time-varying Doppler shifts which depend

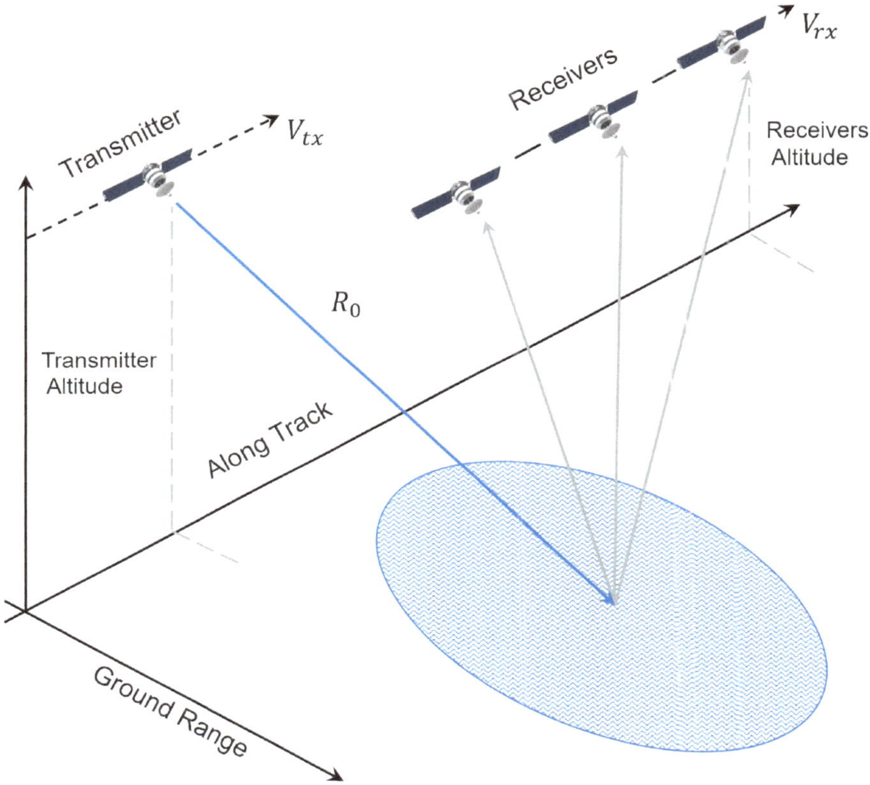

Figure 2.21 Multistatic imaging geometry with spatially separated transmitter and receivers.

on the axis of rotation and speed. This results in the need for a complex motion compensation procedure [45, 46].

2.11 DIGITAL BEAMFORMING

SAR systems can operate in multiple modes and digital beamforming plays a crucial role in this area [47]. Figure 2.24 shows a simplified comparison between analog

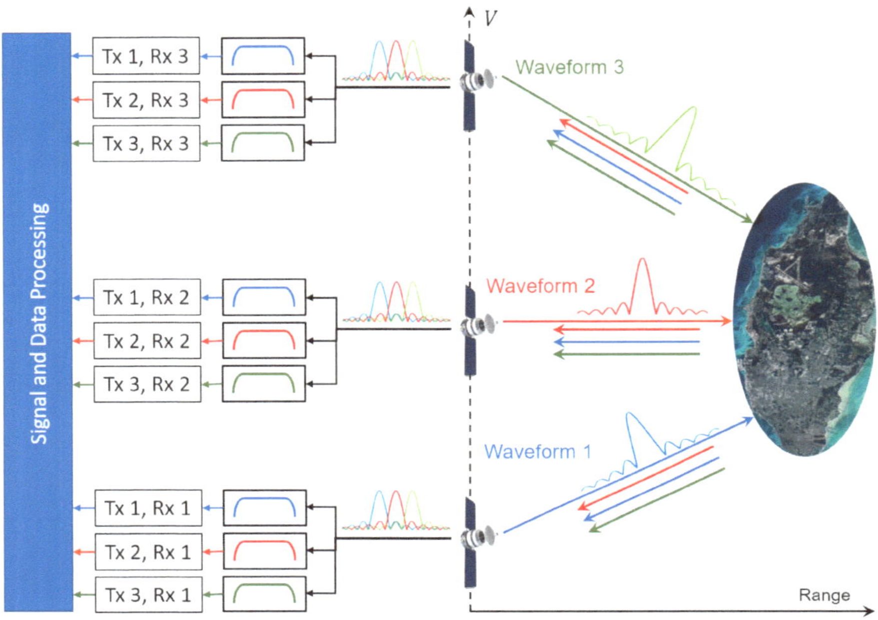

Figure 2.22 Multistatic SAR imaging with orthogonal waveforms.

and digital beamforming. In analog beamforming, the received signal from each antenna element is combined at the radio frequency (RF) level. The formation of beams is accomplished with RF phase shifters and amplifiers associated with each element or subset of elements. The receiver amplifies, filters, downconverts, and digitizes these signals, and delivers those to the radar signal and data processors [2]. In a digital beamforming type architecture, there is a digital receiver at each of the radiating elements or subset of elements. Once the digital signal for each element has been recorded, multiple independent beams are simultaneously formed in software by adding amplitude and phase variations to the digitized signals [48].

Digital beamforming offers many advantages, such as improved dynamic range, simultaneous wide coverage and fine resolution, lower power consumption, and simpler calibration. These advantages come at the expense of increased data rate, increased cost, and complex power budgets. While this places limits on the

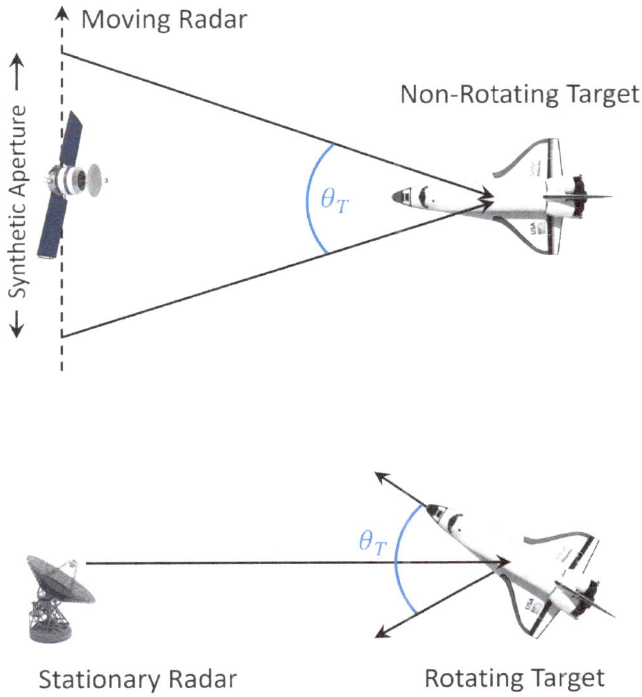

Figure 2.23 Comparison of spotlight SAR to inverse SAR for generating the synthetic aperture with integration angle θ_T. Courtesy NASA.

number of elements in the antenna array, waveform bandwidth, size, and complexity, ongoing research and development are lowering these costs and reducing limitations [47]. Future work in this area offers exciting possibilities for new hybrid SAR imaging modes and ever improving imaging products [49–51].

Most modern SAR systems have multiple operating modes. Tables 2.1–2.4 show the operating modes and capabilities of some modern SAR systems [33, 52–54].

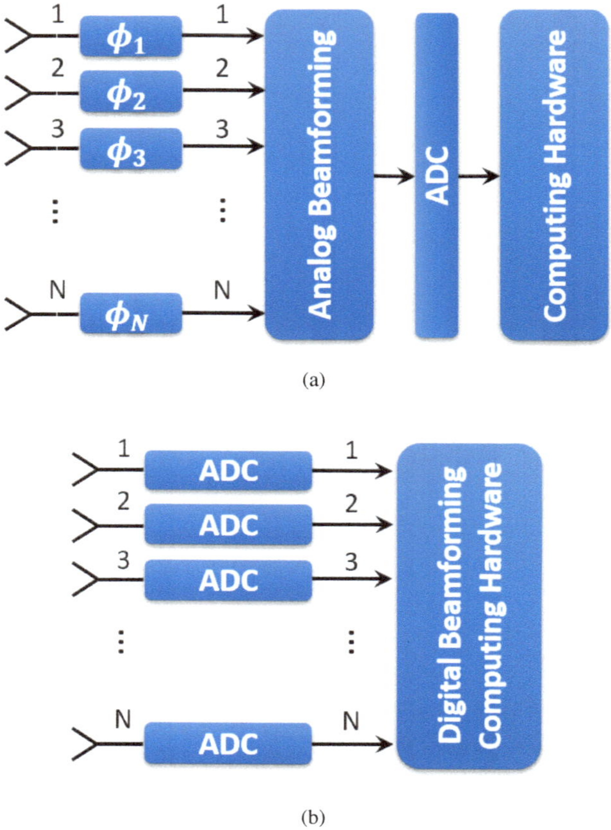

Figure 2.24 Comparison of (a) analog and (b) digital beamforming techniques.

Table 2.1
ICEYE — SAR Imaging Modes

Mode	*Nominal Scene Size (km)*	*Ground-Plane Resolution (m)*	*NESZ (dB)*
Stripmap	30 × 50	3 × 3	−21.5 – −20
Spotlight	5 × 5	1 × 1	−18.5 – −15
SLEA	15 × 15	1 × 1	−18.5 – −15
SCAN	100 × 100	15 × 15	−22.2 – −21.5

Table 2.2
Umbra — SAR Imaging Modes

Mode	*Nominal Scene Size (km)*	*Ground-Plane Resolution (m)*	*Peak Power (W)*
Stripmap	5 to 20 × 50	3 × 3	>500
Spotlight	4 × 4	0.25–2	>500
Extended Dwell	4 × 4	0.25–2	>500
Scanning	Arbitrary Polygon	10 × 10	>500

2.12 EXAMPLES

The sections below illustrate the concepts of this chapter with several Python/MATLAB examples. The examples for this chapter are in the directory *software/python/Chapter2* and the matching MATLAB examples are in the directory *software/matlab/Chapter2*. The reader should consult Chapter 1 for information on how to execute the Python and MATLAB code associated with this book.

2.12.1 Stripmap SAR Range Resolution

This example deals with stripmap SAR range resolution. Using (2.3), the ground-plane range resolution is plotted as a function of signal bandwidth from 1 MHz to

Table 2.3
Capella — SAR Imaging Modes

Mode	*Nominal Scene Size (km)*	*Ground-Plane Resolution (m)*	*NESZ (dB)*
Stripmap	5 × 20	1.2 × 1.2	−20 – −16
Spotlight	5 × 5	0.5 × 0.5	−14 – −10
Site (Sliding Spotlight)	5 × 10	1 × 1	−17 – −14

Table 2.4
RADARSAT Constellation Mission Imaging Modes

Mode	*Swath Width (km)*	*Ground-Plane Resolution (m)*	*NESZ (dB)*
Low Resolution	500	100 m	−22
High Incidence	133	50	−22
Medium Resolution	30–350	16–50	−25 – −22
High Resolution	30	5	−19
Very High Resolution	20	3	−17
Spotlight	20	1	−17
Low Noise	350	100	−25
Quad-Polarization	20	9	−24

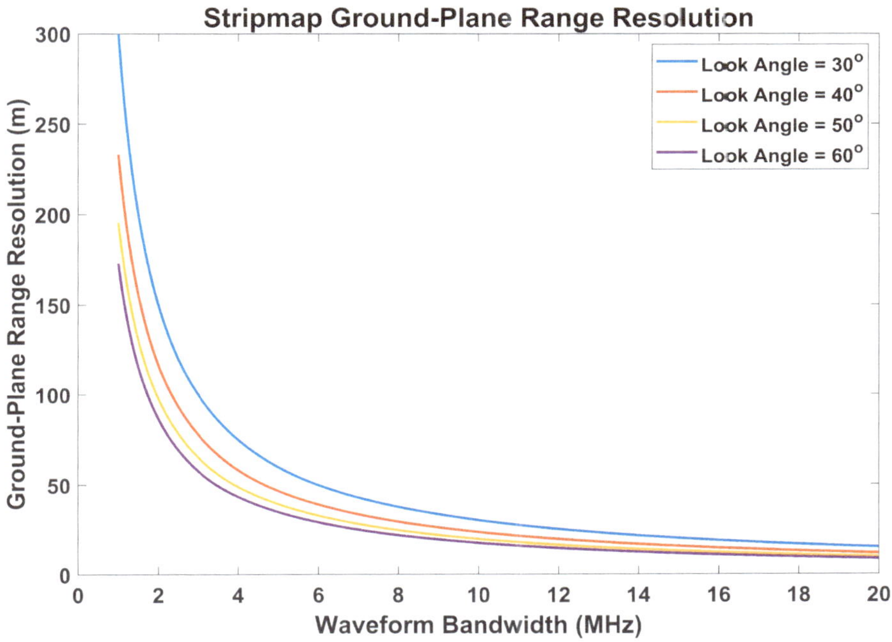

Figure 2.25 Stripmap SAR ground-plane range resolution vs signal bandwidth.

20 MHz. The results are shown in Figure 2.25 for incident angle values of 30°, 40°, 50°, and 60°. As expected, the ground-plane range resolution becomes better with increased signal bandwidth and lower incident angles. The solution to this example is given in the Python notebook *stripmap_range_resolution_example.ipynb* and in the MATLAB live script *stripmap_range_resolution_example.mlx*.

2.12.2 Stripmap SAR Support Band

This example is concerned with the support band for stripmap SAR imaging. Choosing an operating frequency of 10 GHz, the wavelength is calculated as

$$\lambda = \frac{c}{f} \approx \frac{3 \times 10^8}{10 \times 10^9} = 0.03 \qquad \text{(m)}.$$

The imaged area has an along-track distance of 100 m, with a nominal slant range of 10 km, and a cross-range resolution of 1 m. From (2.4), a range resolution of 1 m requires the use of an antenna with a width of $D_a = 2$ m. Substituting the wavelength and antenna width into (2.5) results in a support band of

$$b \in \Big[-\frac{100}{2} - \frac{10\times 10^3 \times 0.03}{4}, \frac{100}{2} + \frac{10\times 10^3 \times 0.03}{4}\Big]$$

$$= \big[-125, 125\big] \qquad \text{(m)}.$$

This means the length of the synthetic aperture must be greater than 250 m. The solution is also given in the Python notebook *stripmap_support_band_example.ipynb* and the MATLAB live script *stripmap_support_band_example.mlx.*

2.12.3 Stripmap SAR Point Spread Function

When studying the spatial resolution of SAR systems, it is important to examine the point spread function. Consider a target located at coordinates (5, 2) m. Figure 2.26 illustrates the point spread function for a system with a signal bandwidth of 20 MHz, an operating frequency of 1 GHz, and a cross-range resolution of 4 m. The solution to this example is given in the Python notebook *stripmap_psf_example.ipynb* and the MATLAB live script *stripmap_psf_example.mlx.*

2.12.4 Stripmap SAR Doppler Bandwidth

The Doppler bandwidth for a stationary target is given in (2.16). For a system operating at S-band, the Doppler bandwidth as a function of platform velocity is shown in Figure 2.27. The results are plotted for antenna beamwidths of 1.0°, 2.5°, and 10°. The solution to this example is given in the Python notebook *stripmap_doppler_bandwidth_example.ipynb* and the MATLAB live script *stripmap_doppler_bandwidth_example.mlx.*

2.12.5 Stripmap SAR Pulse Repetition Frequency

Platform velocity, unambiguous range, and antenna size place an upper and lower limit on the pulse repetition frequency. For this example, the upper and lower

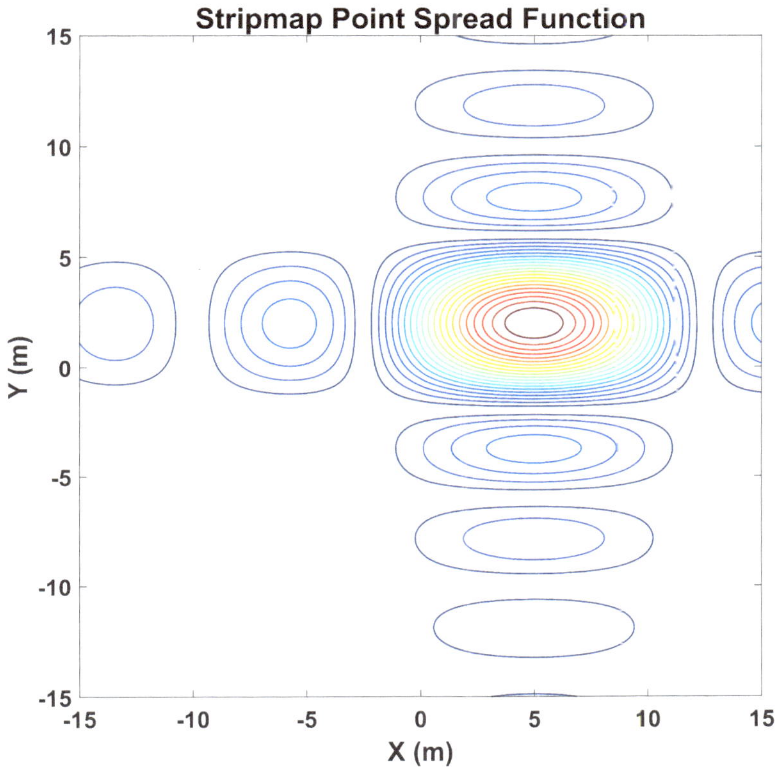

Figure 2.26 Stripmap SAR point spread function with a target location of (5, 2) meters.

pulse repetition frequency limits as a function of platform velocity are calculated using (2.21) and plotted in Figure 2.28. The platform velocity varies from 50 m/s to 2000 m/s. The system has a cross-range resolution of 3.5 m, and the unambiguous range is 200 km. The solution to this example is given in the Python notebook *stripmap_prf_example.ipynb* and the MATLAB live script *stripmap_prf_example.mlx*.

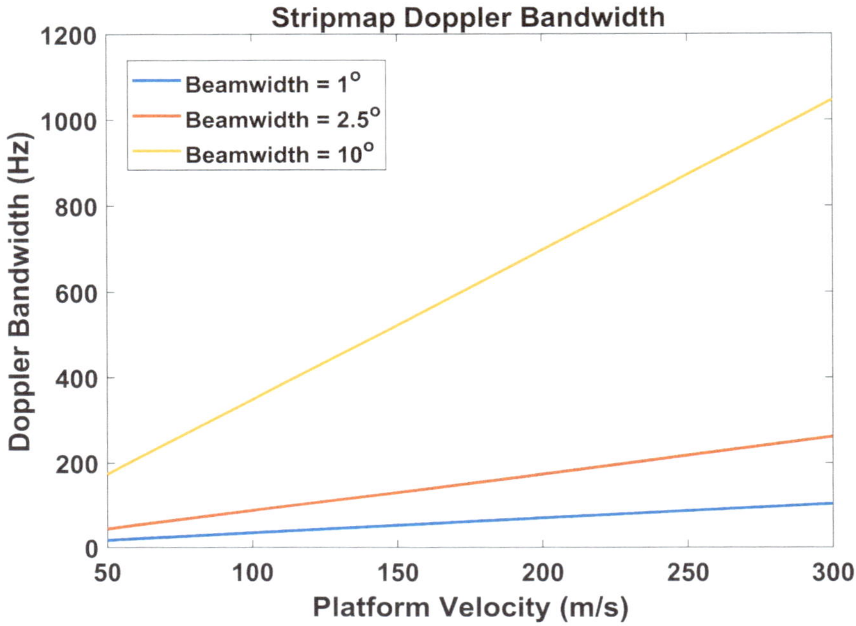

Figure 2.27 Stripmap SAR Doppler bandwidth vs platform velocity.

2.12.6 Stripmap SAR SNR

For stripmap SAR, the SNR is a function of various system and target parameters as well as imaging geometry as given in (2.24). Consider a SAR sensor operating at X-band with a cross-range resolution of 15 m, an average transmitted power of 100 W, and an effective aperture of 5 m^2. The slant range to the target area is 175 km and the target has a radar cross section of (0.05, 1.0, 10.0) m^2. The SNR as a function of velocity from 50 m/s to 500 m/s is shown in Figure 2.29. The solution to this example is given in the Python notebook *stripmap_signal_to_noise_example.ipynb* and the MATLAB live script *stripmap _signal_to_noise_example.mlx*.

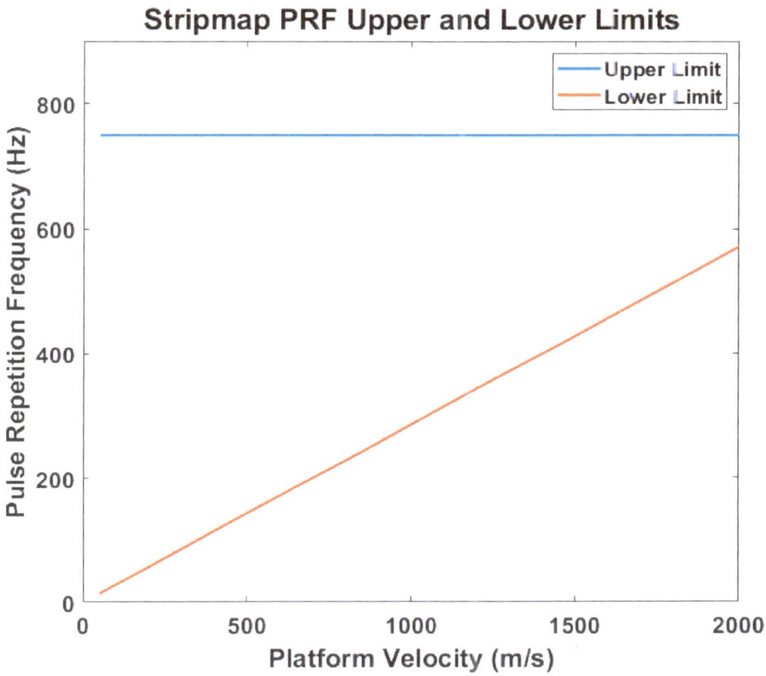

Figure 2.28 Stripmap SAR pulse repetition frequency upper and lower limits vs platform velocity.

2.12.7 Stripmap SAR CNR

For this example, the CNR for a stripmap SAR sensor operating at S-band with a range resolution of 25 m, an average transmitted power of 500 W, and an effective aperture of 1 m^2 is calculated with (2.26). The slant range to the target area is 100 km and the target area has a backscattering coefficient of (0.01, 0.05, 0.25). The CNR as a function of velocity from 25 m/s to 300 m/s is illustrated in Figure 2.30. The solution to this example is given in the Python notebook *stripmap_clutter_to_noise_example.ipynb* and the MATLAB live script *stripmap_clutter_to_noise_example.mlx*.

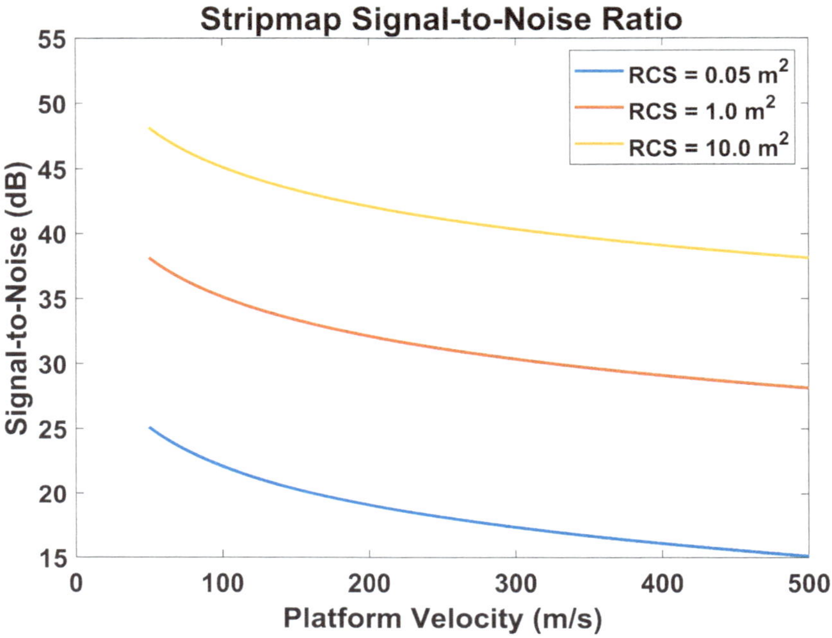

Figure 2.29 Stripmap SAR SNR vs platform velocity.

2.12.8 NESZ

For stripmap SAR imaging of distributed targets, the value of the backscattering coefficient that produces an SNR of one as a function of velocity is found from (2.27). Consider a system capable of operating in multiple modes with range resolutions of (1, 5, 10) m, an average transmitted power of 800 W, and an antenna effective aperture of 3.5 m^2. The radar operates in UHF, and the range to the target area is 200 km. The NESZ is shown in Figure 2.31 for a platform velocity ranging from 10 m/s to 250 m/s. The solution to this example is given in the Python notebook *stripmap_nesz_example.ipynb* and the MATLAB live script *stripmap_nesz_example.mlx*.

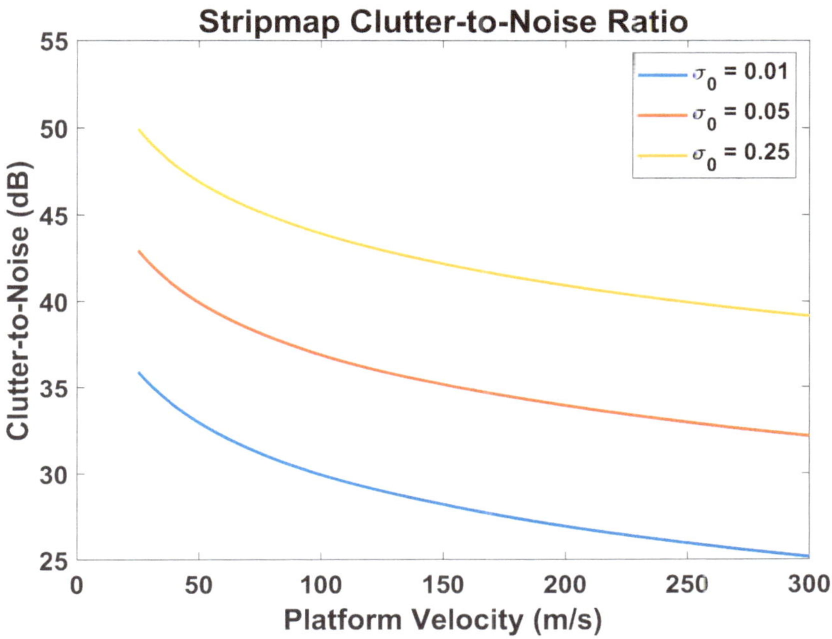

Figure 2.30 Stripmap SAR clutter-to-noise ratio vs platform velocity.

2.12.9 Spotlight SAR Cross-Range Resolution

For a spotlight mode imaging scenario assume there are radars operating at L-, S-, C-, and X-bands at a range of 10 km from the imaging area. Using (2.30), the cross-range resolution is calculated as a function of aperture length and shown in Figure 2.32. The solution to this example is given in the Python notebook *spotlight_resolution_example.ipynb* and the MATLAB live script *spotlight_resolution_example.mlx*.

2.12.10 Spotlight SAR Point Spread Function

The point spread function gives insight to the resolution of the SAR system and is found from (2.24). For this example, a point scatterer is located at (10, 5) m relative

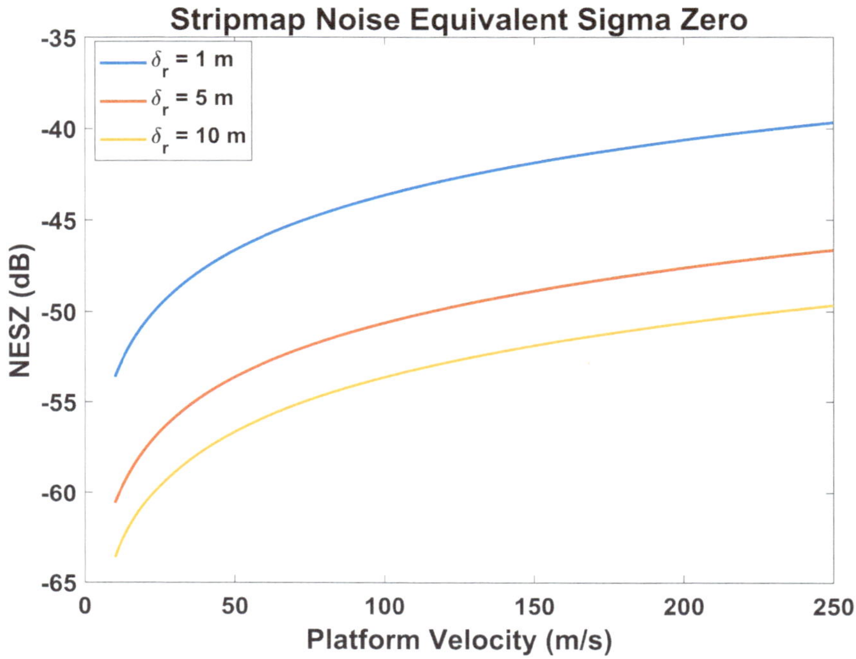

Figure 2.31 Stripmap SAR noise equivalent sigma zero vs platform velocity.

to the center of the imaging area. The range to this area is 100 m, the synthetic aperture length is 1 km, the bandwidth is 50 MHz, and the operating frequency is 1 GHz. Figure 2.33 shows the point spread function for this imaging scenario. The solution to this example is given in the Python notebook *spotlight_psf_example.ipynb* and the MATLAB live script *spotlight_psf_example.mlx*.

2.12.11 Spotlight SAR Doppler Bandwidth

In this example the Doppler bandwidth of a Ku-band spotlight SAR system is studied. Assume the platform has a velocity ranging from 50 m/s to 300 m/s. The synthetic aperture length is 100 m, and the radar makes multiple images from slant ranges of (10, 30, 50) km. The Doppler bandwidth is calculated from (2.38) and is shown in Figure 2.34. The solution to this example is given in the Python

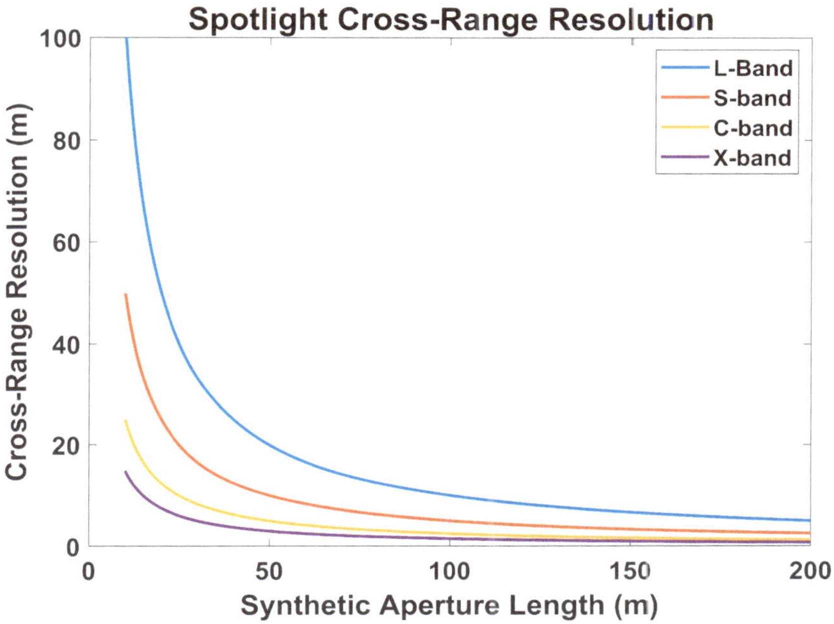

Figure 2.32 Spotlight SAR cross-range resolution vs synthetic aperture length.

notebook *spotlight_doppler_bandwidth_example.ipynb* and the MATLAB live script *spotlight_doppler_bandwidth_example.mlx*.

2.12.12 Spotlight SAR PRF

Platform velocity and unambiguous range place an upper and lower limit on the pulse repetition frequency. For this example, a SAR system platform has a velocity that varies from 50 m/s to 300 m/s and an unambiguous range of 250 km. Using (2.41) with a synthetic aperture length of 350 m and a nominal range of 20 km to the imaging area, the upper and lower limits on the pulse repetition frequency is calculated and plotted in Figure 2.35. The solution to this example is given in the Python notebook *spotlight_prf_example.ipynb* and the MATLAB live script *spotlight_prf_example.mlx*.

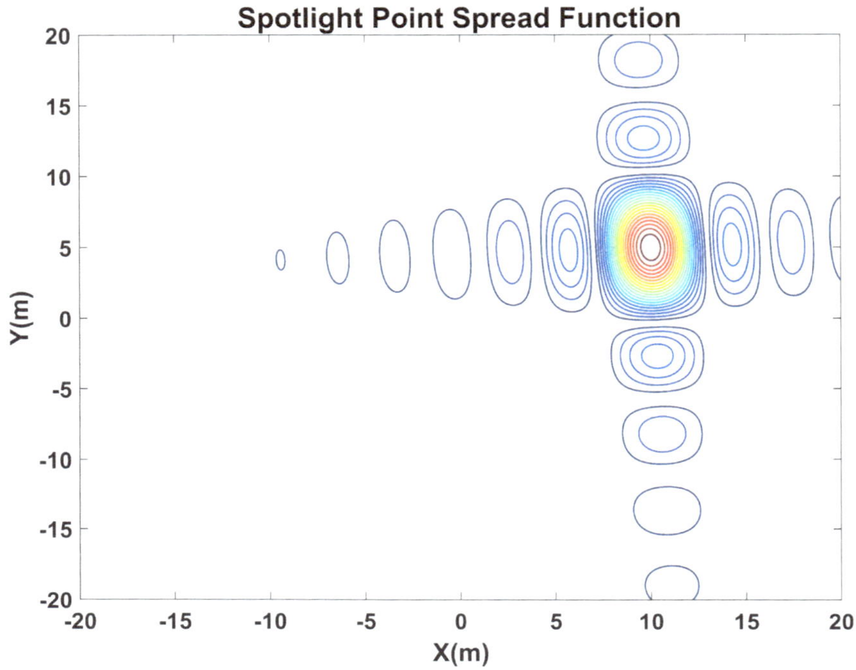

Figure 2.33 Spotlight SAR point spread function for a target located at (10, 5) meters.

2.12.13 Spotlight SAR SNR

In this example, the SNR for spotlight SAR imaging is studied. Assume a SAR system is operating at Ku-band with a bandwidth of 100 MHz, PRF of 10 Hz, pulsewidth is 10 μs, and average transmitted power is 1 kW. The antenna has an effective aperture of 0.5 m^2 and the synthetic aperture length is 500 m. Using (2.42) and (2.43), the SNR before and after signal processing is calculated for slant ranges from 1 to 10 km, and shown in Figure 2.36. The solution to this example is given in the Python notebook *spotlight_snr_example.ipynb* and the MATLAB live script *spotlight_snr_example.mlx*.

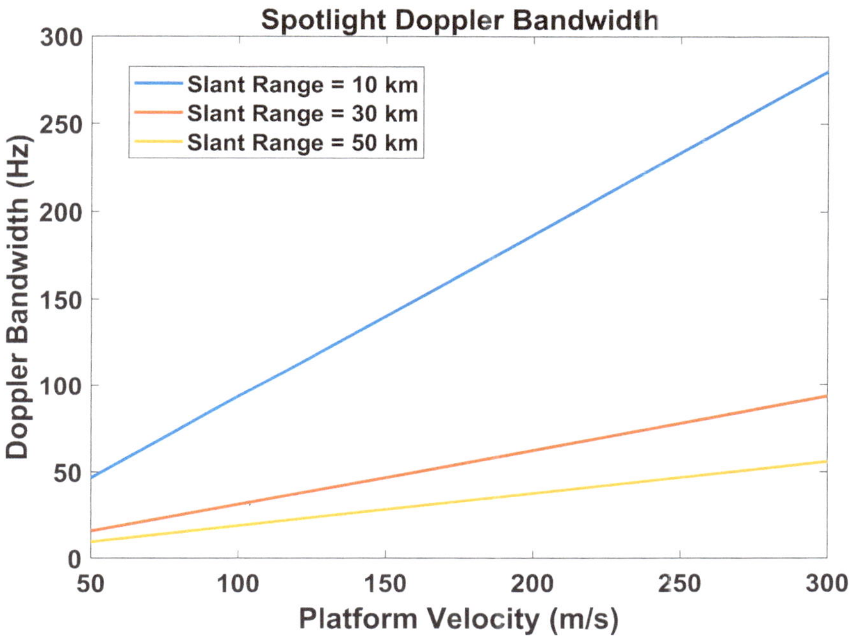

Figure 2.34 Spotlight SAR Doppler bandwidth vs platform velocity.

PROBLEMS

2.1 Describe the advantages and disadvantages of SAR imaging versus electro-optical/infrared (EOIR) imaging.

2.2 For a SAR system in low Earth orbit operating in stripmap mode, what antenna dimensions would be required to produce a swath width of 150 km? Assume the radar operates at L-band.

2.3 Calculate the waveform bandwidth required for a ground-range resolution of 1.5 m. The look angle is 60° and Hanning windowing is applied to the range data.

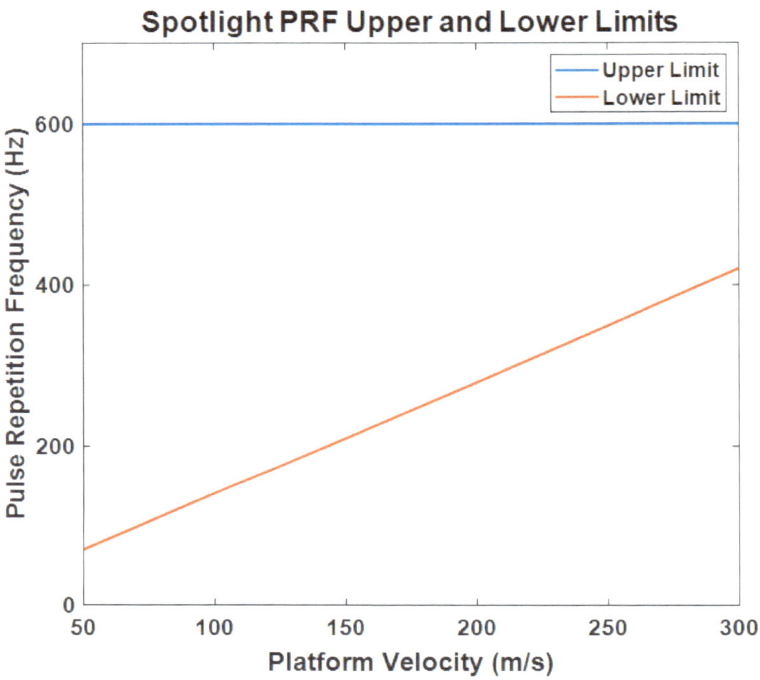

Figure 2.35 Spotlight SAR pulse repetition frequency upper and lower limits vs platform velocity.

2.4 Explain the difference in the support band as the cross-range resolution remains constant and the operating frequency increases.

2.5 Plot the point spread function for a stripmap SAR system with a cross-range resolution of 3 m, and a range resolution of 10 m.

2.6 Compute the Doppler bandwidth for a SAR system operating at K-band, a velocity of 100 m/s, and an antenna beamwidth of 10 degrees.

2.7 Describe the upper and lower pulse repetition frequency limits for stripmap SAR imaging and how those relate to the mapping rate.

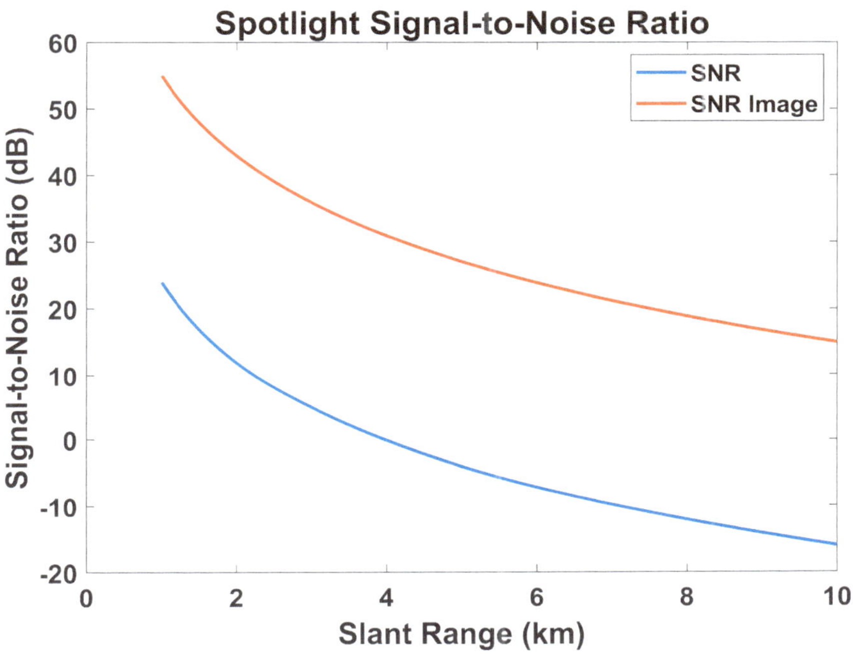

Figure 2.36 Spotlight SAR SNR vs slant range.

2.8 Explain the major differences in the SNR and CNR and how those are dependent on the spatial resolution.

2.9 Calculate the backscattering coefficient that results in an SNR of one for a system operating at C-band, with a velocity of 45 m/s, an average transmitted power of 1 kW, and an effective aperture of 10 m^2. The range to the scene is 10 km and the range resolution is 17 m.

2.10 Describe the advantages and disadvantages of operating a SAR system at a high squint angle.

2.11 Calculate the cross-range resolution for a K-band spotlight SAR system. The range to the imaging area is 60 km, the synthetic aperture length is 200 m.

2.12 Plot the point spread function for spotlight SAR for a target located at (−3, 5) m relative to the scene center. The sensor operates at 10.5 GHz and has a bandwidth of 500 MHz.

2.13 Determine the Doppler bandwidth for spotlight-mode SAR with a platform velocity of 335 m/s. The radar operates at S-band, the synthetic aperture length is 2.5 km, and the range to the imaging area is 10 km.

2.14 What are the major differences in the upper and lower pulse repetition frequency limits for spotlight and stripmap-mode SAR.

2.15 Compute the SNR for spotlight-mode SAR imaging. The transmit power is 600 W, the effective antenna aperture is 20 m^2, and the radar cross section is 7 dB. The radar operates at Ka-band, has a bandwidth of 770 MHz, and a pulsewidth of 100 μs. The range to the imaging area is 50 km. The platform has a velocity of 100 m/s and a PRF of 100 Hz.

2.16 Describe the difference in squint angle imaging between stripmap and spotlight-mode SAR imaging.

2.17 ScanSAR is an imaging mode designed to cover extremely large swaths. How is TOPSAR designed to improve on this concept, and what are the associated advantages?

2.18 Illustrate how digital beamforming could be used to realize the SweepSAR imaging mode and possibly improve upon it.

2.19 Highlight how InSAR has been significant in the remote sensing of the Kilauea volcano in Hawaii.

2.20 Describe the advantages of using multiple polarizations in SAR imagery. Give examples of situations where one polarization may be more useful than another (i.e., HH versus VV).

2.21 In bistatic SAR imaging, why would it be advantageous to use an illuminator of opportunity versus a dedicated transmitting source? What are some of the difficulties that arise when using an illuminator of opportunity?

2.22 Outline a scenario, based on radar-to-target geometry, in which a ground-based radar would not be capable of producing ISAR imaging of an airborne target.

References

[1] A. Moreira, P. Prats-Iraola, M. Younis, G. Krieger, I. Hajnsek, and K. Papathanassiou. "A tutorial on synthetic aperture radar." *IEEE Geoscience and Remote Sensing Magazine*, 1(1):6–43, 2013.

[2] A. Harrison. *Introduction to Radar Using Python and MATLAB*. Artech House, Norwood, 2020.

[3] A. Flores-Anderson, K. Herndon, R. Thapa, and E. Cherrington. The SAR handbook: Comprehensive methodologies for forest monitoring and biomass estimation. Technical report, NASA Space Science and Technology Center, Huntsville, 2019.

[4] The European Space Agency. Sentinel-1 SAR user guide. https://sentinel.esa.int/web/sentinel/technical-guides/sentinel-1-sar. Accessed: 2020-10-03.

[5] Airbus Defence and Space. *TerraSAR-X Image Product Guide*. Airbus, Ottobrunn, 2015.

[6] NASA. NASA's Alaska Satellite Facility Distributed Active Archive Center. http://earthdata.nasa.gov/eosdis/daacs/asf. Accessed: 2020-12-20.

[7] J. Curlander and R. McDonough. *Synthetic Aperture Radar: Systems and Signal Processing*. John Wiley and Sons, New York, 1991.

[8] M. Soumekh. *Synthetic Aperture Radar Signal Processing with MATLAB Algorithms*. John Wiley and Sons, New York, 1999.

[9] M. Born and M. Wolf. *Electromagnetic Theory of Propagation, Interference and Diffraction of Light*. Cambridge University Press, Cambridge, 1999.

[10] L. Rayleigh. Investigations in optics with special reference to the Spectroscope. *Philosophical Magazine and Journal of Science*, 8(49):261–274, 1879.

[11] K. Prabhu. *Window Functions and Their Applications in Signal Processing*. Boca Raton: CRC Press, 2014.

[12] Y. Chan and V. Koo. An introduction to synthetic aperture radar (SAR). *Progress in Electromagnetics Research*, 2:27–60, 2008.

[13] J. Richards. *Remote Sensing With Imaging Radar*. Springer-Verlag, Berlin, 2009.

[14] S. Madsen. Estimating the Doppler centroid of SAR data. *IEEE Transactions on Aerospace and Electronic Systems*, 25(2):134–140, 1989.

[15] J. Curlander, C. Wu, and A. Pang. Automatic processing of spaceborne SAR data. In *International Geoscience and Remote Sensing Symposium*. Zeng, G. 3D Image Reconstruction, 1982.

[16] M. Jin. Optimal Doppler centroid estimation for SAR data from a quasi-homogeneous source. *IEEE Transactions on Geoscience and Remote Sensing*, 24(6):1022–1025, 1986.

[17] F. Li, D. Held, J. Curlander, and C. Wu. Doppler parameter estimation for spaceborne synethic-aperture radars. *IEEE Transactions on Geoscience and Remote Sensing*, 23(1):47–56, 1985.

[18] W. Carrara, R. Majewski, and R. Goodman. *Spotlight Synthetic Aperture Radar: Signal Processing Algorithms*. Artech House, Norwood, 1995.

[19] C. Jakowatz, D. Wahl, P. Eichel, D. Ghiglia, and P. Thompson. *Spotlight-Mode Synthetic Aperture Radar: A Signal Processing Approach*. Springer, New York, 1996.

[20] G. Davidson. *Image Formation From Squint Mode Synthetic Aperture Radar Data*. PhD thesis, The University of British Columbia, 1994.

[21] S. Tang, L. Zhang, and H. So. Focusing high-resolution highly-squinted airborne SAR data with maneuvers. *Remote Sensing*, 10(8):1275, 2018.

[22] F. de Vries. Speckle reduction in SAR imagery by various multi-look techniques. Technical Report ADA349197, TNO Physics and Electronics Laboratory, The Netherlands, 1998.

[23] D. Vavriv and O. Bezvesilniy. Advantages of multi-look SAR processing. In *International Conference on Antenna Theory and Techniques, Odeaas*. 2013.

[24] F. Da Zan and A. Guarnieri. Topsar: Terrain observation by progressive scans. *IEEE Transaction on Geoscience and Remote Sensing*, 44(9):2352–2360, 2006.

[25] A. Freeman, G. Krieger, P. Rosen, M. Younis, W. Johnson, S. Huber, R. Jordan, and A. Moreira. Sweepsar: Beam-forming on receive using a reflector-phased array feed combination for spaceborne SAR. In *IEEE Radar Conference*. Pasadena, 2009.

[26] C. Chuang, S. Shaffer, N. Niamsuwan, S. Li, E. Liao, C. Lim, V. Duong, B. Volain, K. Vines, M. Yang, and K. Wheeler. Nisar L-band digital electronics subsystem: A multichannel system with distributed processors for digital beam forming and mode dependent filtering. In *IEEE Radar Conference*. Philadelphia, 2016.

[27] NASA. NASA-ISRO SAR Mission. http://nisar.jpl.nasa.gov/mission/get-to-know-sar/. Accessed: 2020-09-05.

[28] P. Rosen, S. Hensley, H. Zebker, F. Webb, and E. Fielding. Surface deformation and coherence measurements of Kilauea Volcano, Hawaii, from SIR-C radar interferometry. *Journal of Geophysical Research*, 101:23109–23125, 1996.

[29] D. Massonnet, D. Briole, and A. Amaud. Deflation of Mount Etna monitored by spaceborne radar interferometry. *Nature*, 375(6532):567–570, 1995.

[30] D. Massonnet, M. Rossi, C. Carmona, F. Adragna, G. Peltzer, K. Feigl, and T. Rabaute. The displacement field of the Landers earthquake mapped by radar interferometry. *Nature*, 364(6433):138–142, 1993.

[31] B. Smith-Konter, L. Burkhard, L. Ward, J. Foster, X. Xu, and D. Sandwell. 2020 Kilauea eruption: Line of sight displacement revealed by Sentinel-1 Interferometry. http://pgf.soest.hawaii.edu/Kilauea_insar_2020/. Accessed: 2020-09-12.

[32] NASA Jet Propulsion Laboratory. Space radar image of Kiluchevskoi Volcano, Russia. Accessed: 2020-09-20.

[33] Canadian Space Agency. RADARSAT Constellation Mission. https://www.asc-csa.gc.ca/eng/satellites/radarsat/default.asphttps://www.asc-csa.gc.ca/eng/satellites/radarsat/default.asp, 2022. Accessed: 2022-4-23.

[34] M. Soumekh. Bistatic synthetic aperture radar inversion with application in dynamic object imaging. *IEEE Transactions on Signal Processing*, 39(9):2044–2055, 1991.

[35] J. Glaser. Some results in the bistatic radar cross section (RCS) of complex objects. *Proceedings of the IEEE*, 77(5):639–648, 1989.

[36] R. Eigel, P. Collins, A. Terzuoli, G. Nesti, and J. Fortuny. Bistatic scattering characterization of complex objects. *IEEE Transactions on Geoscience and Remote Sensing*, 38(5):2078–2092, 2000.

[37] C. Yarman, B. Yazc, and M. Cheney. Bistatic synthetic aperture radar imaging for arbitrary flight trajectories. *IEEE Transactions on Geoscience and Remote Sensing*, 17(11):84–93, 2008.

[38] H. Ma, M. Antoniou, A. Stove, J. Winkel, and M. Cherniakov. Maritime moving target localization using passive GNSS-based multistatic radar. *IEEE Transactions on Geoscience and Remote Sensing*, 56(8):4808–4819, 2018.

[39] A. Farina. Tracking function in bistatic and multistatic radar systems. *Proceedings of the IEEE*, 133(7):630–637, 1986.

[40] G. Krieger. Advanced bistatic and multistatic SAR concepts and applications. In *European Conference on Synthetic Aperture Radar (EUSAR)*. Dresden, 2006.

[41] T. Kraus, M. Bachmann, L. Heiderich, G. Krieger, and A. Moreira. Multistatic SAR imaging: Comparison of simulation results and experimental data. In *International Conference on Radar Systems*. Belfast, 2017.

[42] Sandia National Laboratories. Ship dynamics for maritime ISAR imaging. Technical Report SAND2008-1020, Sandia National Laboratories, 2008.

[43] S. Musman, D. Kerr, and C. Bachmann. Automatic recognition of ISAR ship images. *IEEE Transactions on Aerospace and Electronic Systems*, 32(4):1392–1404, 1996.

[44] A. Lazarov. ISAR imaging of a rotating asteroid irradiated by pulsar's electromagnetic emission. *Cybernetics and Information Technologies*, 19(2):38–50, 2019.

[45] V. Chen and H. Ling. *Time-Frequency Transforms for Radar Imaging and Signal Analysis*. Artech House Publishers, Norwood, 2001.

[46] Y. Yan, Z. He, and C. Shi. A review on ISAR motion compensation techniques. In *2nd International Conference on Frontiers of Sensors Technologies (ICFST)*. Shenzhen, 2017.

[47] M. Younis, C. Fischer, and W. Wiesbeck. Digital beamforming in SAR systems. *IEEE Transactions on Geoscience and Remote Sensing*, 41(71):1735–1739, 2003.

[48] S. Talisa, K. O'Haver, T. Comberiate, M. Sharp, and O. Somerlock. Benefits of digital phased array radars. *Proceedings of the IEEE*, 104(3):530–543, 2016.

[49] M. Ludwig, S. D'addio, M. Suess, G. Adamiuk, and A. Suriani. Phased array technology developments for next generation European spaceborne SARs with digital beamforming. In *IEEE International Symposium on Phased Array Systems and Technology*. Waltham, 2019.

[50] S. Ye, T. Yang, and W. Li. A novel algorithm for spaceborne SAR DBF based on sparse spatial spectrum estimation. In *China International SAR Symposium*. Shanghai, 2018.

[51] Q. Xin, Z. Jiang, P. Cheng, and M. He. Signal processing for digital beamforming FMCW SAR. *Mathematical Problems in Engineering*, 2014:10, 2014.

[52] ICEYE. Example SAR data from ICEYE. https://www.iceye.com/downloads/datasets. Accessed: 2021-09-21.

[53] UMBRA. NRO selects Umbra for their strategic commercial enhancements effort. https://umbra.space/blog/nro-selects-umbra-for-their-strategic-commercial-enhancements-effort, 2022. Accessed: 2022-4-23.

[54] Capella Space. SAR Made Easy. https://www.capellaspace.com/. Accessed: 2022-4-5.

Chapter 3

Image Formation

SAR image formation consists of coherently combining raw data from a sensor to produce pixel values in an image. The raw data is termed *phase history data* and is comprised of amplitude and phase information about the energy received by the radar. Phase history data is typically represented by in-phase and quadrature components, often referred to as IQ data [1]. The first SAR images were analog in nature and created by a series of optical lenses. The development of solid-state circuit components in the 1970s made it possible to process the radar return signals digitally. Since then, advances in computing power and radar hardware have led to the research and development of distinctive image formation techniques. This chapter begins with a signal model for the phase history data which is then used to generate one-dimensional range profiles. This is followed by a treatment of a few of the more popular imaging algorithms and SAR data products. The chapter concludes with several Python and MATLAB examples to further illustrate the concepts of SAR image formation. For readability, exp() is used interchangeably with e^x notation in this chapter.

3.1 SIGNAL MODEL

SAR systems commonly use LFM waveforms to achieve fine range resolution while producing SNRs required for imaging. An LFM waveform may be written as [1]

$$s(t) = \exp\left[j2\pi\left(f_0 t + \frac{\beta}{2}t^2\right)\right] \quad 0 \le t \le \tau, \tag{3.1}$$

where

f_0 = radar operating frequency (Hz),
β = frequency slope (B/τ) (Hz/s),
τ = pulsewidth (s),
$s(t)$ = LFM signal.

The amplitude of the LFM waveform is kept constant during transmission. The frequency of the LFM waveform varies linearly during transmission at a rate of βt, which results in a signal bandwidth of $\beta\tau$. The return signal from a point-target in the range dimension is then

$$s_r(t) = \mathrm{A}\exp\left[j2\pi\left(f_0(t - t_0) + \frac{\beta}{2}(t - t_0)^2\right)\right], \tag{3.2}$$

where $t_0 = 2R_0/c$, is the time delay to a point-target located at R_0, and the factor, A, is related to the target's radar cross section, antenna gain, path loss, and other factors. The instantaneous frequency is the time derivative of the phase, which is written as

$$f_i = \frac{1}{2\pi}\frac{d\phi(t)}{dt} \qquad \text{(Hz)}. \tag{3.3}$$

The phase in (3.2) is given by

$$\phi(t) = 2\pi\left[f_0(t - t_0) + \frac{\beta}{2}(t - t_0)^2\right] \qquad \text{(rad)}. \tag{3.4}$$

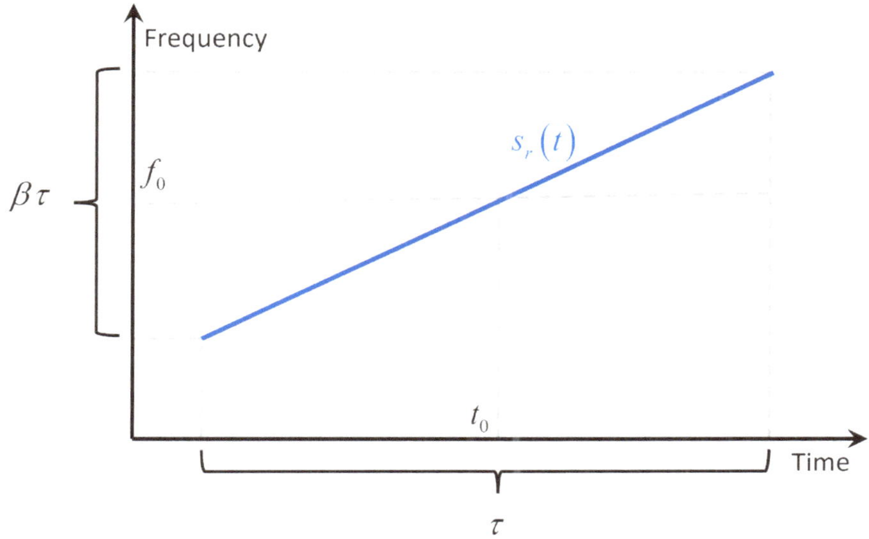

Figure 3.1 Instantaneous frequency of the return LFM signal.

Using (3.4) in (3.3), the instantaneous frequency is then

$$f_i = f_0 + \beta(t - t_0) \qquad \text{(Hz)}. \tag{3.5}$$

Figure 3.1 depicts the instantaneous frequency as a function of time for the return signal.

For a SAR platform moving with a constant velocity, the slant range to a point-target located at R_0 is expressed as

$$r(t) = \sqrt{R_0^2 + (vt)^2} \qquad \text{(m)}, \tag{3.6}$$

where

$$
\begin{aligned}
R_0 &= \text{range at closest approach (m)},\\
v &= \text{platform velocity (m/s)},\\
t &= \text{time (s)},\\
r(t) &= \text{slant range to the target (m)}.
\end{aligned}
$$

If $vt/R_0 \ll 1$, the binomial approximation may be used, and (3.6) is then written as

$$r(t) = R_0 + \frac{(vt)^2}{2R_0} \qquad \text{(m)}. \tag{3.7}$$

Figure 3.2 illustrates the parabolic variation of the slant range for a constant velocity platform. For a point-target located at $r(t)$, the azimuth signal may be represented as

$$s_a(t) = \mathrm{A}\exp\left[j\frac{4\pi}{\lambda}r(t)\right] = \mathrm{A}\exp\left[j\frac{4\pi}{\lambda}\left(R_0 + \frac{(vt)^2}{2R_0}\right)\right]. \tag{3.8}$$

The phase of the return signal in the cross-range direction is causally related to the slant range to the target by

$$\phi(t) = \frac{4\pi r(t)}{\lambda} = \frac{4\pi}{\lambda}\left[R_0 + \frac{(vt)^2}{2R_0}\right] \qquad \text{(rad)}. \tag{3.9}$$

As before, the instantaneous frequency is found from the time derivative of the phase as

$$f_i = \frac{v^2 t}{\lambda R_0} \qquad \text{(Hz)}. \tag{3.10}$$

The instantaneous frequency is shown in Figure 3.3, and the Doppler bandwidth is then written as

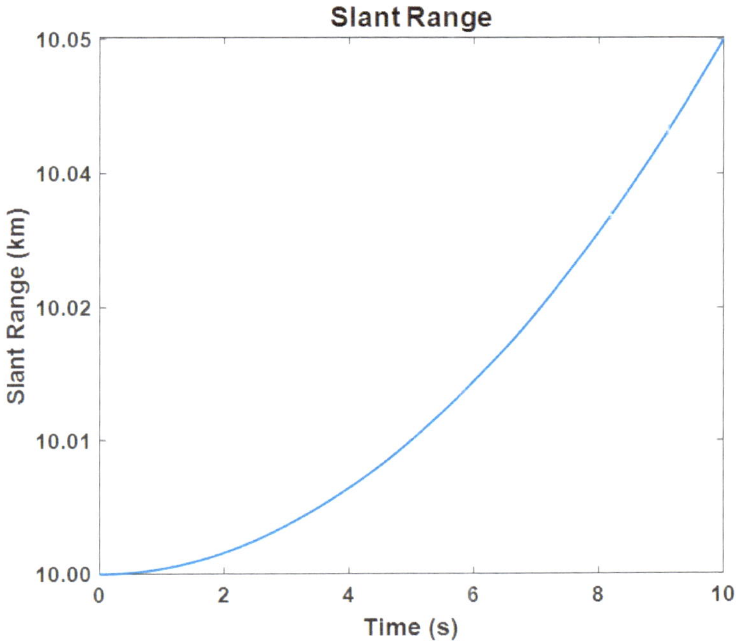

Figure 3.2 Slant range as a function of time for a constant velocity platform.

$$f_D = \frac{v^2 T_{sa}}{\lambda R_0} \qquad \text{(Hz)}. \tag{3.11}$$

where T_{sa} is the integration time along the synthetic aperture.

The signal model will be used in the following sections to illustrate various image formation algorithms. While these algorithms use the information in the signal in differing ways to create the images, the basic signal model with all the phase information will remain the same.

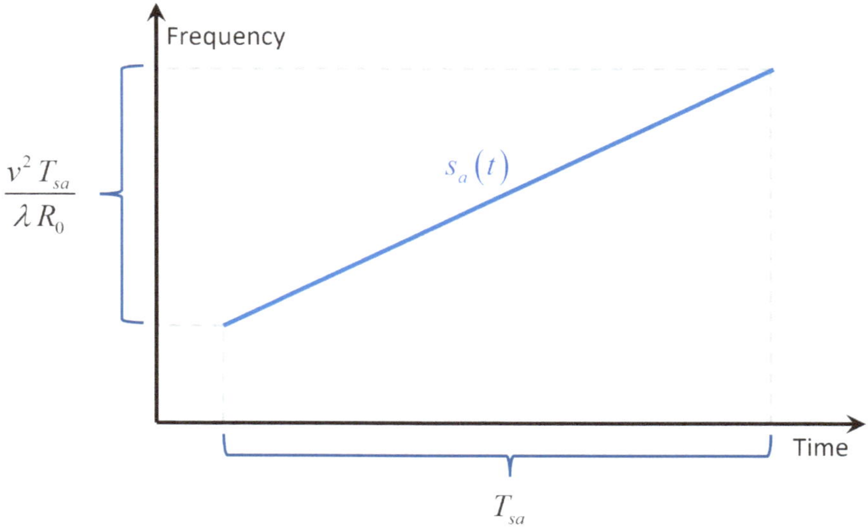

Figure 3.3 Instantaneous frequency in the cross-range dimension.

3.2 ONE-DIMENSIONAL RANGE PROFILES

Range profiles represent the reflectivity of a target as a function of slant range and can be thought of as one-dimensional images of the target. Chapter 2 introduced the range resolution of LFM waveforms. If the range resolution is fine enough, individual scatterers on the target can be resolved. This leads to identification of target geometry, which aids in target classification and discrimination, detailed mapping, and improved environmental monitoring. Radars generate range profiles by pulse compression of the return signal which is often accomplished through use of a matched filter, or in the case of large bandwidth waveforms, a stretch processor.

3.2.1 Matched Filter

Consider the block diagram of a matched filter as illustrated in Figure 3.4. The input signal to the matched signal is the received signal given in (3.2) plus noise, $n_r(t)$. If the filter response, $h(t)$, is linear, the filter output is written as

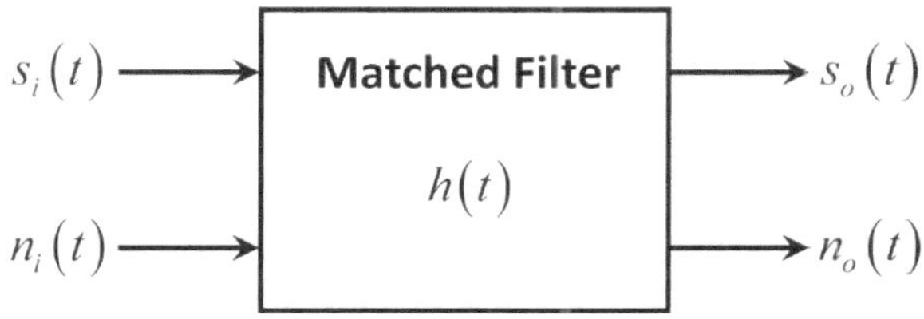

Figure 3.4 Block diagram of a matched filter.

$$s_o(t) = s_r(t) * h(t), \tag{3.12}$$

$$n_o(t) = n_r(t) * h(t), \tag{3.13}$$

and the symbol, $(*)$, represents linear convolution. The filter response, $h(t)$, is the conjugate of a scaled, time reversed, and time shifted version of the transmitted signal, $s(t)$. This correspondence of the filter response to the transmitted signal gives rise to the name *matched filter* [1]. Employing the frequency domain form of (3.12) allows the output of the matched filter to be written as

$$s_o(t) = \mathcal{F}^{-1}\Big\{\mathcal{F}\{h(t)\} \times \mathcal{F}\{s_r(t)\}\Big\}, \tag{3.14}$$

where

$s_r(t)$	=	matched filter input signal (received signal),
$h(t)$	=	matched filter impulse response,
$\mathcal{F}\{\cdot\}$	=	Fourier transform operator,
$s_o(t)$	=	matched filter output.

The matched filter may be implemented with the Fourier transform and inverse Fourier transform from the SciPy module *fftpack* [2]. This module includes various functions for computing discrete and fast Fourier transforms along with several helper functions. The expression in (3.14) is implemented in the code shown in Listing 3.1. The target is located at a time delay of 0.2 seconds, and the output is shown in Figure 3.5. As seen in the figure, the peak of the matched filter output occurs at $t = 0.2$ seconds, corresponding to the target location. The first null is located at $\pm 1/B$ seconds from the peak, where B is the bandwidth. Therefore, targets will be unambiguous in time if separated by at least $1/B$ seconds. With the round-trip delay time of $2R/c$ seconds, the range resolution for a single LFM pulse is $c/2B$ meters.

Listing 3.1 Calculation of the Matched Filter Output

```
from scipy.fftpack import fft, ifft, fftshift
from scipy import conj, linspace, exp
from scipy.constants import pi
from matplotlib import pyplot as plt

# Pulsewidth (s)
t = linspace(-1, 1, 1024)

# Time delay to the target (s)
t0 = 0.2

# Set up the transmit and receive signals (s)
st = exp(1j * 10 * pi * t ** 2)
sr = exp(1j * 10 * pi * (t - t0) ** 2)

# Impulse response and matched filtering
Hf = fft(conj(st))
Si = fft(sr)
so = fftshift(ifft(Si * Hf))

# Plot the matched filter output
plt.figure(1)
plt.plot(t, abs(so) / max(abs(so)))
plt.title('Matched Filter Output')
plt.xlabel('Time Delay (s)')
plt.ylabel('Relative Amplitude')
plt.show()
```

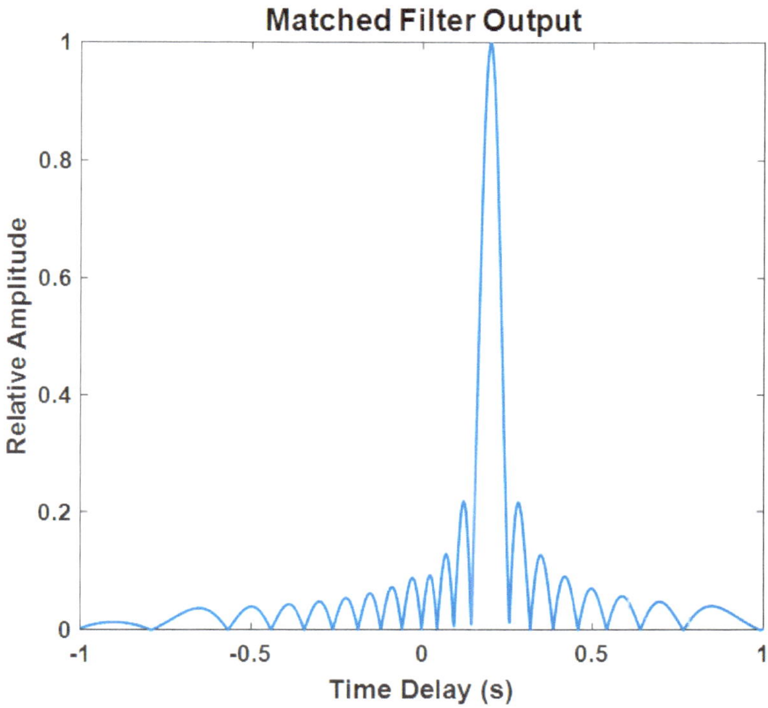

Figure 3.5 Matched filter output for Listing 3.1.

3.2.2 Stretch Processor

Stretch processing is a compression technique used for processing large bandwidth waveforms employed in high range-resolution systems such as SAR. Consider the block diagram of the stretch processor given in Figure 3.6. In general, stretch processing consists of the following steps:

1. Mix the return signal with a reference signal.
2. Low-pass filter the mixed signals.

3. Convert the filtered signals to the digital domain with an analog-to-digital converter.
4. Fourier transform the digital signals.

There are requirements on the pulsewidth of the reference signal and range window size that may be processed [1, 3, 4]. The output of the mixing step is written as

$$s_o(t) = \mathrm{A}\exp\left[j2\pi f_0\big(t_0 - t_{ref}\big) + j2\pi\beta\big(t_0 - t_{ref}\big) + j\pi\beta\big(t_{ref}^2 - t_0^2\big)\right], \quad (3.15)$$

where t_{ref} is the time delay of the reference signal. Recall the instantaneous frequency is the time derivative of the phase as expressed in (3.3), which leads to

$$f_i = \beta\big(t_0 - t_{ref}\big) \qquad \text{(Hz)}. \quad (3.16)$$

This allows the time delay of the target to be written as

$$t_0 = \frac{f_i}{\beta} + t_{ref} \qquad \text{(s)}. \quad (3.17)$$

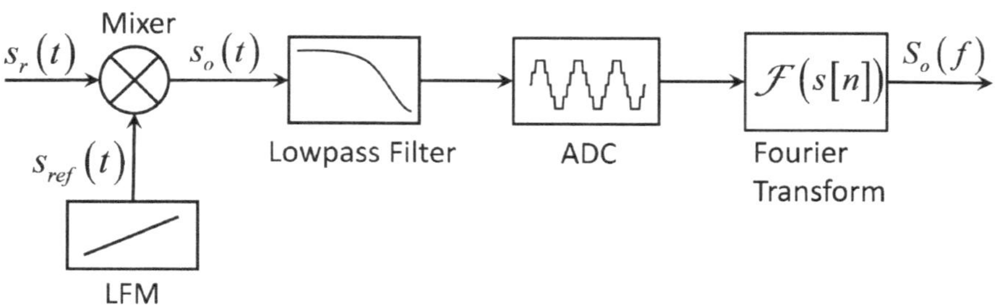

Figure 3.6 Stretch processor block diagram.

Therefore, the target range is related to the instantaneous frequency as

$$R_0 = \frac{c}{2}\left[\frac{f_i}{B}\tau_p + t_{ref}\right] \qquad \text{(m)}. \tag{3.18}$$

The expression in (3.16) indicates the resulting instantaneous frequency is constant and related to the range of the target, as illustrated in Figure 3.7. Listing 3.2 illustrates these concepts in Python code, with the output given in Figure 3.8. To study the resolution associated with the stretch processor, consider the situation where there are two point-targets present at ranges R_1 and R_2, with associated time delays of $t_1 = 2R_1/c$ and $t_2 = 2R_2/c$. Using (3.16), the instantaneous frequencies are

$$f_{i,1} = \beta\big(t_1 - t_{ref}\big) \qquad \text{(Hz)}, \tag{3.19}$$

$$f_{i,2} = \beta\big(t_2 - t_{ref}\big) \qquad \text{(Hz)}. \tag{3.20}$$

The frequency resolution of the stretch processor is written as

$$\delta f = \frac{1}{\tau_p} \qquad \text{(Hz)}. \tag{3.21}$$

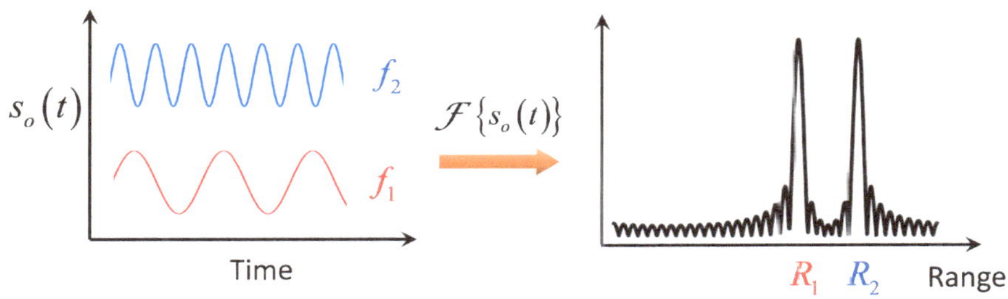

Figure 3.7 Stretch processor instantaneous frequency related to range.

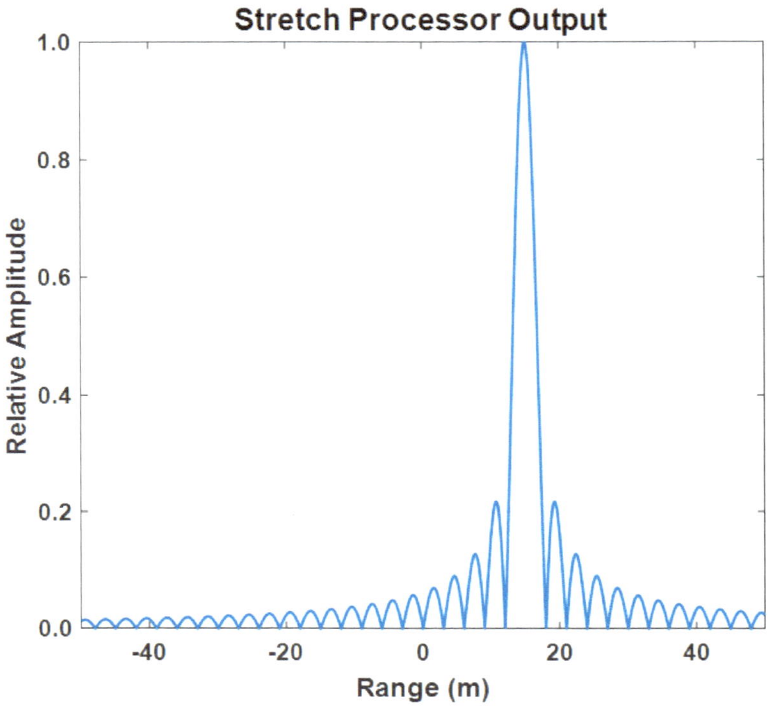

Figure 3.8 Stretch processor output for Listing 3.2.

The two point-targets may be resolved if the associated ranges are such that the instantaneous frequencies are greater than the frequency resolution of the stretch processor. This is expressed as

$$f_{i,1} - f_{i,2} = \Big[\beta\big(t_1 - t_{ref}\big) - \beta\big(t_2 - t_{ref}\big)\Big] \geq \frac{1}{\tau_p} \qquad \text{(Hz)}. \tag{3.22}$$

This leads to the following time resolution

$$\big(t_2 - t_1\big) \geq \frac{1}{B} \qquad \text{(s)}, \tag{3.23}$$

Listing 3.2 Calculation of the Stretch Processor Output

```
from scipy.fftpack import fft, fftshift, fftfreq
from numpy import ceil, linspace, exp
from scipy.constants import pi, c
from matplotlib import pyplot as plt

# Bandwidth (Hz)
bandwidth = 50e6

# Pulsewidth (s)
pulsewidth = 1e-3

# Range window length +/- (m)
range_window_length = 50.0

# Target range (m)
target_range = 15.0

# Time delay to target (s)
t0 = 2.0 * target_range / c

# Number of samples
number_of_samples = int(ceil(4 * bandwidth *
                      range_window_length / c))

# Time sampling
t, dt = linspace(-0.5 * pulsewidth, 0.5 * pulsewidth,
                      number_of_samples, retstep=True)

# Sampled signal after mixing
so = exp(1j * 2.0 * pi * bandwidth / pulsewidth * t0 * t)

# Fourier transform
so = fftshift(fft(so, 512))

# FFT frequencies
frequencies = fftshift(fftfreq(512, dt))

# Range window
range_window = 0.5 * frequencies * c * pulsewidth / bandwidth
```

```

# Plot the stretch processor output
plt.figure(figsize=(12,9))
plt.plot(range_window, abs(so) / max(abs(so)))
plt.tick_params(labelsize=18)
plt.grid(linestyle=':', linewidth=0.5)
plt.title('Stretch Processor Output', size=24)
plt.xlabel('Range (m)', size=20)
plt.ylabel('Relative Amplitude', size=22)
plt.show()
```

which gives a range resolution of

$$\delta R = \big(R_2 - R_1\big) = \frac{c}{2B} \qquad \text{(m)}. \tag{3.24}$$

This is the same as the range resolution for the matched filter, as shown in Section 3.2.1. For more detailed discussions on both stretch processing and matched filtering, the reader is referred to [1, 3, 5].

3.3 TWO-DIMENSIONAL IMAGING

In the previous section, one-dimensional range profiles were studied. These profiles provide target information such as relative range between scatterers, apparent length, and lead/trail scatterers. Two-dimensional imaging can achieve high resolution in the azimuth direction and higher SNRs due to the larger aperture and coherent integration of several pulses. This leads to detailed imagery with applications in environmental monitoring, moving target indication, interferometry, intelligence, surveillance, and reconnaissance.

There are various methods for forming two-dimensional SAR imagery from collected radar data, each having unique advantages and disadvantages. The following sections cover a few of the more popular methods, citing some of the strengths and weaknesses, as well as scenarios where one method may be better suited than others. For a more comprehensive comparison of algorithms, the reader is referred to the following excellent sources [6–8].

3.3.1 Range Doppler Algorithm

The range-Doppler algorithm (RDA) is perhaps the most widely used image formation algorithm for processing SAR data [9, 10]. It was originally developed in the late 1970s for processing Seasat data, and later used to process spaceborne SAR data [11]. RDA operates on the range and azimuth domain data separately as one-dimensional processes. This separation of range and azimuth is an approximation, and some form of range cell migration correction must be incorporated. The basic steps in RDA are:

1. Range compression.
2. Conversion to range-Doppler domain.
3. Range cell migration correction.
4. Azimuth compression.

The first step in RDA is range compression and this step may be performed via matched filtering as discussed in Section 3.2. A sensitivity time control attenuator is often added to correct for the signal's amplitude variation due to large differences in range [1].

Conversion to the range-Doppler domain is accomplished by taking a one-dimensional FFT along the azimuth direction and is often performed on blocks of data covered by the length of the azimuth matched filter. At this point, the signal from a point-target would follow a hyperbolic path in the range-Doppler domain which can cross several range cells, see Example 3.4.3. The signal must be aligned in range to capture all the energy from the target to then be passed to the azimuth compression step, otherwise the target will be blurred in the final SAR image.

Before range cell migration correction (RCMC) is performed, the Doppler centroid must be estimated. This is a key step in processing SAR data as errors in the estimate result in degraded SNR, sidelobe levels, and ambiguity levels. As referenced in Section 2.2.5, several techniques have been developed for this purpose. A rather simple technique for illustration is the average phase change method [12]. This method works by taking the phase change between two azimuth cells for each range cell and finding the mean as

$$f_{DC} = \frac{\sum_{i=1}^{N-1} B_i}{2\pi N PRF} \qquad \text{(Hz)}, \tag{3.25}$$

where N is the number of azimuth angles,

$$B_i = \angle\left\{\sum_{j=1}^{M} s_{i,j} \times s_{i+1,j}^*\right\} \qquad \text{(rad)}, \tag{3.26}$$

M is the number of range cells, and $s_{i,j}$ is the range-Doppler data. The correction for range migration at every Doppler frequency is then

$$R_{\mathit{offset}}^{m,n} = \frac{R_m}{\sqrt{1 - \left(\frac{\lambda \Delta f\, n}{2v_{\mathit{eff}}}\right)^2}} - R_m \qquad \text{(m)}, \tag{3.27}$$

where

$$\Delta f = \frac{PRF}{N} + f_{DC} \qquad \text{(Hz)}. \tag{3.28}$$

Similar to the range-Doppler conversion, RCMC may also be performed in a blockwise manner.

The final step is compression of the data in the azimuth direction, which is accomplished with a matched filter. This requires the creation of an azimuth reference function and makes use of the estimated Doppler centroid in (3.25). As was shown in Chapter 2, the Doppler frequency in the azimuth direction is an LFM function. The frequency slope for the azimuth reference function is given by

$$\beta_{az} = \frac{2v_{\mathit{eff}}^2}{\lambda R_s} \qquad \text{(Hz)}, \tag{3.29}$$

where R_s is the squinted range for each image cell location. This is expressed as

$$R_s^{m,n} = \frac{R_m}{\sqrt{1 - \left(\frac{\lambda\,\Delta f\,n}{2v_{\mathit{eff}}}\right)^2}} \qquad \text{(m)}. \tag{3.30}$$

Therefore, each range cell requires a unique reference function, as shown in Figure 3.9.

3.3.2 Polar Format Algorithm

The polar format algorithm (PFA) is perhaps one of the more straightforward and easy-to-understand image formation techniques. PFA was originally developed for ISAR imaging [13], but will be treated here in a more generic sense.

Recall the signal model given in (3.2) with the instantaneous frequency given in (3.5). By collecting data at several azimuth angles and a constant range (spotlight mode), the signal's wavenumber space may be written as

$$k_x = k\cos\left(\theta_n\right) \qquad \text{(rad/m)}, \tag{3.31}$$

$$k_y = k\sin\left(\theta_n\right) \qquad \text{(rad/m)}, \tag{3.32}$$

where θ_n are the azimuth sample angles and

$$k = \frac{2\pi}{\lambda} = \frac{2\pi}{c}\left[f_0 + \beta(t - t_0)\right] \qquad \text{(rad/m)}. \tag{3.33}$$

The wavenumber is the signal's spatial frequency and is measured in cycles per unit distance. From (3.31)–(3.33), the wavenumber space is filled in a polar raster as shown in Figure 3.10. If the data filling the wavenumber space were continuous, an image could be formed by taking the inverse Fourier transform as

$$s(x,y) = \int_{-\infty}^{\infty}\int_{-\infty}^{\infty} S\left(k_x\,x, k_y\,y\right)\exp\left[j\left(k_x\,x + k_y\,y\right)\right]dk_x\,dk_y. \tag{3.34}$$

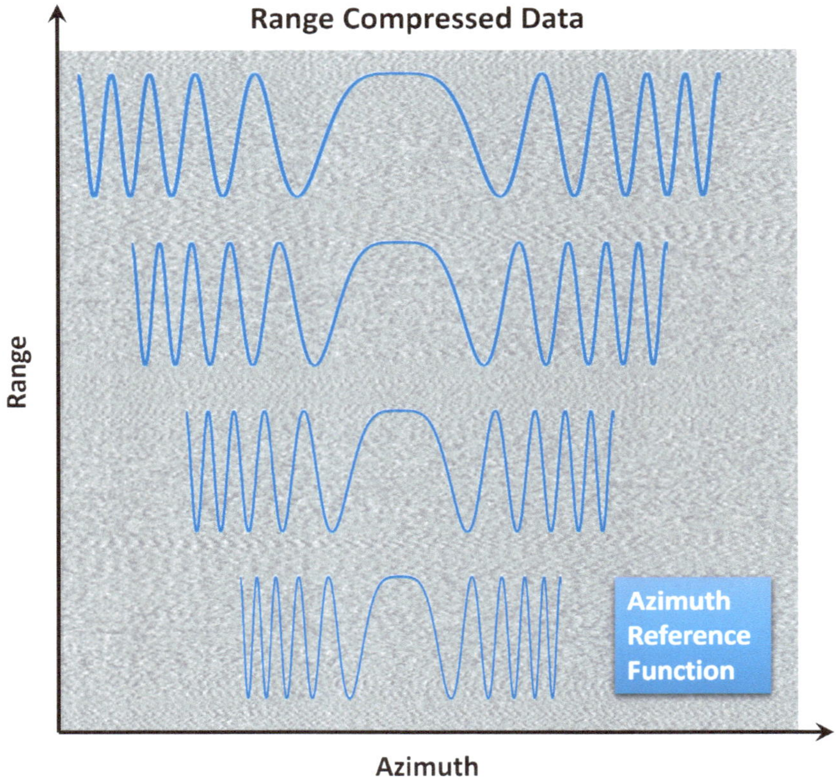

Figure 3.9 Azimuth reference function variation versus range.

Since modern radar systems make use of digital signal processing, the two-dimensional inverse discrete Fourier transform is employed, and (3.34) is replaced with

$$s(x, y) = \sum_{m-1}^{M} \sum_{n=1}^{N} S\big(k_x^{m,n}\, x, k_y^{m,n}\, y\big) \exp\Big[j\big(k_x\, x + k_y\, y\big)\Big], \tag{3.35}$$

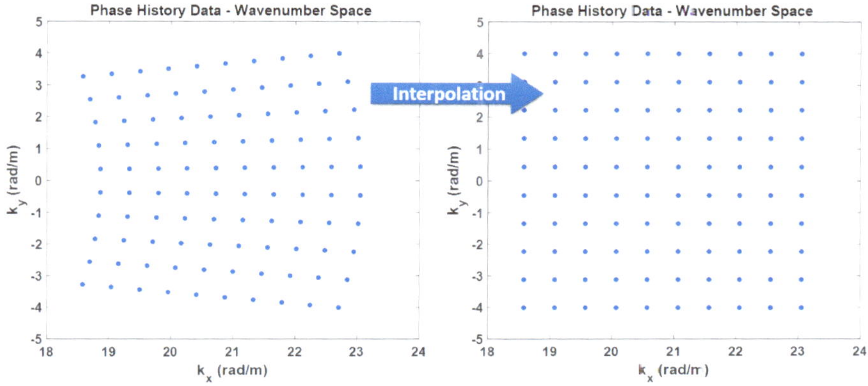

Figure 3.10 Spotlight SAR phase history data in wavenumber space.

where the superscripts m and n represent the samples in the range and azimuth directions, respectively. Implementing (3.35) results in a perfectly acceptable image. However, performing the double summation for every (x, y) location is computationally expensive. The kernel in (3.35) is not suited for fast Fourier transform (FFT) methods as these work on regular grids [14]. If the bandwidth is small compared to the center frequency and the span of angles is also small,

$$\beta T \ll 2\pi f_0 \qquad \text{(Hz)}, \tag{3.36}$$

$$\Delta\theta \ll 1 \qquad \text{(deg)}, \tag{3.37}$$

then the wavenumber space may be approximated as Cartesian. For high-resolution systems, the approximation is invalid and a method for reformatting the data onto a Cartesian grid is necessary. This is commonly accomplished through two one-dimensional interpolations. The first interpolation is performed in the range direction and the second along the azimuth direction. Interpolation of complex data is not as simple as it may seem and depends on the nature of the data itself. A poor interpolation scheme will add artifacts in the final image. Also, interpolation processing dominates the processing times for classical implementations of PFA. The reader is referred to the following sources for more information about complex data interpolation [15–17]. This results in the wavenumber sample locations shown in Figure 3.10. Note k_x only depends on the index m (range), and k_y only depends on the index n (azimuth). The inverse Fourier transform of (3.35) is now separable, and for performance may be calculated by two one-dimensional transforms as

$$s(x, y) = \sum_{n=1}^{N} \left\{ \sum_{m=1}^{M} S\big(k_x^m, k_y^n\big) \exp\left[jk_x^m\, x\right] \right\} \exp\left[jk_y^n\, y\right]. \tag{3.38}$$

While an image may be formed in this manner, the high-frequency data are further apart than the low-frequency data, resulting in image degradation [1]. Also, it was assumed the wavefront is planar and residual phase errors were ignored. These approximations require further post processing to mitigate image artifacts [18]. Therefore, other image formation techniques are desired.

3.3.3 Backprojection Algorithm

The image formation algorithms presented in the previous sections make certain assumptions about range cell migration, platform motion, and homogeneity of the point spread function [19]. The backprojection imaging algorithm (BPA) has far fewer assumptions and therefore offers several advantages over other imaging methods. BPA is the same regardless of the imaging mode, while other methods typically need modification depending on the imaging mode and geometry. Also, BPA is well suited for the bistatic and multistatic imaging geometries discussed in Section 2.9. Unambiguous aliasing, often called spectral wrapping, is problematic for many imaging methods, while BPA is unaffected [20]. Since BPA effectively produces a matched filter on a pixel-by-pixel basis, higher order effects such as antenna gain compensation can be included on individual pulses. Convolution methods use the same response across the entire processing window and can only roughly address higher order effects. Many image formation techniques require a process known as orthorectification to map slant plane images to the ground plane and effectively remove effects of tilt and relief [21]. BPA forms images in the ground plane making orthorectification significantly simpler. With BPA there is no fundamental limit on the size of the imaging swath nor on the regularity of the relative velocity between the sensor and the swath.

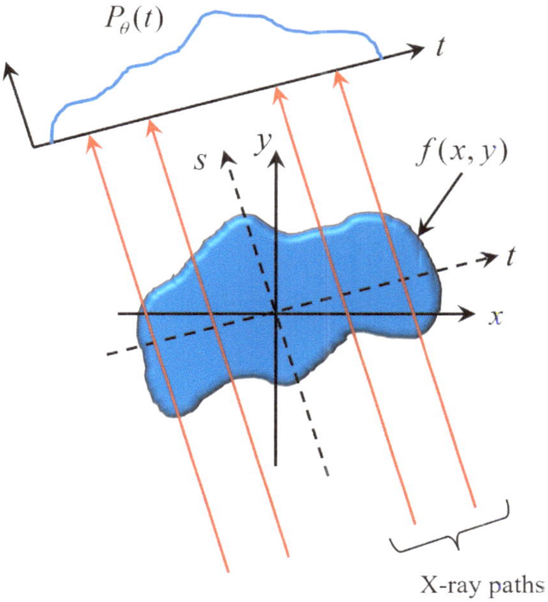

Figure 3.11 Basic geometry for two-dimensional X-ray imaging.

3.3.3.1 Line Integrals and Projections

Consider the basic geometry for two-dimensional X-ray tomography shown in Figure 3.11, where the object's two-dimensional density function is given by $f(x, y)$. $P_\theta(t)$ is the value of the Radon transform of the density function, $f(x, y)$, and is found by forming a set of line integrals. Mathematically, the Radon transform is expressed as [22]

$$P_\theta(t) = \int_l f(x, y) ds$$

$$= \int_{-\infty}^{\infty} \int_{-\infty}^{\infty} f(x, y)\delta(x \cos\theta + y \sin\theta - t)\, dx\, dy. \tag{3.39}$$

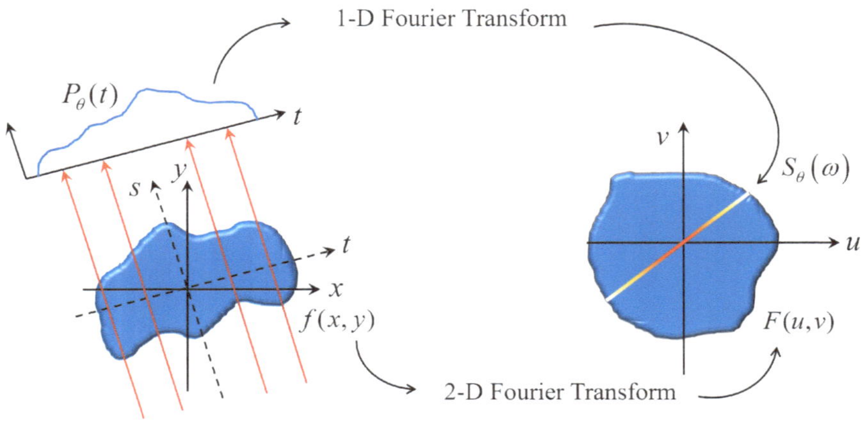

Figure 3.12 Illustration of the Fourier slice theorem.

Once the Radon transform is found, a method for reconstructing the original density function is needed. Image reconstruction is accomplished with the use of the Fourier slice theorem and the filtered backprojection algorithm.

3.3.3.2 Fourier Slice Theorem

The Fourier slice theorem, illustrated in Figure 3.12, states the one-dimensional Fourier transform of a parallel projection is equal to a slice of the two-dimensional Fourier transform of the original function [23]. To demonstrate the Fourier slice theorem, start by defining the two-dimensional Fourier transform of the object's density function as

$$\mathcal{F}(u, v) = \int_{-\infty}^{\infty}\int_{-\infty}^{\infty} f(x, y) \exp\Big[-j2\pi\big(ux + vy\big)\Big]\, dx\, dy, \tag{3.40}$$

and the one-dimensional Fourier transform of the projection function, $P_\theta(t)$, as

$$S_\theta(\omega) = \int_{-\infty}^{\infty} P_\theta(t) \exp\left[-j2\pi\omega t\right] dt. \tag{3.41}$$

Since no assumption has been made about the object's orientation, θ is set to 0 without loss of generality. This leads to $v = 0$, and allows the two-dimensional Fourier transform to be written as

$$\mathcal{F}(u, 0) = \int_{-\infty}^{\infty}\int_{-\infty}^{\infty} f(x, y) \exp\left[-j2\pi ux\right] dx\, dy. \tag{3.42}$$

Note the phase term is now a function of x only, and the double integral may be written as

$$\mathcal{F}(u, 0) = \int_{-\infty}^{\infty} \left[\int_{-\infty}^{\infty} f(x, y) dy\right] \exp\left[-j2\pi ux\right] dx. \tag{3.43}$$

The term inside the brackets is an expression for the projection along lines of constant x. This may be written as

$$P_{\theta=0}(x) = \int_{-\infty}^{\infty} f(x, y) dy. \tag{3.44}$$

Substituting (3.44) into (3.43) results in

$$\mathcal{F}(u, 0) = \int_{-\infty}^{\infty} P_{\theta=0}(x) \exp\left[-j2\pi ux\right] dx. \tag{3.45}$$

Recalling the definition of the one-dimensional transform in (3.41), the relationship between the one-dimensional projection function and the two-dimensional Fourier transform of the object's density function is

$$\mathcal{F}(u,0) = S_{\theta=0}(\omega). \tag{3.46}$$

As stated earlier, no assumption was made about the object's orientation; therefore, the general expression of the Fourier slice theorem is

$$\mathcal{F}(u,v) = S_{\theta}(\omega). \tag{3.47}$$

While Radon's original work showed the mathematical procedure for reconstruction of a density function from projections, the inverse Radon transform cannot be solved analytically. Some of the earliest reconstruction approaches were algebraic and those will be covered in the following sections. Researchers began using Fourier transforms in the solution, leading to a method known as filtered backprojection [24]. It is the most used reconstruction technique in medical imaging and is an extremely attractive method for SAR imaging [25, 26]. It is computationally efficient and employs FFTs. It is highly parallelizable as each projection's contribution and each pixel's value may be computed independently. Also, image reconstruction may begin as soon as the first projection function is obtained, rather than waiting for all projections to be collected. Finally, the filtered backprojection algorithm does not depend on uniform angular sampling in the projection data, and multiple rotations or platform passes can be used.

3.3.3.3 Filtered Backprojection

To illustrate the filtered backprojection algorithm, begin by taking the inverse Fourier transform of the object's two-dimensional spatial frequency domain representation as

$$f(x,y) = \int_{-\infty}^{\infty}\int_{-\infty}^{\infty} \mathcal{F}(u,v)\exp\Big[j2\pi\big(ux+vy\big)\Big]\,du\,dv. \tag{3.48}$$

Since the projection functions fill in the spatial frequency domain in a polar fashion, make the following substitutions

$$u = \omega \cos\theta, \tag{3.49}$$

$$v = \omega \sin\theta, \tag{3.50}$$

$$du\,dv = \omega\,d\omega\,d\theta \tag{3.51}$$

which results in

$$f(x,y) = \int_0^{2\pi}\int_0^{\infty} \mathcal{F}(\omega,\theta)\exp\Big[j2\pi\omega\,(x\cos\theta + y\sin\theta)\Big]\omega\,d\omega\,d\theta. \tag{3.52}$$

Splitting (3.52) into two integrals gives

$$f(x,y) = \int_0^{\pi}\int_0^{\infty} \mathcal{F}(\omega,\theta)\exp\big[j2\pi\omega t\big]\omega\,d\omega\,d\theta$$

$$+ \int_0^{\pi}\int_0^{\infty} \mathcal{F}(\omega,\theta+\pi)\exp\big[j2\pi\omega t'\big]\omega\,d\omega\,d\theta, \tag{3.53}$$

where $t = x\cos\theta + y\sin\theta$ and $t' = x\cos(\theta+\pi) + y\sin(\theta+\pi)$. Making use of the relation, $\mathcal{F}(\omega,\theta+\pi) = \mathcal{F}(-\omega,\theta)$, allows (3.53) to be written as

$$f(x,y) = \int_0^{pi}\int_0^{\infty} \mathcal{F}(\omega,\theta)\exp\Big[j2\pi\omega\,(x\cos\theta + y\sin\theta)\Big]|\omega|\,d\omega\,d\theta. \tag{3.54}$$

The inner integral represents the convolution of the projection with a filter as

$$P_\theta(t) * h(t) = \int_{-\infty}^{\infty} \mathcal{F}(\omega,\theta)\exp\big[j2\pi\omega(x\cos\theta + y\sin\theta)\big]|\omega|\,d\omega. \tag{3.55}$$

The filtered projection function in (3.55) is then projected onto the image space along a line in the same direction as the projection was obtained. This is expressed as

$$f(x, y) = \int_0^\pi P_\theta(t) * h(t)\, d\theta. \tag{3.56}$$

The backprojection imaging algorithm is as follows:

1. Construct the spatial domain target density array $f(x_i, y_j)$.
2. Pulse-compress the return signal (one-dimensional range profiles) and perform filtering.
3. Upsample the return signal for improved image quality. Upsample factor can be large ($>$100).
4. For a given pixel point, (x_i, y_j), find the nearest upsampled time point (interpolate) and add the value of the filtered data to the array component $f(x_i, y_j)$.
5. Repeat for all pixel locations (x_i, y_j) and all projections $P_\theta(t)$.

When processing the reflected waveform of a radar system as direct transductions of the spatial frequencies of a target's reflectivity function, the spatial frequency formulation makes the SAR imaging problem analogous to X-ray medical imaging [27]. Consider the imaging scenario where energy from the radar illuminates an area that contains a target to be imaged, as shown in Figure 3.13. Assume the target structure across the illuminated region is described by a two-dimensional reflectivity such that the radar returns from all scatterers on a constant range line are received simultaneously by the radar. In the previous section, the projection, $P_\theta(t)$, represented a projection of the X-ray attenuation function of an object. For SAR imaging, the projection represents a projection of the reflectivity function of a target. The projections may now be collected at several angles, as illustrated in Figure 3.14. Using the Fourier slice theorem as before, the projection data is used to fill in the reflectivity function's two-dimensional spatial frequency domain. The filtered backprojection algorithm is then used to reconstruct the reflectivity function from this two-dimensional spatial frequency data.

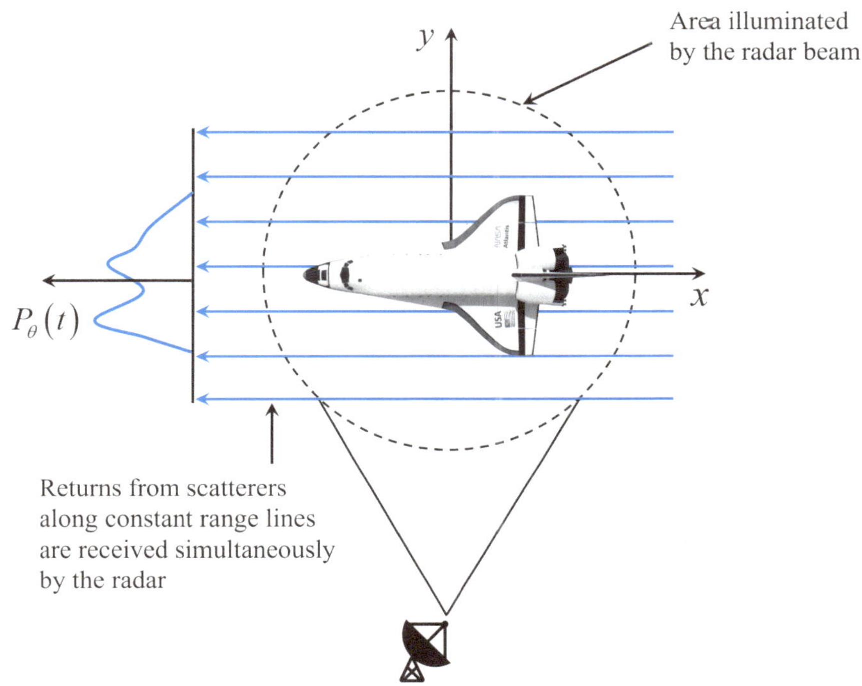

Figure 3.13 Basic geometry for tomographic SAR imaging. Courtesy NASA.

Watch this animation of backprojection.

3.3.4 Algebraic Methods

Another approach for SAR image formation is to assume the image consists of an array of unknowns, set up algebraic equations for the unknowns in terms of the measured data, and solve for these unknowns in an iterative manner [28, 29]. Methods in this class are known as algebraic reconstruction techniques (ART). Conceptually, ART is simpler than the transform-based methods in Sections 3.3.1 and 3.3.2. However, the transform-based methods are simpler to implement. As with

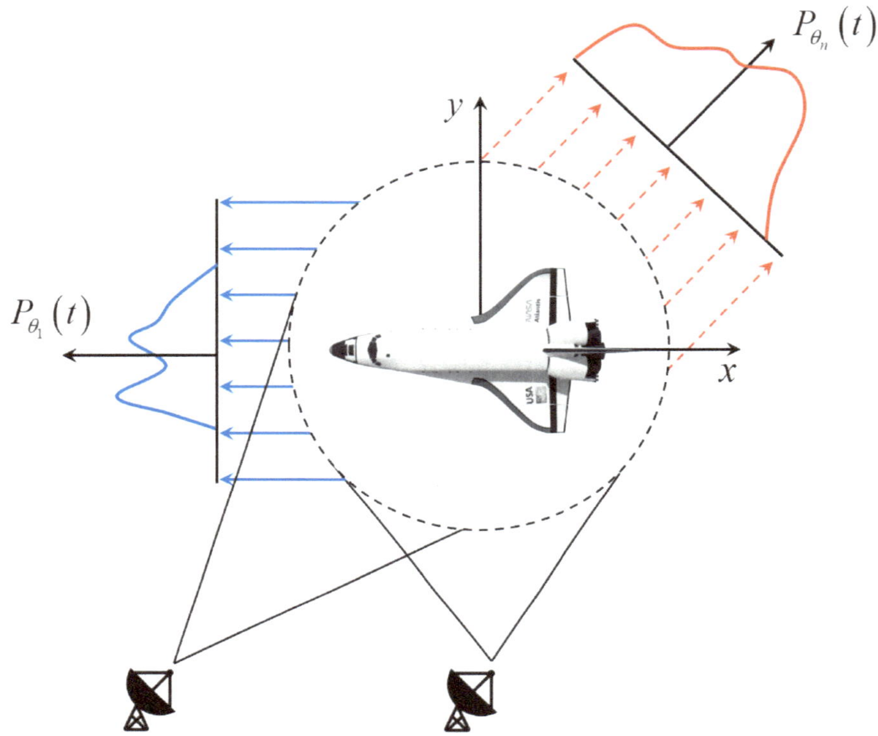

Figure 3.14 Projections collected at various angles due to target or sensor motion. Courtesy NASA.

BPA, ART inherently handles irregular sampling along the synthetic aperture and arbitrary image shapes. ART is also an attractive method when the transmitted and received energy undergoes refraction due to the atmosphere [28]. In these cases, it is also possible to combine ART with ray tracing methods to first create an image ignoring refraction, then use ray tracing to come up with a more accurate set of algebraic equations [30, 31]. However, scenarios dealing with refraction often have effects from diffraction as well.

Figure 3.15 shows the basic geometry for tomographic SAR imaging with a rectangular grid superimposed onto the image $f(x, y)$. The measurements are stacked such that

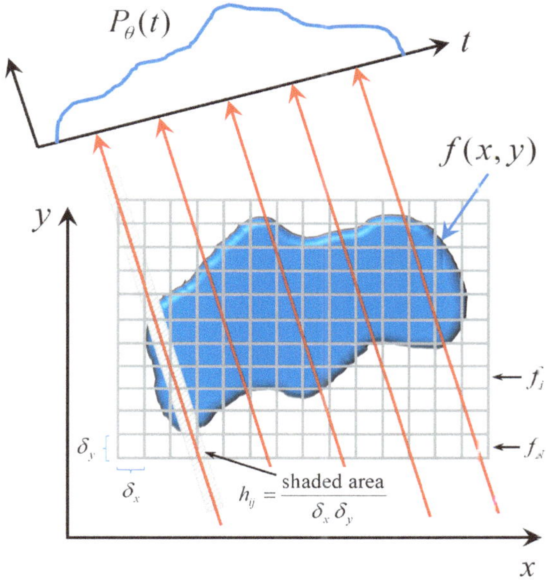

Figure 3.15 Basic geometry for tomographic SAR imaging with grid overlay for ART.

$$\vec{p} = \mathbf{H}\vec{f}, \tag{3.57}$$

where f_j are the pixels in the image, p_i are the measurements and $H_{i,j}$ represents the contribution of the *j*th pixel to the *i*th measurement. Note $f : \mathbb{R}^N$, $p : \mathbb{R}^M$, and $\mathbf{H} : \mathbb{R}^{MXN}$. The element $H_{i,j}$ is the fractional area of the *j*th pixel intercepted by the *i*th measurement, as shown in Figure 3.15. Note, $\mathbf{H}$ is sparse since a small number of pixels contribute to a line integral.

For small number of pixels, N, and small number of measurements, M, conventional matrix theory methods could be used to invert (3.57). In practice, M and N are far too large any direct matrix inversion. Thus, iterative methods must be employed for solving (3.57). Iterative methods to solve this problem were originally developed for computed tomography (CT) and electron microscopy. These methods are based on the method of projections and were first proposed in [32] and later covered in detail in [33]. These methods result in a least-squares solution meeting the constrained minimization

$$\min_{i} \|\vec{f}_i\|_2^2 \qquad \text{such that} \qquad \mathbf{H}\vec{f} = \vec{p}. \tag{3.58}$$

The ART algorithm proceeds by initializing the image pixel values $\vec{f}$. If no a priori information is available, then set $\vec{f}^{\,0} = 0$. The iterative step is then computed by

$$f_j^{k+1} = f_j^k + \frac{p_i - \langle h_i, f^k \rangle}{\|h_i\|} H_{i,j}, \tag{3.59}$$

where f_j^k is the *j*th element of $\vec{f}$ at the *k*th iteration, h_i is the *i*th row of $\mathbf{H}$, and $H_{i,j}$ is the *(i,j)*th element of $\mathbf{H}$.

As stated above, an image $\vec{f}$ with N pixels has N degrees of freedom. The image may be considered as a single point in N-dimensional space. Writing out each equation in (3.57) gives a representation of a hyperplane. If a unique solution exists, the intersection of all the hyperplanes will be a point. If the hyperplanes could be made orthogonal to each other, then the solution would be found with a single iteration. While such an orthogonalization is possible, it is not computationally feasible.

If $M > N$, then the system of equations is said to be overdetermined. For this case, no unique solution exists, and the iterative solution does not converge, but rather oscillates about the intersection of the hyperplanes [28].

If $M < N$, no unique solution exists, and an infinite number of solutions are possible. For example, with (3.57), the solution can be anywhere on the hyperplane. The best attempt would be drawing a projection from $\vec{f}^{\,0}$ onto this hyperplane. The solution is then the point on the hyperplane closest to the initial guess. It has been shown that the iterative approach would then converge to a solution $\vec{f}_s$, such that $|\vec{f}^{\,0} - \vec{f}_s$ is minimized [33].

3.3.4.1 Simultaneous Iterative Reconstruction Technique

Simultaneous iterative reconstruction technique (SIRT) uses the same iterative correction as given in (3.59) to compute the change in the jth pixel due to the ith equation. The major difference is the value of the jth pixel is not modified until all the measurements have been processed. At that point, the pixel value is modified and is effectively the average of all the calculated changes for that pixel. This establishes a single iteration in the SIRT algorithm. Each successive iteration goes back through all the measurements and a new average change is computed. SIRT has the expense of slower convergence but tends to produce higher quality images [29, 34].

3.3.4.2 Simultaneous Algebraic Reconstruction Technique

The simultaneous algebraic reconstruction technique (SART) approach is a variation of the algebraic reconstruction techniques in Sections 3.3.4 and 3.3.4.1 and combines the best aspects of ART and SIRT [35]. Specifically, this reconstruction method produces quality images and numerical accuracy in a single iteration. Errors in the calculation of the measurement integrals by finite summation is replaced with bilinear elements. To reduce noise in the measured data, SART applies the correction term for all measurements in one projection simultaneously. Also, the quality of the reconstructions is further enhanced through heuristic procedures [28, 35, 36].

3.3.4.3 Multiplicative Algebraic Reconstruction Technique

The minimum norm regularization in ART, given in (3.58), is quite common but not the only regularization used to obtain a unique solution to the SAR imaging problem. Multiplicative algebraic reconstruction technique (MART) is a maximum entropy solution satisfying

$$\min_{f} \sum_{i} \vec{f} \ln \vec{f_i} \quad \text{such that} \quad \mathbf{H}\vec{f} = \vec{p}. \tag{3.60}$$

Maximum entropy methods are based on information theory and treats the image as a probability density function. Prior to use in SAR imaging, maximum entropy methods were suggested for use in both radio astronomy and CT image reconstruction [37, 38]. Maximum entropy methods tend to have uniform intensity distributions and work well in situations with large areas of weak scattering or strong scattering.

The MART algorithm begins by initializing the image pixel values as $\vec{f}^0 = \exp(-1)$. The updates are then calculated by

$$f_j^{k+1} = f_j^k \left[\frac{p_i}{\langle h_i, x^k \rangle} \right]^{\lambda, h_{i,j}}, \tag{3.61}$$

where λ is a relaxation value. There are constraints on the form of (3.61) for the algorithm and a few others, such as Elfving MART, Willis MART, which have been developed [39].

3.3.5 Nontraditional Methods

In addition to the image formation techniques presented in the preceding sections, various other algorithms have been researched. These include omega-k, overlapped subaperture, multiple signal classification (MUSIC), and amplitude and phase estimation of a sinusoid (APES). These are beyond the scope of this book and the reader is referred to the following sources for more information on these and other techniques for SAR image formation [40–42].

3.4 EXAMPLES

The sections below illustrate the concepts of this chapter with several Python/MATLAB examples. The examples for this chapter are in the directory *software/python/Chapter3* and the matching MATLAB examples are in the directory *software/matlab/Chapter3*. The reader should consult Chapter 1 for information on how to execute the Python and MATLAB code associated with this book.

3.4.1 Range Profile — Point Targets

As a first example, consider an LFM waveform transmitted with a bandwidth of 15 Hz incident upon a point-target located at a time delay of 3.5 s. The real and imaginary part of the return signal is shown in Figure 3.16 and the associated range profile is given in Figure 3.17. The time delay is evident in both these figures. The instantaneous frequency is given in Figure 3.18, and the frequency spectrum is shown in Figure 3.19. Comparing Figure 3.18 and Figure 3.19, the bandwidth and instantaneous frequency are consistent. The above example is given in the Python notebook *range_profile_point_target_example.ipynb* and the MATLAB live script *range_profile_point_target_example.mlx*.

3.4.2 Range Profile — Backhoe

As a more realistic example, consider the backhoe shown in Figure 3.20. There are many scattering mechanisms associated with this type of complex target contributing to the total backscattered field. The scattering data was simulated at X-band with a high-frequency electromagnetic scattering code and provided by the U.S. Air Force Sensor Data Management System (SDMS) [43]. Range profiles for the backhoe are shown in Figure 3.21. The bandwidth is 3 GHz, corresponding to a range resolution of approximately 5 cm, and VV, HH, and HV polarizations are shown. The copolarized signals (VV, HH) typically have higher intensity and are due to specular, surface, and volume scattering. Cross-polarized signals (HV, VH) usually have weak intensity and are associated with multiple bounce type scattering and doubly curved surfaces and are very dependent on the orientation of the targets. For comparison, Figure 3.22 shows the range profiles at a bandwidth of 300 MHz, which corresponds to a range resolution of 1 m. Figures 3.21 and 3.22 illustrate the dependency of range resolution on waveform bandwidth, as much more target detail is discernable at the higher bandwidth. This leads to better performance for classification and discrimination algorithms. Figures 3.21 and 3.22 were created at an elevation angle of 0°. Figure 3.23 shows the range profiles at an elevation angle of 30° and a bandwidth of 3 GHz. Note the shorter apparent length of the target in Figure 3.23 due to the range profile being created in the slant plane. The above example is given in the Python notebook *range_profile_backhoe_example.ipynb* and the MATLAB live script *range_profile_backhoe_example.mlx*.

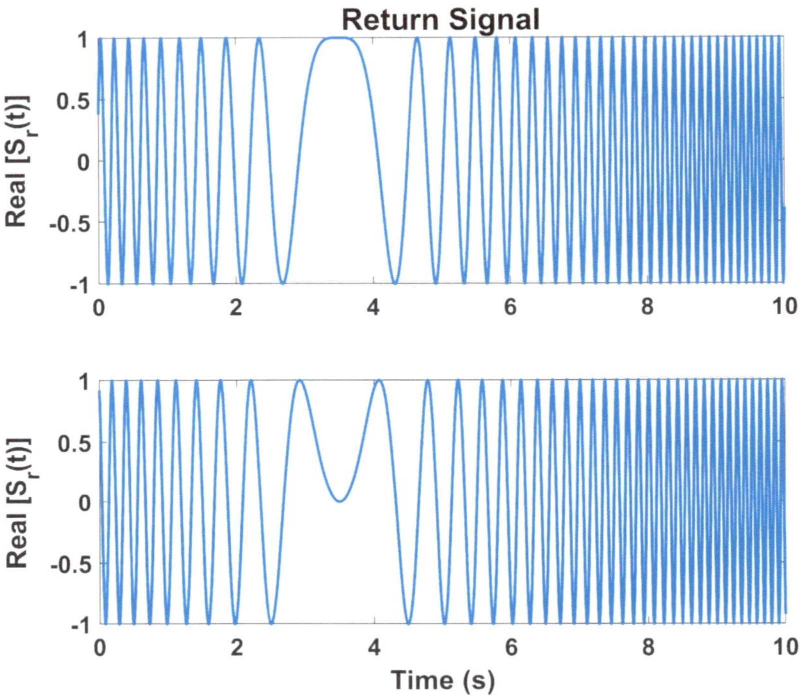

Figure 3.16 Real and imaginary components of the LFM return signal from a point target with a delay of 3.5 seconds.

3.4.3 Range Doppler Algorithm — Point Targets

Consider an imaging scene consisting of four point-targets of varying intensity. The point-targets are located at coordinates $x = [-20, 0, 16, 24]$ and $y = [-1, 8, -6, 16]$ meters relative to the scene center. The radar cross section associated with each target is $\sigma = [15, 8, 10, 15]$ m^2. The operating frequency is 10 GHz, the bandwidth is 100 MHz, and the pulsewidth is 200 μs. The synthetic aperture length is 200 m and the range to the scene center is 10 km. The range-compressed data is shown in Figure 3.24(a) and the range cell migration corrected data is given

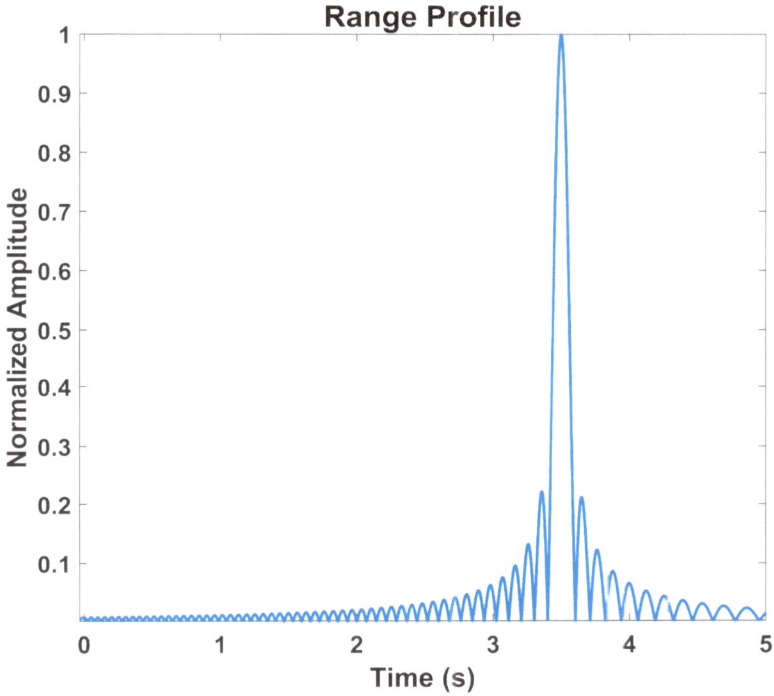

Figure 3.17 One-dimensional range profile of a point-target with a delay of 3.5 seconds.

in Figure 3.24(b). Azimuth compression completes the process, and the final focused image is shown in Figure 3.25. The above example is given in the Python notebook *range_doppler_point_target_example.ipynb* and the MATLAB live script *range_doppler_point_target_example.mlx*.

3.4.4 Range Doppler Algorithm — Backhoe

In this example, the range-Doppler algorithm is used to create two-dimensional imagery of the backhoe shown in Figure 3.20, and described in Section 3.4.2. Figure 3.26 shows the image obtained from RDA with an azimuth span of 10° and a waveform bandwidth of approximately 5 GHz. The figure also displays an

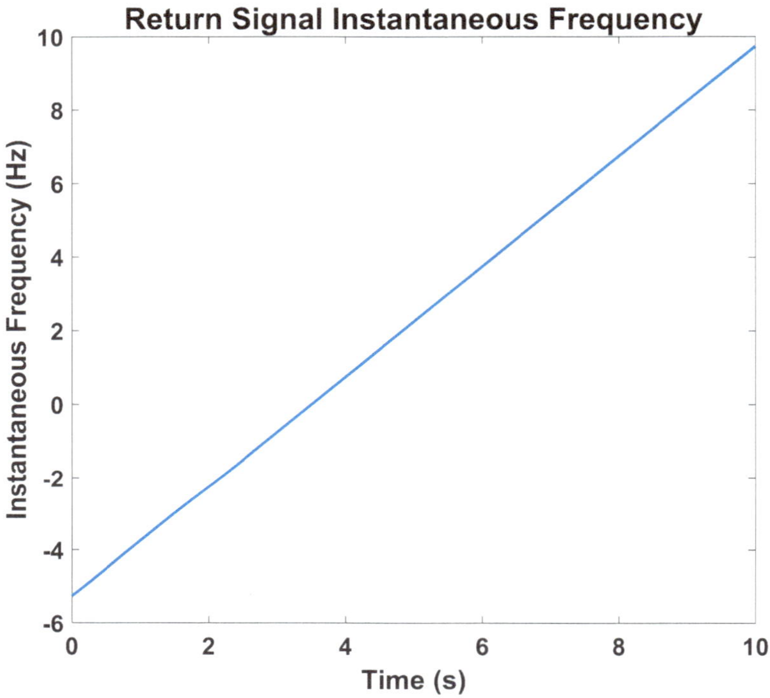

Figure 3.18 Instantaneous frequency of the LFM return signal from a point target with a delay of 3.5 seconds.

overlay of the target facet file used to simulate the data. The major scattering structures such as the front and rear wheels, front and rear buckets, and connecting arm framework are evident in the imagery. The above example is given in the Python notebook *range_doppler_example.ipynb* and the MATLAB live script *range_doppler_example.mlx.*

3.4.5 Polar Format Algorithm — Point Targets

This example examines PFA imaging for a scene consisting of four point-targets with different scattering intensities. The point targets are located at coordinates

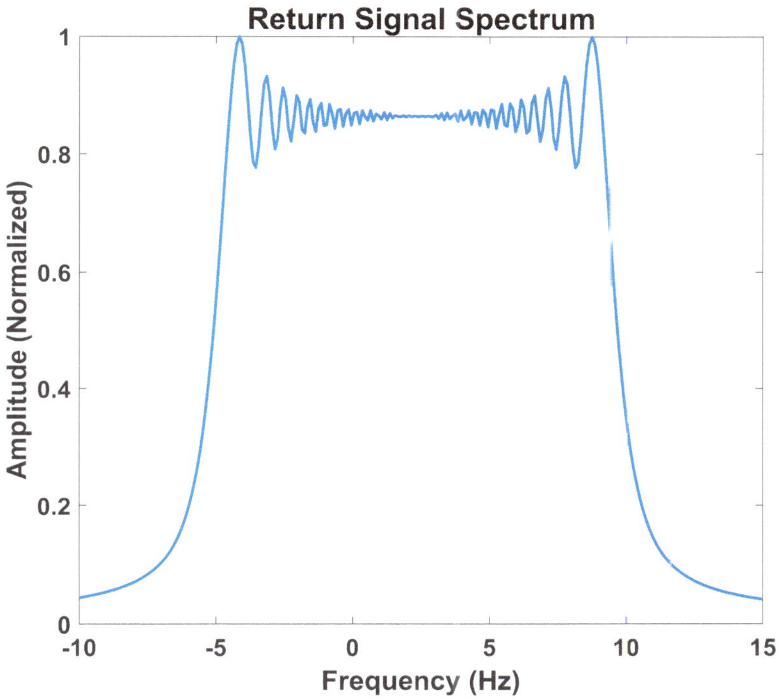

Figure 3.19 Frequency spectrum of the LFM return signal from a point target with a delay of 3.5 seconds.

of $x = [10, -5, 2.5, -2.5]$ and $y = [10, -5, -1, 8]$ meters relative to the scene center, and the range to the scene center is 1 km. The radar cross section of the point targets is $\sigma = [10, 10, 7, 5]$ m^2. The operating frequency is 5 GHz and the waveform bandwidth is 250 MHz. The azimuth span of the synthetic aperture is 6°. The raw phase history data for this scene is shown in Figure 3.27(a). The raw phase history data is then interpolated onto a Cartesian grid as shown in Figure 3.27(b). A one-dimensional inverse Fourier transform is computed in the range direction, followed by a one-dimensional inverse Fourier transform in the azimuth direction to give the results shown in Figure 3.28. The above example is given in the Python

Figure 3.20 CAD model of a backhoe used for one-dimensional range profiles. Courtesy of Walton C. Gibson.

notebook *polar_format_point_targets_example.ipynb* and the MATLAB live script *polar_format_point_targets_example.mlx*.

3.4.6 Polar Format Algorithm — Learjet

In this example, PFA is used to create two-dimensional SAR imagery of a Learjet, shown in Figure 3.29. The target was simulated at S-band using the high frequency radar cross-section code, Lucernhammer MT, from Tripoint Industries [44]. Figure 3.30 displays the image obtained from PFA with an azimuth span of 10° and a waveform bandwidth of 1 GHz. This figure also shows an overlay of the target facet file used to generate the simulated data. The major scattering structures are

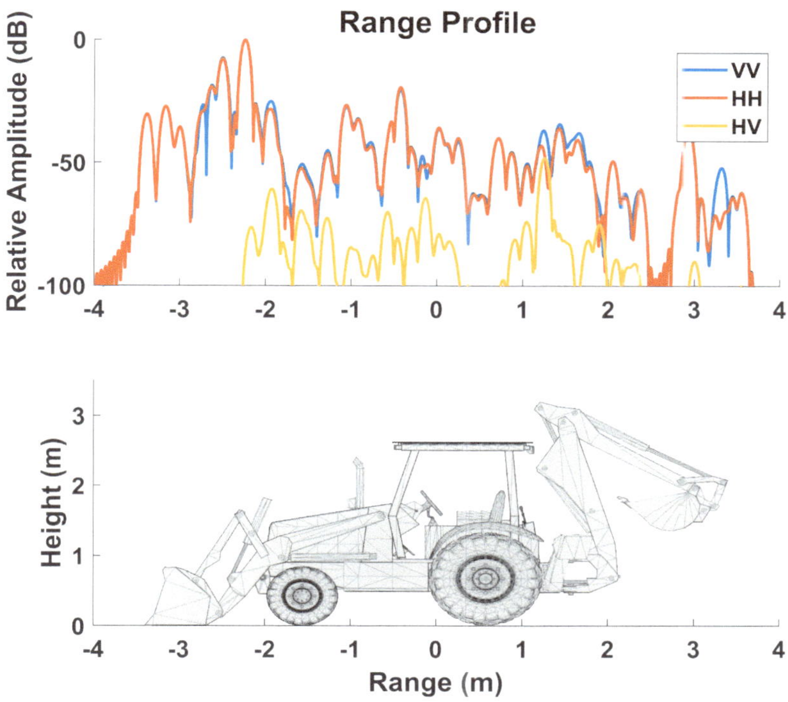

Figure 3.21 X-band range profiles for VV, HH, and HV polarizations at a bandwidth of 3 GHz.

evident in the SAR imagery. The above example is given in the Python notebook *polar_format_example.ipynb* and the MATLAB live script *polar_format_example.mlx*.

3.4.7 Backprojection Algorithm — Point Targets

For this example, consider an imaging scene consisting of five point-targets with different scattering intensities. The point targets are located at coordinates of $x = [3, 0, -3, -5, 8]$ and $y = [-3, 0, 8, 6, 4]$ meters relative to the scene center, and the range to the scene center is 50 km. The radar cross section of the point targets is $\sigma = [10, 7, 18, 15, 10]$ m^2. The operating frequency is 5 GHz, and the waveform bandwidth is 300 MHz. Figure 3.31(a) shows the image after a single pulse has

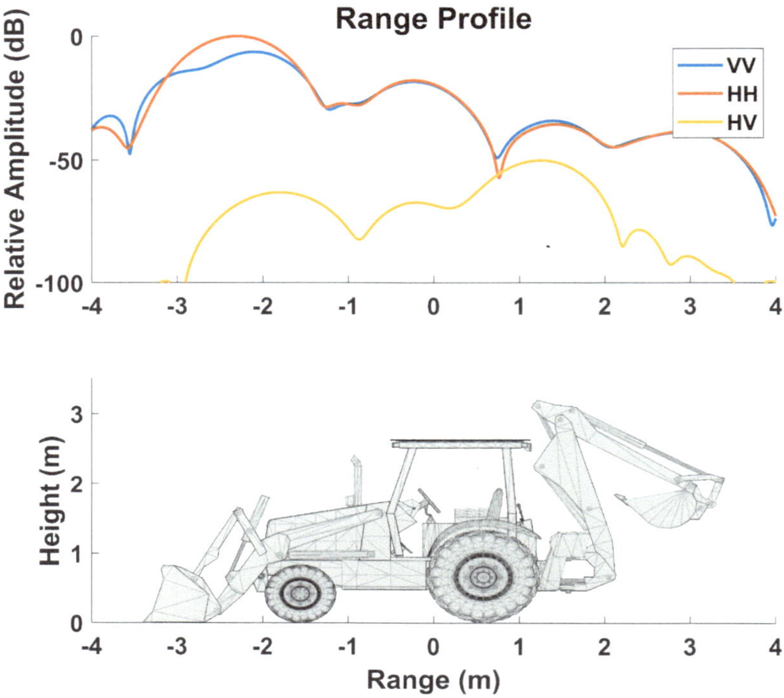

Figure 3.22 X-band range profiles for VV, HH, and HV polarizations at a bandwidth of 300 MHz.

been backprojected. Note the image is composed on range lines that are constant in the cross-range direction. Figure 3.31(b) shows the image after 100 pulses and the image is now starting to take form. Figure 3.32(a) shows the image after 500 pulses and Figure 3.32(b) shows the image after all pulses have been backprojected and the image is now fully created. The above example is given in the Python notebook *backprojection_point_targets_example.ipynb* and the MATLAB live script *backprojection_point_targets_example.mlx*.

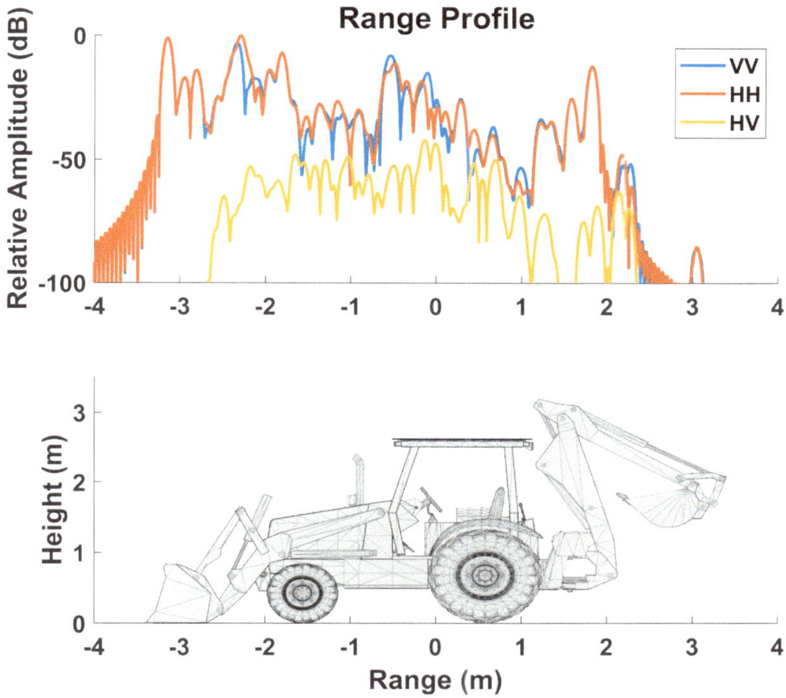

Figure 3.23 X-band range profiles for VV, HH, and HV polarizations at a bandwidth of 3 GHz and elevation angle of 30°.

3.4.8 Backprojection Algorithm — Toyota Avalon

For this example, consider the two-dimensional SAR imaging of a Toyota Avalon. The target was simulated at X-band with a high-frequency electromagnetic scattering code and provided by the U.S. Air Force SDMS [43]. Figure 3.33(a) shows the image obtained by backprojecting the data for a 6° azimuth span. As seen, little information about the target structure is present in this image. Since backprojection is well suited for large azimuth spans, images were created with 90°, 180°, and 360° azimuth spans. The results are given in Figure 3.33(b), Figure 3.34(a), and Figure 3.34(b) respectively. With increasing azimuth spans, more target detail is

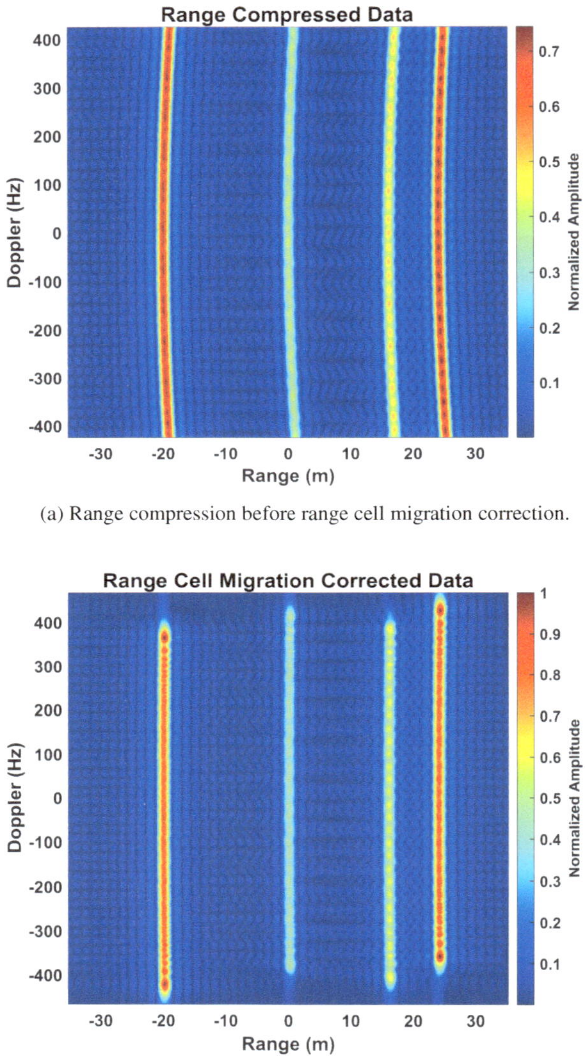

(a) Range compression before range cell migration correction.

(b) Range compression after range cell migration correction.

Figure 3.24 Range compression after range cell migration correction.

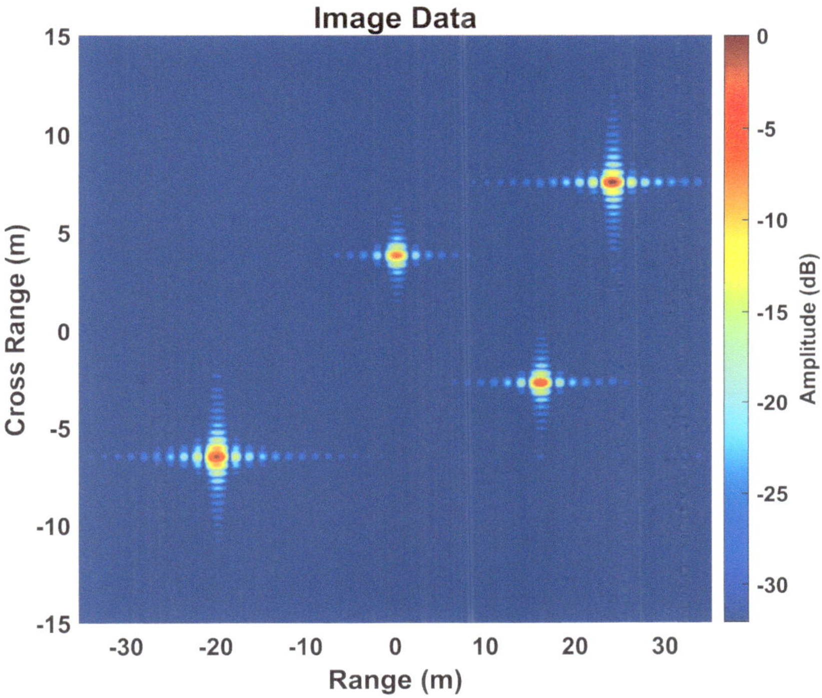

Figure 3.25 Focused image for range-Doppler imaging of point targets.

visible in the images, providing valuable information for mapping, environmental monitoring, classification, and discrimination. The above example is given in the Python notebook *backprojection_example.ipynb* and the MATLAB live script *backprojection_example.mlx*.

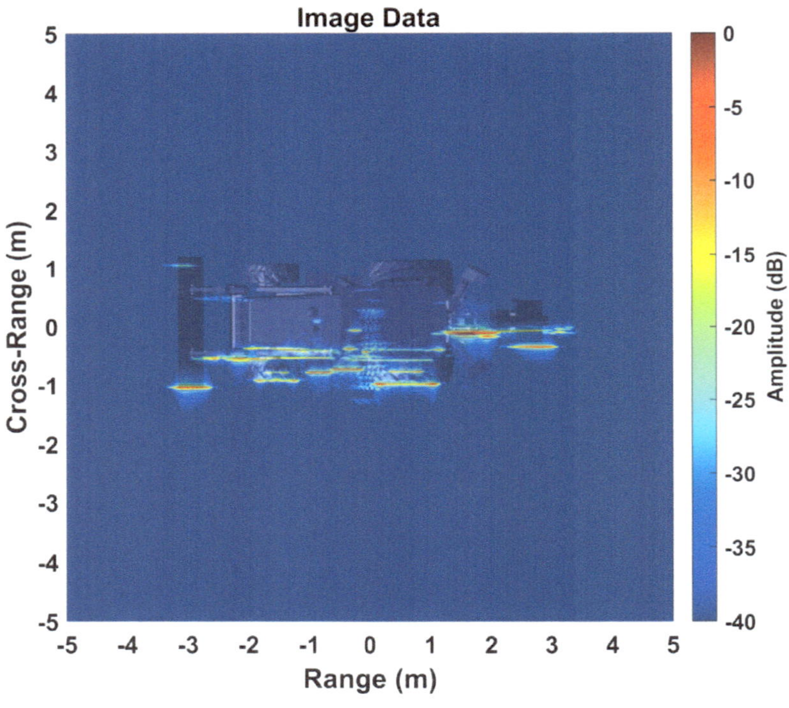

Figure 3.26 Range-Doppler image formation of a backhoe at X-band.

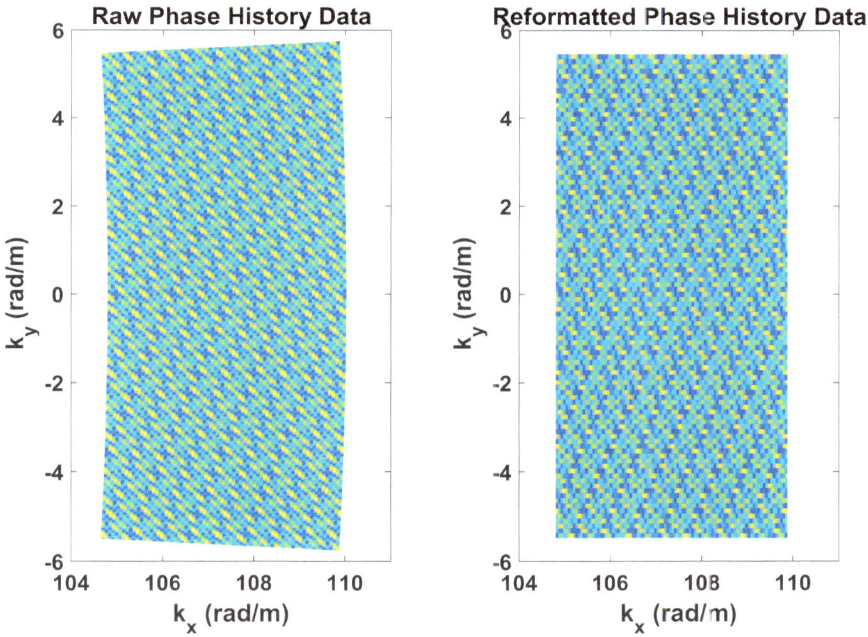

(a) Raw data. (b) Reformatted data.

Figure 3.27 Raw phase history data reformatted onto a Cartesian grid for polar format imaging.

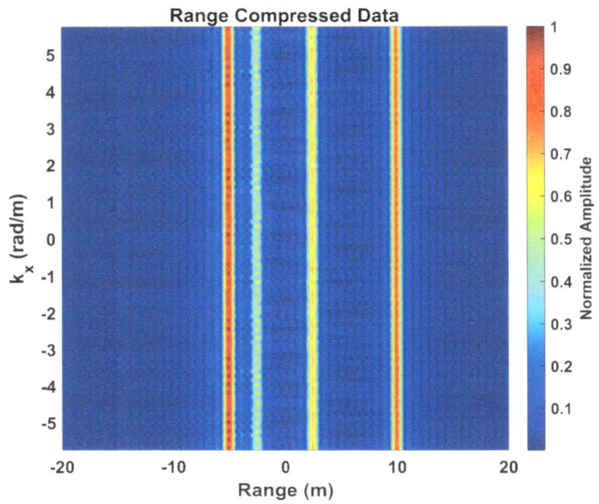

(a) PFA range compressed data for point targets.

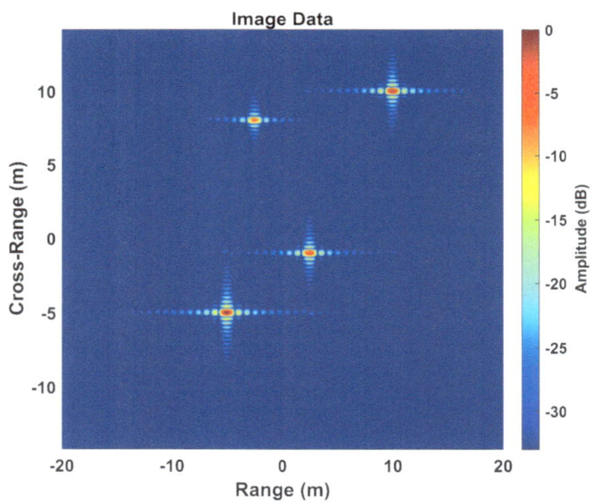

(b) PFA focused image of point targets.

Figure 3.28 Range and azimuth compression to form the focused image using the polar format algorithm.

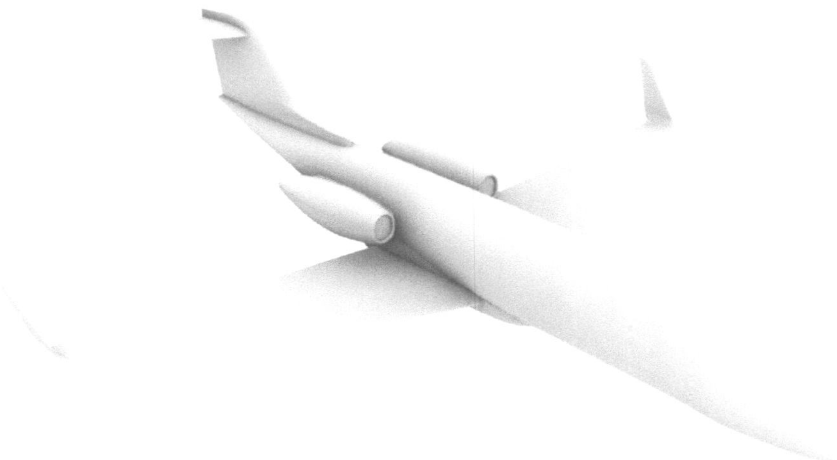

Figure 3.29 CAD model of a Learjet used for polar format imaging. Courtesy of Walton C. Gibson.

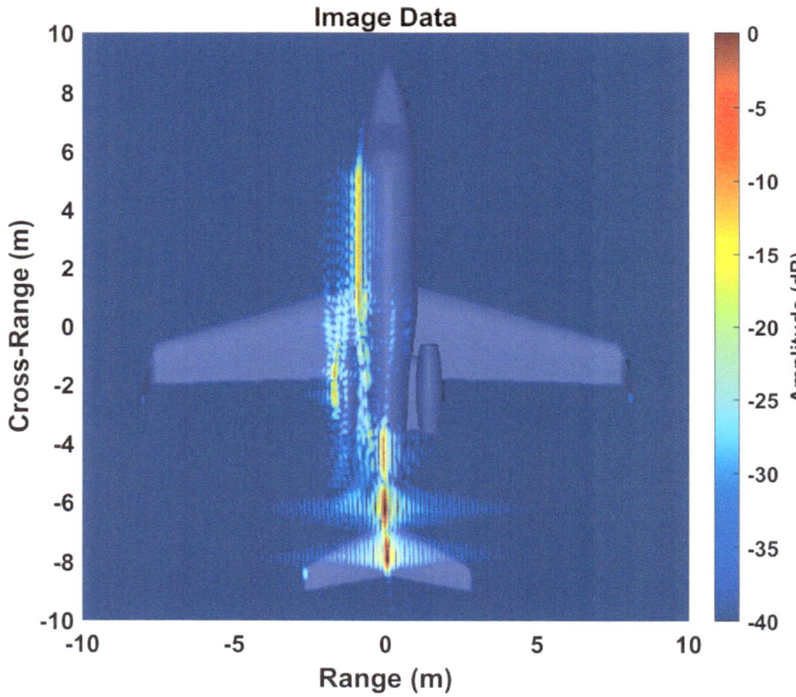

Figure 3.30 Polar format image formation of a Learjet at S-band.

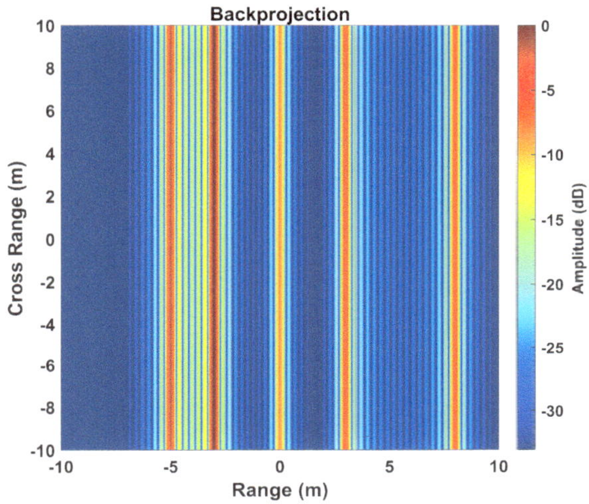

(a) Backprojected image after one pulse.

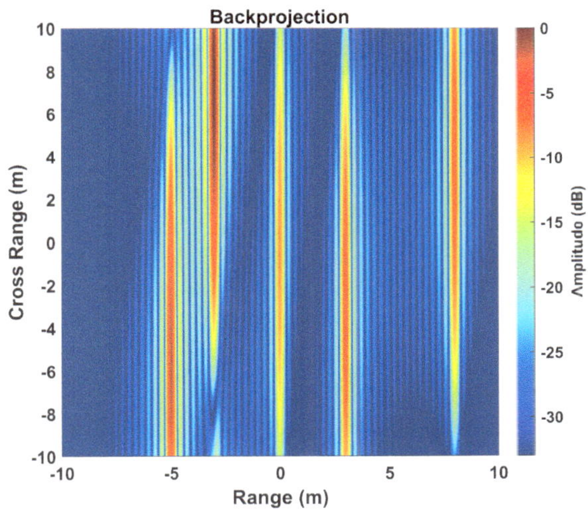

(b) Backprojected image after 100 pulses.

Figure 3.31 Backprojection image reconstruction of point targets.

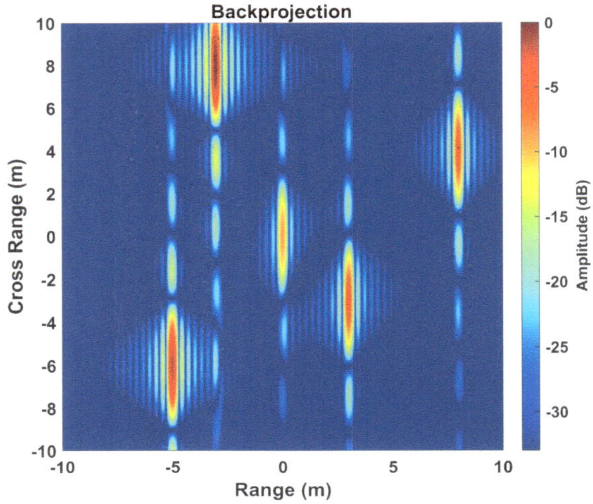

(a) Backprojected image after 500 pulses.

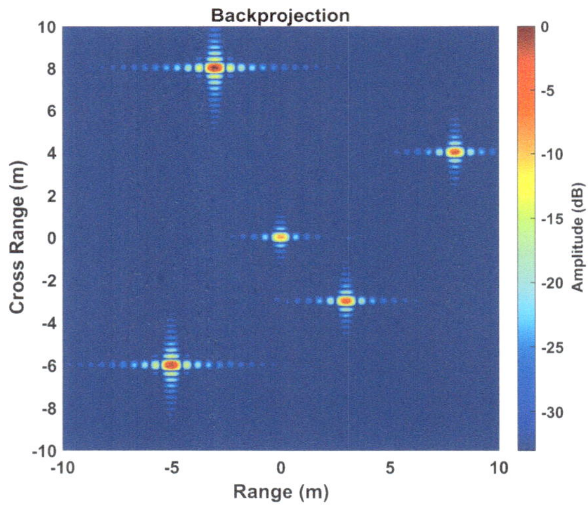

(b) Backprojected image with all pulses.

Figure 3.32 Backprojection image reconstruction of point targets.

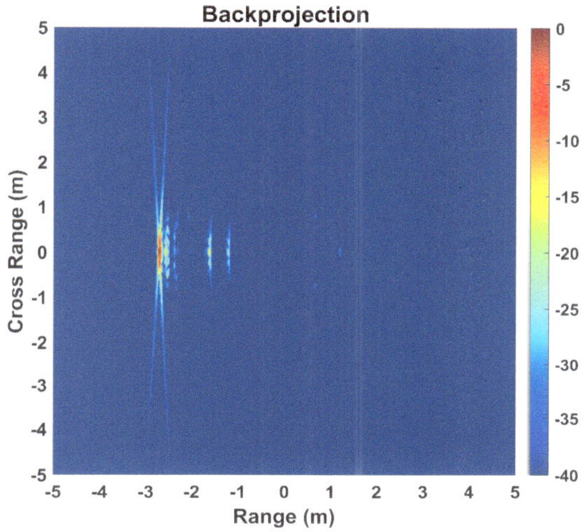

(a) Backprojected image with 6° azimuth span.

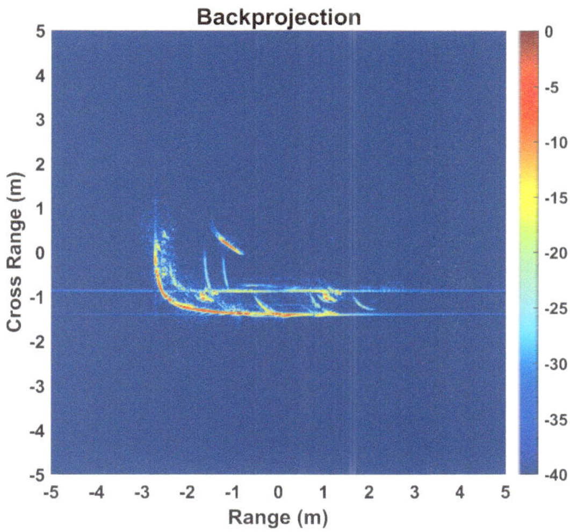

(b) Backprojected image 90° azimuth span.

Figure 3.33 Backprojection image reconstruction of a Toyota Avalon at X-band.

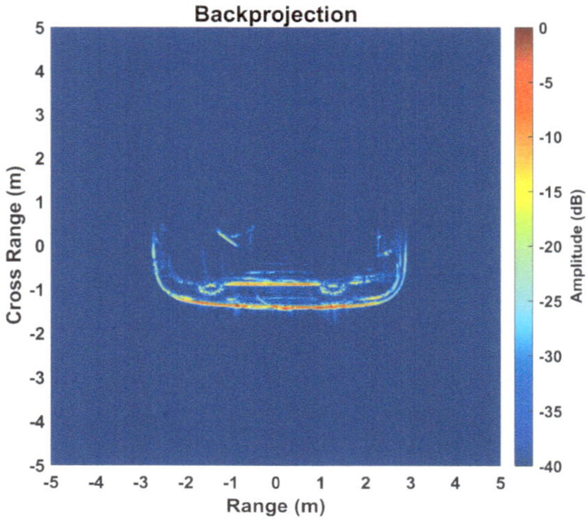

(a) Backprojected image with 180° azimuth span.

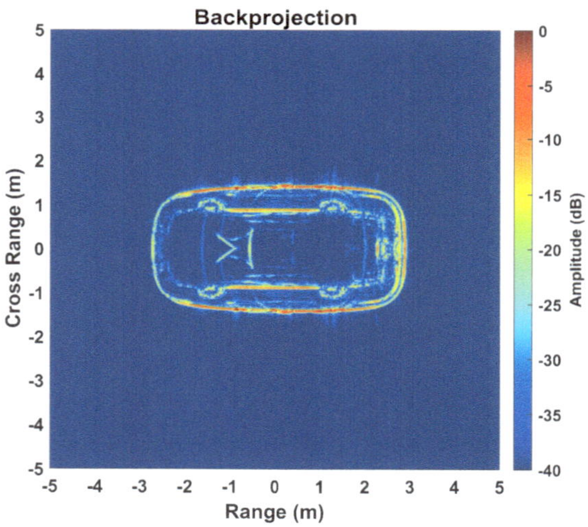

(b) Backprojected image with 360° azimuth span.

Figure 3.34 Backprojection image reconstruction of a Toyota Avalon at X-band.

PROBLEMS

3.1 Using the signal model given in (3.2), plot the return signal for three point-targets located at ranges of 100 m, 150 m, and 210 m. The radar cross-section for each target is 10 dBsm, 12 dBsm, and 7 dBsm. The waveform bandwidth is 100 MHz and the pulsewidth is 0.5 ms. Also, compare the matched filter output versus the stretch processor output for the three point-targets.

3.2 For a linear frequency modulated waveform with a bandwidth of 500 MHz and a pulsewidth of 0.5 ms, find the instantaneous frequency for stretch processing when the target is at a range of 100 km and the reference signal time delay is 0.6 ms.

3.3 Describe the major advantages of stretch processing and give practical examples of when stretch processing should be employed over matched filtering.

3.4 Explain the difference in resolution and SNR when using matched filtering versus stretch processing.

3.5 What is the effect of waveform bandwidth, pulsewidth, and polarization on one-dimensional range profiles?

3.6 Illustrate an imaging scenario where one-dimensional imaging may be preferred over two-dimensional imaging.

3.7 For range-Doppler imaging, what image artifacts are associated with error in the estimation of the Doppler centroid?

3.8 Calculate and plot the range cell migration correction given in (3.26) for Example 3.5.3 above.

3.9 Characterize the bandwidth and azimuth span limitations in (3.36) and (3.37) for polar format imaging. Using the code given in the Python notebook *range_doppler_example.ipynb* or the MATLAB live script *range_doppler_example.mlx*, explain the image artifacts that arise when these limits are exceeded.

3.10 Describe the advantages and disadvantages of the backprojection algorithm versus the polar format algorithm. Give practical examples of situations where one method is preferred over the other.

3.11 What is the role of the Radon transform in the development of modern systems employing backprojection imaging?

3.12 Explain the differences between SIRT, SART, and MART. Give examples of situations where one may outperform the others.

3.13 Compare the sampling requirements, including irregular sampling, between algebraic reconstruction methods and transform-based methods.

3.14 Expound on the convergence of algebraic reconstruction methods, including the number of iterations, condition number of the system matrix, existence of solutions, and constrained minimizations.

3.15 Briefly describe some nontraditional imaging techniques and applications.

3.16 What is a situation where a researcher would want access to the raw radar measurements in addition to standard SAR data products such as single look complex images?

3.17 For target classification and discrimination, what SAR data products would be most beneficial?

3.18 Research specialized SAR data products from various sensors including Sentinel-1's ocean swell spectra, ocean wind fields, and surface radial velocities.

References

[1] A. Harrison. *Introduction to Radar Using Python and MATLAB*. Artech House, Norwood, 2020.

[2] E. Jones et al. SciPy: Open source scientific tools for Python. `http://www.scipy.org/`. Accessed: 2021-06-23.

[3] B. Mahafaza. *Radar Systems Analysis and Design Using MATLAB*, 3rd ed. Chapman and Hall/CRC, New York, 2015.

[4] M. Soumekh. *Synthetic Aperture Radar Signal Processing with MATLAB Algorithms*. John Wiley and Sons, New York, 1999.

[5] M. Skolnik. *Radar Handbook*, 3rd ed. McGraw-Hill, New York, 2008.

[6] C. Jakowatz, D. Wahl, D. Yocky, B. Bray, W. Bow, and J. Richards. Comparison of algorithms for use in real-time spotlight-mode SAR image formation. In *SPIE Defense and Security Symposium*. Orlando, 2004.

[7] C. Jakowatz and N. Doren. Comparison of polar formatting and back-projection algorithms for spotlight-mode SAR image formation. In *SPIE Defense and Secuirty Symposium*. Orlando, 2006.

[8] M. Horvath and B. Rigling. A comparison of SAR imaging algorithms for high-squint angle trajectories. *SPIE Proceedings, Algorithms for Synthetic Aperture Radar Imagery XVIII*, 8051, 2011.

[9] J. Curlander and R. McDonough. *Synthetic Aperture Radar: Systems and Signal Processing*. John Wiley and Sons, New York, 1991.

[10] I. Cumming and J. Bennett. Digital processing of SEASAT SAR data. In *IEEE International Conference on Acoustics, Speech and Signal Processing*. Washington, 1979.

[11] N. Dastgir. *Processing SAR Data using Range Doppler and Chirp Scaling Algorithms*. Royal Institute of Technology, Stockholm, 2007.

[12] L. Cumming, F. Wong, and B. Hawkins. RADARSAT-1 Doppler centroid estimation using phase-based estimators. Toulouse, 2000.

[13] J. Walker. Range-Doppler imaging of rotating objects. *IEEE Transactions on Aerospace and Electronic Systems*, 16(1):23–51, 1980.

[14] G. Bergland. A guided tour of the fast Fourier transform. *IEEE Spectrum*, pages 41–52, 1969.

[15] A. Doerry, E. Bishop, J. Miller, V. Horndt, and D. Small. Designing interpolation kernels for SAR data. In *SPIE Defense, Security and Sensing Symposium*. Baltimore, 2012.

[16] D. Wang, M. Ercegovac, and Y. Xiao. Complex function approximation using two-dimensional interpolation. *IEEE Transactions on Computers*, 63(12):2948–2960, 2014.

[17] J. Zhang, X. Huang, and H. Suzuki. Phase-shifted interpolation of complex signals. *IEEE Communications Letters*, 16(9):1466–1469, 2012.

[18] A. Doerry. Basics of polar-format algorithm for processing synthetic aperture radar images. Technical Report SAND2012-3369, Sandia National Laboratories, Albuquerque, New Mexico 87185, 2012.

[19] R. Lewitt. Reconstruction algorithms: Transform methods. *Proceedings of the IEEE*, 71:390–408, 1983.

[20] M. Duersch. *Backprojection for Synthetic Aperture Radar*. PhD thesis, Brigham Young University, Provo, 2013.

[21] E. Slonecker, B. Johnson, and J. McMahon. An automated imagery orthorectification pilot. *Journal of Remote Sensing*, 3(1):1–16, 2009.

[22] J. Radon and P. Parks (translator). On the determination of functions from their integral values along certain manifolds. *IEEE Transactions on Medical Imaging*, 5(4):170–175, 1986.

[23] R. Bracewell. Strip integration in radio astronomy. *Australian Journal of Physics*, 9:198–217, 1956.

[24] R. Bates and T. Peters. Towards improvements in tomography. *New Zealand Journal of Science*, 14:883–896, 1971.

[25] G. Hounsfield. Computerized transverse axial scanning (tomography): Part 1. *British Journal of Radiology*, 46:1016–1022, 1973.

[26] D. Munson, J. O'Brien, and W. Jenkins. A tomographic formulation of spotlight-mode synthetic aperture radar. *Proceedings of the IEEE*, 71(8):917–925, 1983.

[27] C. Jakowatz, D. Wahl, P. Eichel, D. Ghiglia, and P. Thompson. *Spotlight-Mode Synthetic Aperture Radar: A Signal Processing Approach*. Springer, New York, 1996.

[28] A. Kak and M. Slaney. *Principles of Computerized Tomographic Imaging*. IEEE Press, 1988.

[29] G. Herman. *Fundamentals of Computerized Tomography: Image Reconstruction from Projections, Radar Handbook* 2nd ed. Springer-Verlag, New York, 2009.

[30] A. Andersen. *The Application of Ray Tracing Towwards a Correction for Refracting Effects in Computed Tomography with Diffracting Sources*. PhD thesis, School of Electrical Engineering, Purdue University, Lafayette, 1984.

[31] A. Andersen and A. Kak. Digital ray tracing in two-dimensional refractive fields. *Journal of the Acoustical Society of America*, 72:1593–1606, 1982.

[32] S. Kaczmarz. Angenaherte auflosung von systemen linearer gleichungen. *Bulletin of the Polish Academy of Sciences*, 6:355–357, 1937.

[33] K. Tanabe. Projection method for solving a singular system. *Numerische Mathematik*, 17:203–214, 1971.

[34] J. Trampert and J. Leveque. Simulatneous iterative reconstruction technique: Pyhsical interpretation based on the generalized least squares solution. *Journal of Geophysical Research*, 95(559):12, 1990.

[35] A. Andersen and A. Kak. Simultaneous algebraic reconstruction technique (SART): A superior implementation of ART. *Ultrasonic Imaging*, 6(1):81–94, 1984.

[36] M. Jiang and G. Wang. Convergence of the simultaneous algebraic reconstruction technique (SART). *IEEE Transactions on Image Processing*, 12(8):957–961, 2003.

[37] S. Gull and G. Daniell. Image reconstruction from incomplete and noisy data. *Nature*, 272:686–690, 1978.

[38] R. Gordon, R. Bender, and G. Herman. Algebraic reconstruction technique (ART) for three-dimensional electron microscopy and x-ray photography. *Journal of Theoretical Biology*, 29:471–481, 1970.

[39] M. Willis. *Algebraic Reconstruction Algorithms for Remote Sensing Image Enhancement*. PhD thesis, Brigham Young University, Provo, 2000.

[40] M. Tolman. *A Detailed Look at the Omega-K Algorithm for Processing Synthetic Aperture Radar Data*. PhD thesis, Brigham Young University, Provo, 2008.

[41] S. Hamidi, M. Nezhad-Ahmadi, and S. Safavi-Naeini. Sar imaging for a 79GHz FMCW radar based on MUSIC method. In *18th International Symposium on Antenna Technology and Applied Electromagnetics*, pages 1–2, 2018.

[42] J. Li and M. Palsetia. Using APES for interferometric SAR imaging. *Algorithms for Synthetic Aperture Radar Imagery III*, 2757:70–81, 1996.

[43] U. S. Air Force. Sensor data management system (SDMS). https://www.sdms.afrl.af.mil, 2021. Accessed: 2020-12-15.

[44] Tripoint Industries. lucernhammer : Electromagnetic signature / radar cross section prediction. http://lucernhammer.tripointindustries.com/. Accessed: 2021-12-05.

Chapter 4

Three-Dimensional Imaging

Three-dimensional SAR imagery reveals information about the scene in the range, azimuth, and elevation directions and is far more discernable than two-dimensional imagery. The information in the added dimension gives more details and insight into the size, shape, and orientation of objects within a scene. It can also solve some of the imaging issues found in two-dimensional imagery, such as layover [1]. The chapter begins with an overview of the history and development of three-dimensional SAR imaging. The extension of the point spread function to three dimensions and an analysis of the range, azimuth, and elevation resolution is then presented. This is followed by the development of the three-dimensional polar format algorithm, which results in a generic framework for three-dimensional imaging. The linear trace version of the Fourier slice theorem is given along with the derivation of the three-dimensional backprojection algorithm. The chapter concludes with a few Python and MATLAB examples to further illustrate the concepts of three-dimensional SAR imaging. For readability, exp() is used interchangeably with e^x notation in this chapter.

4.1 DEVELOPMENT

Three-dimensional SAR imaging algorithms and sensors have been developed for numerous applications including through-wall surveillance, buried object detection and classification, persistent building monitoring, digital elevation model development, environmental survey, and ecological process modeling [1–6]. The subsequent sections will review the development of some of the more common systems and applications.

4.1.1 Airborne and Spaceborne Systems

Three-dimensional SAR imaging from airborne and spaceborne platforms can broadly be categorized into two groups with their own timelines. The first began in the 1960s with the use of InSAR and stereoscopic SAR (StereoSAR) to obtain elevation information about the scene [7–10]. InSAR and StereoSAR use two or more two-dimensional complex SAR images to extract topographic information about the terrain [7, 11–14]. This is accomplished by registering the images and then forming an interferogram by multiplying one image with the complex conjugate of the other. Phase unwrapping is then employed to construct a digital elevation model of the scene [2, 15]. An example of such a sensor is the TanDEM-X, which has been acquiring global digital elevation model (DEM) data since 2010 [16].

Since InSAR uses phase information to determine the relative distance in path length to scatterers between two images, scatterer position information is retrieved while scatterer intensity is not. Thus, researchers began investigating methods to determine both scatterer location and intensity. In 1995, the U.S. Naval Research Lab proposed a different concept for using a three-dimensional synthetic aperture to form three-dimensional images [17]. This technique is commonly referred to as tomographic SAR (TomoSAR) and is an extension of conventional two-dimensional SAR imaging principles to three-dimensions and allows for the determination of both the position and intensity of scatterers in the scene [18]. Research on this topic has been conducted in many countries and several sensors have been developed [19–24]. As an example, European Space Agency has planned the 2023 launch of the BIOMASS Earth observing satellite, which is based on a P-band polarimetric SAR sensor. The mission will use TomoSAR techniques, in addition to others, to provide complete information about the vertical structure, scattering nature, and dynamic processes of the world's forests [25, 26].

While significant research is being conducted around the world, InSAR and TomoSAR remain the two main imaging modes in use for airborne and spaceborne three-dimensional SAR missions [27–30]. For airborne and spaceborne sensors operating in TomoSAR mode, the aperture size in the elevation direction is typically much smaller than the azimuth aperture and is often sparsely sampled, requiring super resolution methods. While beyond the scope of this book, the reader is referred to these excellent sources on the subject [3, 30–32].

Figure 4.1 illustrates a generic data collection geometry for three-dimensional SAR imaging from airborne and spaceborne platforms. Here, the aperture in the azimuth direction is synthesized by the flight path of the platform, while the aperture in the elevation direction in synthesized by taking multiple passes at varying altitudes.

Watch this animation of a 3-D SAR collection geometry.

4.1.2 Through-Wall Systems

Through-wall radar imaging systems allow for gathering of information about the location, classification, and movement of targets from behind walls, barriers, and other obstructions. This information can be critical to military and law enforcement conducting operations in urban areas, as well as search and rescue efforts. These systems typically employ UWB waveforms, also referred to as impulse or carrier-free waveforms and current systems have bandwidths as large as 3 GHz [4]. Note, the FCC defines ultrawideband to be a waveform with a bandwidth greater than 25% of the center frequency. Research and development of UWB systems increased rapidly with the FCC's Report and Order in 2002 which allows unlicensed UWB operation of imaging, vehicle radar, and communication systems [33, 34]. UWB systems have very fine range resolution arising from short pulse duration. Also, the low spectral density of these waveforms results in low health hazard to humans.

The earliest UWB through-wall imaging system was publicly announced in 2003 by Time Domain Corporation [4]. Since then, there has been much research and development performed by groups such as DARPA, the Canadian Department of National Defense, Eureka Aerospace, CyTerra Corporation, and Camero-Tech. Advancements in hardware, computing, and imaging algorithms have led to sophisticated systems such as the Xaver™ 800, which can display real-time three-dimensional imagery for intelligence, surveillance, and reconnaissance [5]. The aperture for through-wall systems is synthesized by using an antenna array with individual elements placed in a grid layout. Each antenna element transmits

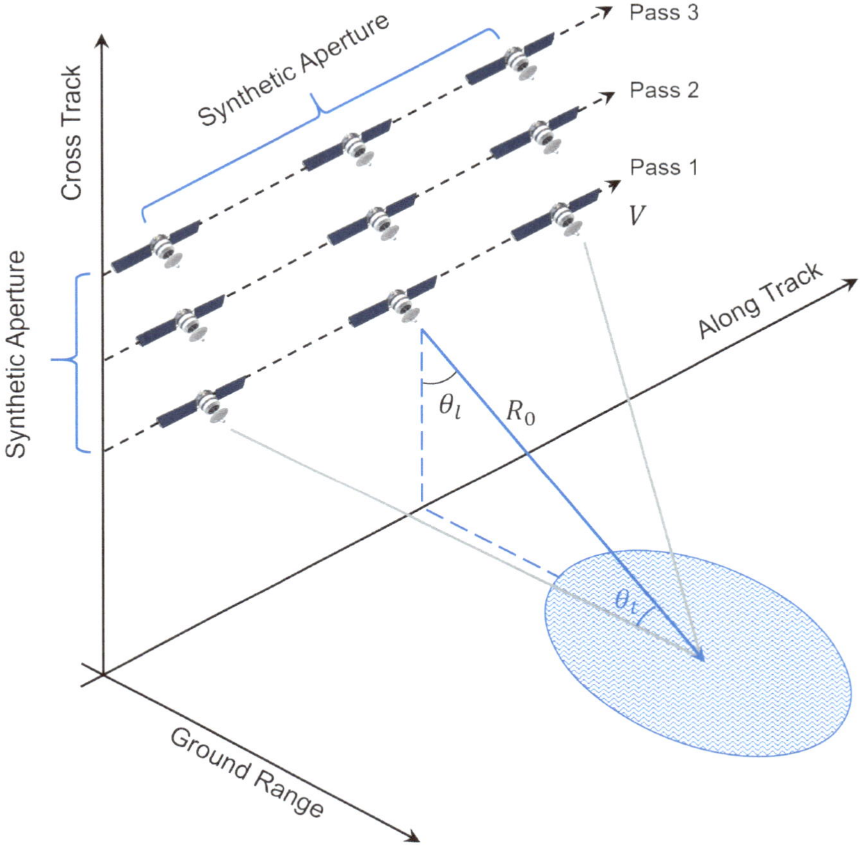

Figure 4.1 Basic three-dimensional data collection geometry for airborne and spaceborne platforms.

and receives the radar signal thus creating the aperture. This application of three-dimensional radar imaging continues to see much attention from commercial, government, and academic communities, and the reader is referred to several excellent sources on the subject [4, 35–38].

4.1.3 ISAR

There has been considerable research into three-dimensional ISAR imaging over the past two decades [39–42]. However, most of this research has been conducted with either simulated data or under ideal lab conditions. Some experiments have been performed with civilian aircraft and drone measurements [41, 42]. As with airborne and spaceborne three-dimensional imaging, InSAR and TomoSAR are the two main imaging modes. Nearfield methods dealing with wavefront curvature have also been investigated [40, 43]. There are several issues due to unknown target motion that are still to be solved before this is a reliable and robust imaging technique [39, 44].

4.2 POINT SPREAD FUNCTION

The resolution and point spread function for spotlight mode imaging was investigated in Section 2.3.1. Extending this to three dimensions, the resolution along the slant range and ground range directions is the same as those given in (2.2) and (2.3), and the resolution in the azimuth direction is given in (2.30) and (2.31). Analogous to the azimuth direction, the resolution in the elevation direction may be expressed as

$$\delta_e = \frac{\lambda}{4\sin\left(\dfrac{\phi_s}{2}\right)} \qquad \text{(m)}, \tag{4.1}$$

where ϕ_s is the angle subtended by the synthetic aperture in the elevation direction. As before, the small angle approximation is used in (4.1) to give

$$\delta_e = \frac{\lambda}{2\phi_s} \qquad \text{(m)}. \tag{4.2}$$

The expressions in (2.28) and (4.1) are based on approximating the target spectral support region as a cuboid, illustrated in Figure 4.2. This region is defined by the span

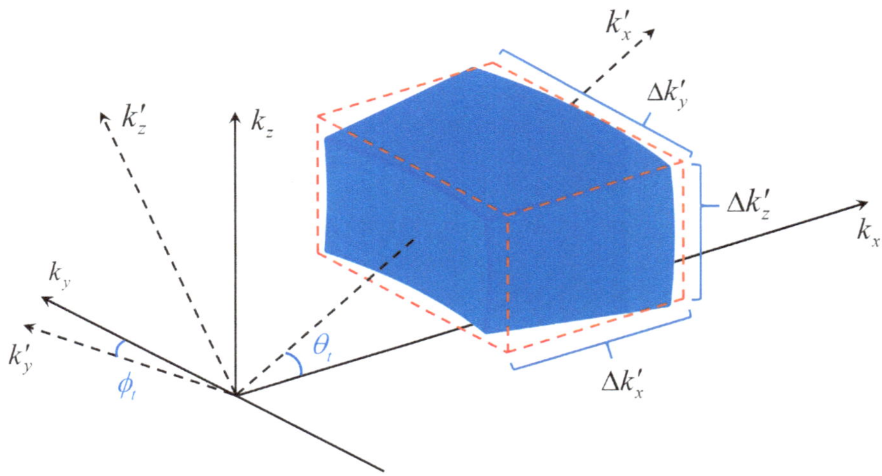

Figure 4.2 Three-dimensional spectral support for a point target located at (θ_i, ϕ_i).

$$\Delta k'_x = 2\big(k_{max} - k_{min}\big) \qquad \text{(rad/m)}, \tag{4.3}$$

$$\Delta k'_y = 2k_c\Big[\sin\big(\theta_t^+\big) - \sin\big(\theta_t^-\big)\Big] \qquad \text{(rad/m)}, \tag{4.4}$$

$$\Delta k'_z = 2k_c\Big[\sin\big(\phi_t^+\big) - \sin\big(\phi_t^-\big)\Big] \qquad \text{(rad/m)}, \tag{4.5}$$

where $\theta_t^\pm$ is the azimuth angle to the point target from the endpoints of the synthetic aperture in the azimuth direction, $\phi^\pm$ is the elevation angle to the point target from the endpoints of the synthetic aperture in the elevation direction, and k_c is the wavenumber at the center frequency. The point spread function is written as

$$psf = \text{sinc}\left(\frac{\Delta k'_x x'}{2\pi}\right)\text{sinc}\left(\frac{\Delta k'_y y'}{2\pi}\right)\text{sinc}\left(\frac{\Delta k'_z z'}{2\pi}\right). \tag{4.6}$$

The resulting range and cross-range resolutions are calculated by substituting (4.3), (4.4), and (4.5) into

$$\delta_{x'} = \frac{2\pi}{\Delta k_{x'}} \quad \text{(m)}, \tag{4.7}$$

$$\delta_{y'} = \frac{2\pi}{\Delta k_{y'}} \quad \text{(m)}, \tag{4.8}$$

$$\delta_{z'} = \frac{2\pi}{\Delta k_{z'}} \quad \text{(m)}. \tag{4.9}$$

This results in

$$\delta_{x'} = \frac{c}{2B} \quad \text{(m)}, \tag{4.10}$$

$$\delta_{y'} = \frac{\lambda R_t}{2L_a \cos(\theta_t)} \quad \text{(m)}, \tag{4.11}$$

$$\delta_{z'} = \frac{\lambda R_t}{2L_a \cos(\phi_t)} \quad \text{(m)}, \tag{4.12}$$

where R_t is the range to the target, L_a is the aperture length in the azimuth direction, and L_e is the aperture length in the elevation direction. The expressions above are a good approximation when the bandwidth is much smaller than the radar center frequency and the angular span of the synthetic aperture is small. However, the approximation does not hold for wide bandwidth systems where the bandwidth is comparable to the radar center frequency or synthetic apertures with wide angular spans. For those systems, the point spread function is shift-varying and its shape resembles a three-dimensional funnel. Figure 4.3 illustrates the three-dimensional point spread function and associated resolutions.

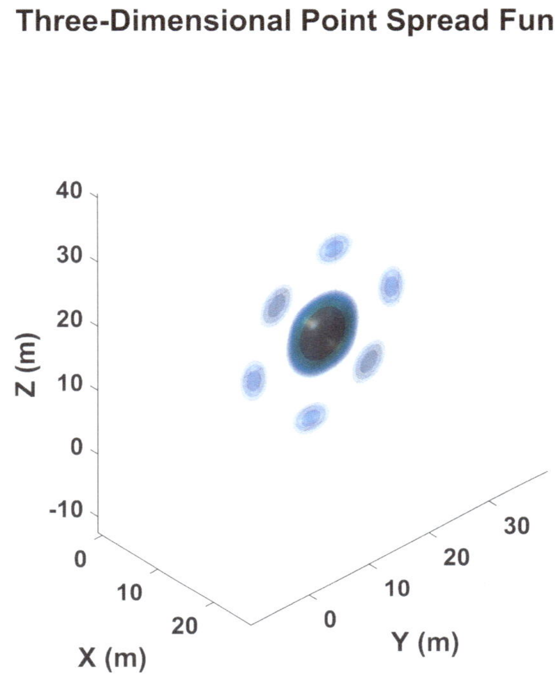

Figure 4.3 Three-dimensional point spread function for a target location of (13, 15, 14) meters.

4.3 POLAR FORMAT IMAGING

Recall the signal model given in (3.2) with the instantaneous frequency given in (3.5). By collecting data at several azimuth and elevation angles and a constant range (spotlight mode), the signal's wavenumber space may be written as

$$k_x = k\cos\left(\theta_n\right)\cos\left(\phi_p\right) \qquad \text{(rad/m)}, \tag{4.13}$$

$$k_y = k\sin\left(\theta_n\right)\cos\left(\phi_p\right) \qquad \text{(rad/m)}, \tag{4.14}$$

$$k_z = k\sin\left(\phi_p\right) \qquad \text{(rad/m)}, \tag{4.15}$$

where θ_n are the azimuth sample angles, ϕ_p are the elevation sample angles, and the wavenumber, k, is given in (3.33). From (4.13)–(4.15), and (3.33), the wavenumber space is filled in a spherical raster as illustrated in Figure 4.4(a). If the data filling the wavenumber space were continuous, an image could be formed by taking the inverse Fourier transform as

$$s(x, y, z) = \int_{-\infty}^{\infty}\int_{-\infty}^{\infty}\int_{-\infty}^{\infty} S(k_x, k_y, k_z)\times \\ \exp\left[j(k_x x + k_y y + k_z z)\right] dk_x\, dk_y\, dk_z. \tag{4.16}$$

The three-dimensional inverse discrete Fourier transform is employed, and (4.16) is replaced with

$$s(x, y, z) = \sum_{m=1}^{M}\sum_{n=1}^{N}\sum_{p=1}^{P} S\left(k_x^{m,n,p}, k_y^{m,n,p}, k_z^{m,n,p}\right)\times \\ \exp\left[j\left(k_x^{m,n,p} x + k_y^{m,n,p} y + k_z^{m,n,p} z\right)\right], \tag{4.17}$$

where the superscripts m, n, and p represent the samples in the range azimuth, and elevation dimensions, respectively. Implementing (4.17) results in a perfectly acceptable image. However, performing the triple summation for every location is computationally expensive. The kernel in (4.17) is not suited for FFT methods as these work on regular grids [45]. If the bandwidth is small compared to the center frequency and the span of the angles is also small,

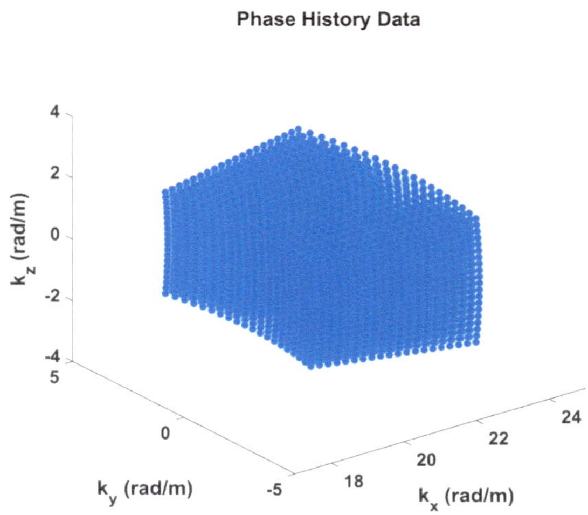

(a) Raw phase history data spherical raster.

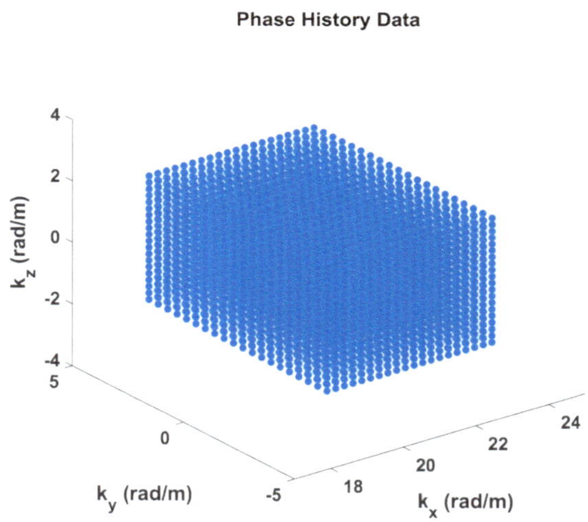

(b) Phase history data after rectangular reformatting.

Figure 4.4 Spotlight SAR phase history data in wavenumber space.

$$\beta T \ll 2\pi f_0 \qquad \text{(Hz)}, \tag{4.18}$$

$$\Delta\theta \ll 1 \qquad \text{(deg)}, \tag{4.19}$$

$$\Delta\phi \ll 1 \qquad \text{(deg)}, \tag{4.20}$$

then the wavenumber space may be approximated as Cartesian. For high-resolution systems, the approximation is invalid and a method for reformatting the data onto a Cartesian grid is necessary. This is commonly accomplished through three one-dimensional interpolations. The first interpolation is performed in the range direction, the second along the azimuth direction, and the third along the elevation direction. The resulting wavenumber sample locations are shown in Figure 4.4(b). Note k_x only depends on the index m (range), k_y only depends on the index n (azimuth), and k_z only depends on the index p (elevation). The inverse discrete Fourier transform of (4.17) is now separable, and for performance may be calculated by three one-dimensional transforms as

$$s(x, y, z) = \sum_{p=1}^{P} \left\{ \sum_{n=1}^{N} \left\{ \sum_{m=1}^{M} S\left(k_x^m, k_y^n, k_z^p\right) e_x \right\} e_y \right\} e_z, \tag{4.21}$$

where $e_x = \exp\left[jk_x^m\, x\right]$, $e_y = \left[jk_y^n\, y\right]$, and $e_z = \exp\left[jk_z^p\, z\right]$. Similar to the two-dimensional case, the high-frequency data are further apart than the low-frequency data, resulting in image degradation [46]. Also, it was assumed the wavefront is planar and residual phase errors were ignored. These approximations require further post processing to mitigate image artifacts [47]. Due to the limitations of polar format on waveform bandwidth and angular span, this method is limited for three-dimensional imaging and is presented here for illustrative purposes. However, there has been recent research into using a combination of polar formatting and L_1 regularization [48].

4.4 BACKPROJECTION IMAGING

In 1917, Johann Radon demonstrated that it was possible to reconstruct an original density function from projection data [49]. In two dimensions, the Radon transform represents the projection of a density function defined in a plane to the space of lines in the plane. The value of the Radon transform for a particular line is found by integrating the density function over that line. This is used as a starting point to two-dimensional SAR imaging using the filtered backprojection algorithm in Section 3.3.3. Radon further included formulas for the transform in three dimensions, in which the integral is taken over planes rather than lines as in two dimensions. The Radon transform has since been generalized to higher dimensional Euclidean spaces. The complex analog of the Radon transform is known as the Penrose transform [50–52].

When extending the Fourier slice theorem to three dimensions, two different versions may be derived. The first is the planar slice version, which relates planar projection functions to planar Fourier transform slices, as originally shown by Radon [46, 49, 53]. The planar slice theorem states the Fourier transform of the projection of a three-dimensional density function onto a two-dimensional subspace is equal to a two-dimensional slice of the three-dimensional Fourier transform of the density function. The slice through the three-dimensional frequency space is parallel to two-dimensional projection as illustrated in Figure 4.5. This version is commonly used in infrared, optical, and X-ray imaging applications. For example, backprojection has been used with data from a middle wave infrared camera to thermally and chemically characterize plasma torches [54]. Light emission tomography uses backprojection to perform three-dimensional imaging of small animals for biological and medical research [55]. The Zeiss X-ray series of scanners aid in the inspection and analysis of various production components such as automotive assemblies, consumer electronics, and medical equipment [56]. Finally, GE Healthcare employs backprojection with X-ray measurements to create three-dimensional imaging products to detect and analyze cardiac, oncological, cerebral, and metal artifacts [57].

The second version of the Fourier slice theorem in three-dimensions is the linear trace version, which relates one-dimensional projection functions to traces of the three-dimensional Fourier transform [46, 58–60]. This is the version used for radar applications as pulse compressed radar data is one-dimensional, as seen in

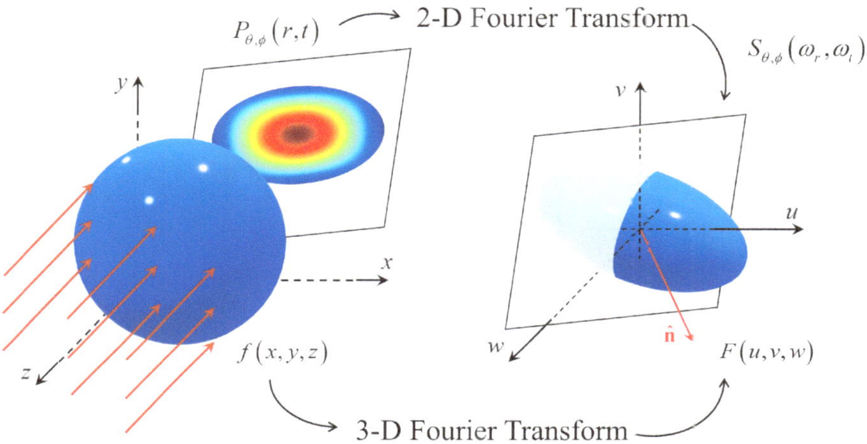

Figure 4.5 Three-dimensional spatial frequency domain from two-dimensional projection functions.

Section 3.2. The subsequent sections present the linear trace theorem followed by the filtered backprojection algorithm in three-dimensions.

4.4.1 Linear Trace Theorem

To show the linear trace theorem, start by defining the three-dimensional Fourier transform of the target's reflectivity function as

$$F(u,v,w) = \int_{-\infty}^{\infty}\int_{-\infty}^{\infty}\int_{-\infty}^{\infty} f(x,y,z)e^{-j2\pi(u\,x+v\,y+w\,z)}\,dx\,dy\,dz, \tag{4.22}$$

and the one-dimensional Fourier transform of the projection, $P_{\theta,\psi}(t)$, as

$$S_{\theta,\psi}(\omega) = \int_{-\infty}^{\infty} P_{\theta,\psi}(t)e^{-j2\pi\omega t}dt. \tag{4.23}$$

As in the two-dimensional case, no assumption has been made about the target's orientation, allowing $\theta = 0$ and $\psi = \pi/2$ without loss of generality. This leads to $v = 0$ and $w = 0$, and the three-dimensional Fourier transform is

$$F(u,0,0) = \int_{-\infty}^{\infty}\int_{-\infty}^{\infty}\int_{-\infty}^{\infty} f(x,y,z)e^{-j2\pi ux}\, dx\, dy\, dz. \tag{4.24}$$

Note, the phase factor is now a function of x only, and the triple integral may be written as

$$F(u,0,0) = \int_{-\infty}^{\infty}\left[\int_{-\infty}^{\infty}\int_{-\infty}^{\infty} f(x,y,z)\, dy\, dz\right] e^{-j2\pi ux}\, dx. \tag{4.25}$$

The term inside the brackets is an expression for the projection along planes of constant x. This may be expressed as

$$P_{\theta=0,\psi=\pi/2}(x) = \int_{-\infty}^{\infty}\int_{-\infty}^{\infty} f(x,y,z)\, dy\, dz. \tag{4.26}$$

Substituting (4.26) into (4.25) results in

$$F(u,0,0) = \int_{-\infty}^{\infty} P_{\theta=0,\psi=\pi/2}(x)e^{-j2\pi ux}\, dx. \tag{4.27}$$

Recalling the definition of the one-dimensional Fourier transform in (4.23), the relationship between the one-dimensional projection function and the three-dimensional Fourier transform of the object's density function is

$$F(u,0,0) = S_{\theta=0,\psi=\pi/2}(\omega). \tag{4.28}$$

As in the two-dimensional case in Section 3.3.3.2, no assumption has been made about the object's orientation; therefore, the general expression of the linear trace theorem is

$$F(u, v, w) = S_{\theta,\psi}(\omega). \tag{4.29}$$

A collection of projections at pairs of angles $[(\theta_1, \psi_1), (\theta_i, \psi_j), \ldots (\theta_M, \psi_N)]$ fills in the three-dimensional spatial frequency domain of the object's density function, as illustrated in Figure 4.6, and the image may be reconstructed with the backprojection algorithm.

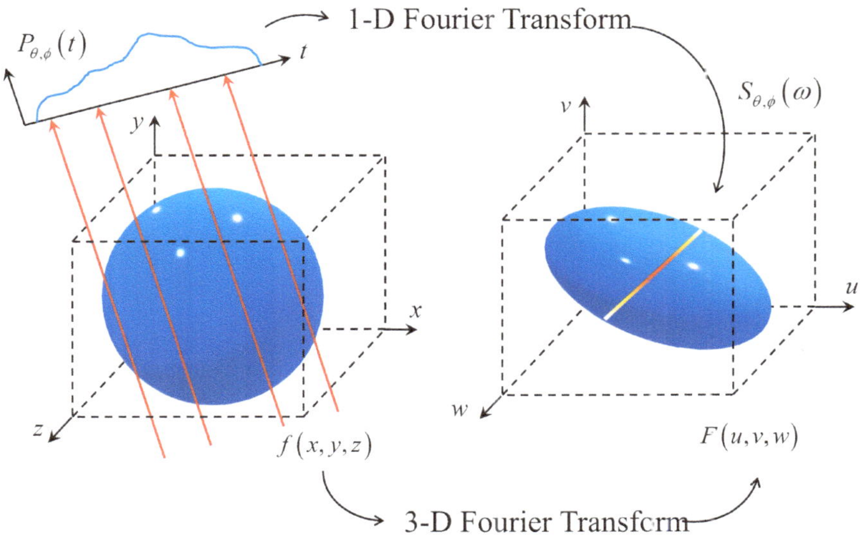

Figure 4.6 Three-dimensional spatial frequency domain from one-dimensional projection functions.

4.4.2 Filtered Backprojection

For the three-dimensional filtered backprojection algorithm, begin by taking the inverse Fourier transform of the object's three-dimensional spatial frequency domain representation as

$$f(x, y, z) = \int_{-\infty}^{\infty}\int_{-\infty}^{\infty}\int_{-\infty}^{\infty} F(u, v, w)\, e^{j2\pi(xu+yv+zw)}\, du\, dv\, dw. \tag{4.30}$$

Since the projection functions fill in the spatial frequency domain in a spherical fashion, making the following substitutions

$$u = \omega \cos\theta \sin\psi, \tag{4.31}$$

$$v = \omega \sin\theta \sin\psi, \tag{4.32}$$

$$w = \omega \cos\psi, \tag{4.33}$$

$$du\, dv\, dw = \omega^2 \sin\psi\, d\omega\, d\theta\, d\psi, \tag{4.34}$$

results in

$$f(x, y, z) = \int_{0}^{\pi}\int_{0}^{2\pi}\int_{0}^{\infty} F(\omega, \theta, \psi)\, e^{j2\pi\omega t}\, \omega^2 \sin\psi\, d\omega\, d\theta\, d\psi, \tag{4.35}$$

where

$$t = x\cos\theta\sin\psi + y\sin\theta\sin\psi + z\cos\psi \qquad \text{(rad)}. \tag{4.36}$$

The expression in (4.35) may be written as

$$f(x, y, z) = \int_{0}^{\pi}\int_{0}^{2\pi}\left[\int_{0}^{\infty} F(\omega, \theta, \psi)\, e^{j2\pi\omega t}\, \omega^2 d\omega\right] \sin\psi\, d\theta\, d\psi. \tag{4.37}$$

The inner integral represents the convolution of the projection function with a filter as

$$P_{\theta,\psi}(t) * h(t) = \int_0^\infty F(\omega,\theta,\psi)\, e^{j2\pi\omega t} \omega^2 \, d\omega. \tag{4.38}$$

The filtered projection function in (4.38) is then projected onto the image space along a line in the same direction as the projection was obtained. This is expressed as

$$f(x,y,z) = \int_0^\pi \int_0^{2\pi} P_{\theta,\psi}(t) * h(t) \sin\psi \, d\theta \, d\psi. \tag{4.39}$$

The backprojection imaging process proceeds as in the two-dimensional case in Section 3.3.3.3. The difference is the spatial target reflectivity array, $f(x_i, y_j, z_k)$, is three-dimensional, and the time point is based on the three-dimensional pixel point, (x_i, y_j, z_k).

4.5 EXAMPLES

The sections that follow illustrate the concepts of three-dimensional SAR imaging with a few Python/MATLAB examples. The Python examples for this chapter are in the directory *software/python/Chapter4* and the matching MATLAB examples are in the directory *software/matlab/Chapter4*. The reader should consult Chapter 1 for information on how to execute the Python and MATLAB code associated with this book.

4.5.1 Polar Format Algorithm — Point Targets

This example examines polar format imaging for a scene consisting of four point-targets with different scattering intensities. The point targets are located at coordinates of $x = [-10, 5, -8, 8]$, $y = [10, -3, -8, 12]$, and $z = [5, 0, -5, 11]$ meters relative to the scene center, and the range to the scene center is 1 km. The radar cross section of the point targets is $\sigma = [10, 10, 7, 5]$ m^2. The operating frequency is 5 GHz, and the waveform bandwidth is 100 MHz. The azimuth span of the synthetic aperture is 1°, and the elevation span of the synthetic aperture is also 1°. Figure 4.7 shows the three-dimensional imagery resulting from the polar format algorithm. For completeness, Figures 4.8–4.10, shows the imagery viewed in the x–y, x–z, and y–z planes, respectively. This example is given in the Python notebook *polar_format_point_targets_example.ipynb* and the MATLAB live script *polar_format_point_targets_example.mlx*.

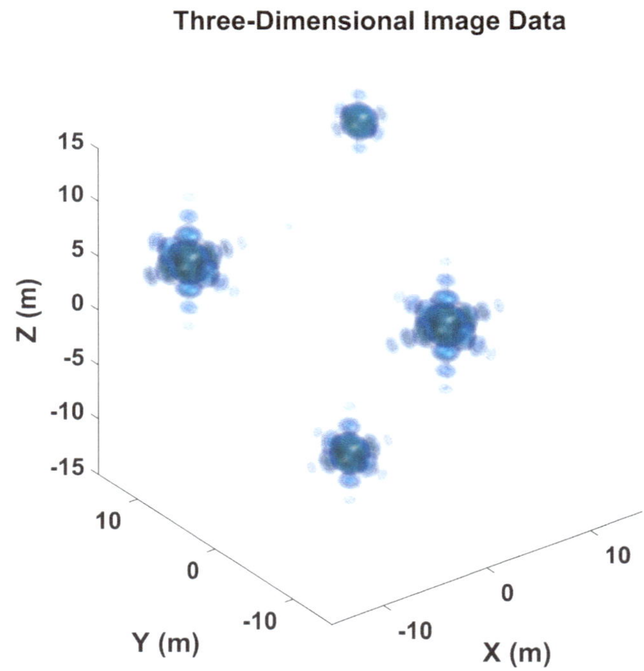

Figure 4.7 Three-dimensional imagery of point targets generated with the polar format algorithm.

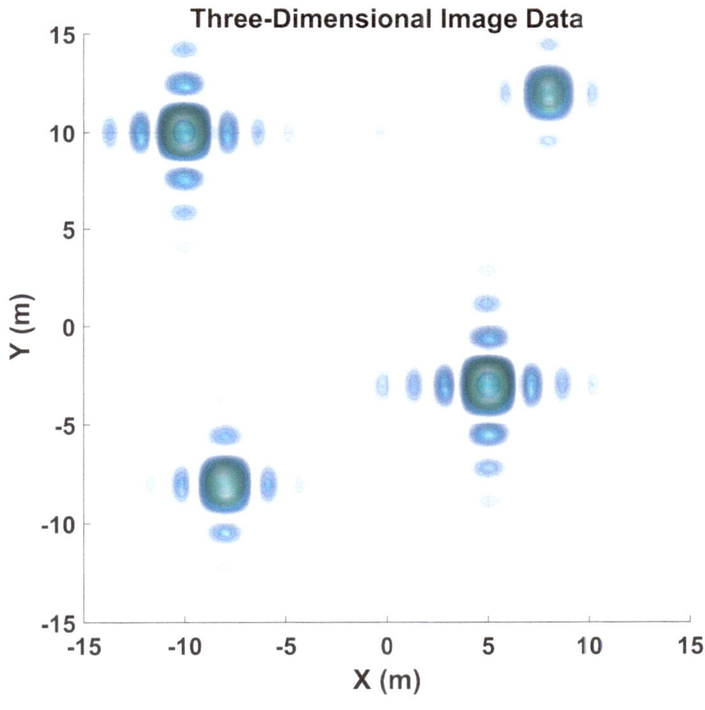

Figure 4.8 Three-dimensional imagery of point targets generated with the polar format algorithm as viewed in the x–y plane.

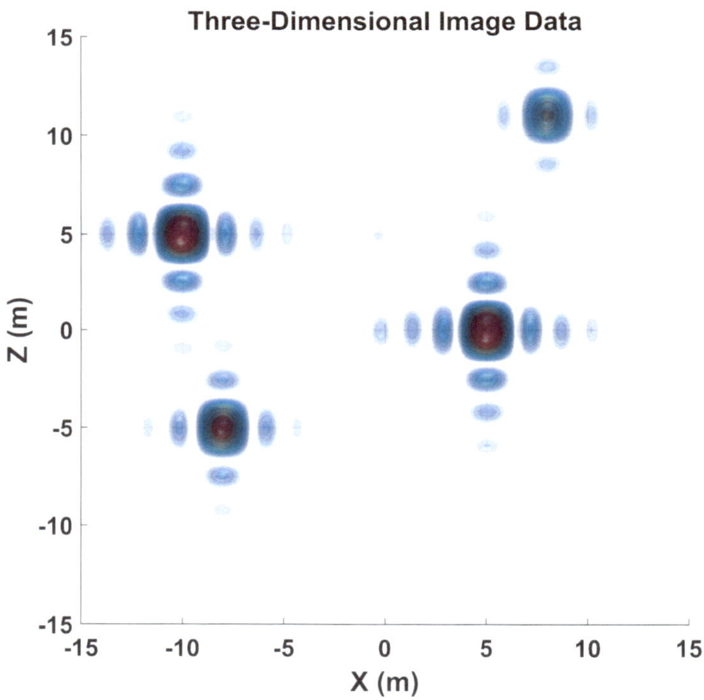

Figure 4.9 Three-dimensional imagery of point targets generated with the polar format algorithm as viewed in the x–z plane.

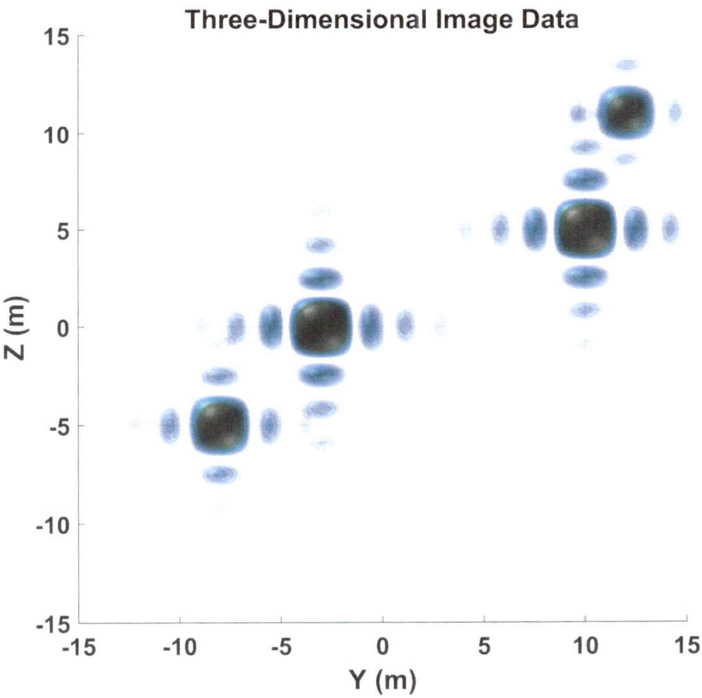

Figure 4.10 Three-dimensional imagery of point targets generated with the polar format algorithm as viewed in the y–z plane.

4.5.2 Backprojection Algorithm — Learjet

For this example, the backprojection algorithm is used to create three-dimensional SAR imagery of a Learjet, shown in Figure 3.30. The target was simulated at S-band with an operating bandwidth of 1 GHz, using the high frequency radar cross-section code, Lucernhammer MT, from Tripoint Industries [61]. Since the backprojection algorithm does not require continuous sampling, data was used from three azimuth spans of $[-15°, 15°]$, $[75°, 105°]$, and $[255°, 285°]$. The elevation span is $[-15°, 15°]$ for each of the azimuth spans. Figure 4.11 shows the resulting three-dimensional imagery for the Learjet. Note the overall shape of the aircraft is visible with greater geometry detail for the empennage. The width of the aircraft is also discernable. Figures 4.12–4.14 are given to illustrate other features visible in the SAR imagery, such as the shape and placement of the engines. This example is given in the Python notebook *backprojection_example.ipynb* and the MATLAB live script *backprojection_example.mlx*.

Watch this animation of 3-D SAR imagery of the Learjet.

4.5.3 Backprojection Algorithm — Backhoe

This example deals with the backprojection imaging of a backhoe illustrated in Figure 3.20. The target was simulated at C-band with an operating bandwidth of 1GHz using the high frequency radar cross-section code, Lucernhammer MT, from Tripoint Industries [61]. Unlike the previous example, the backhoe is only illuminated from broadside and the elevation span is $[0°, 30°]$, and the azimuth span is $[75°, 105°]$. Figure 4.15 and Figure 4.16 show the resulting three-dimensional imagery from the backprojection algorithm for the backhoe from alternative views, while Figure 4.17 is given to show the SAR imagery as viewed in the x–z plane. Note the shape and orientation of the backhoe is quite visible in the imagery. Also, there are many more scattering features, such as corners, edges, and multibounce locations than the Learjet, giving rise to more discernable imagery. This example is given in the Python notebook *backprojection_example.ipynb* and the MATLAB live script *backprojection_example.mlx*.

Watch this animation of 3-D SAR imagery of the backhoe.

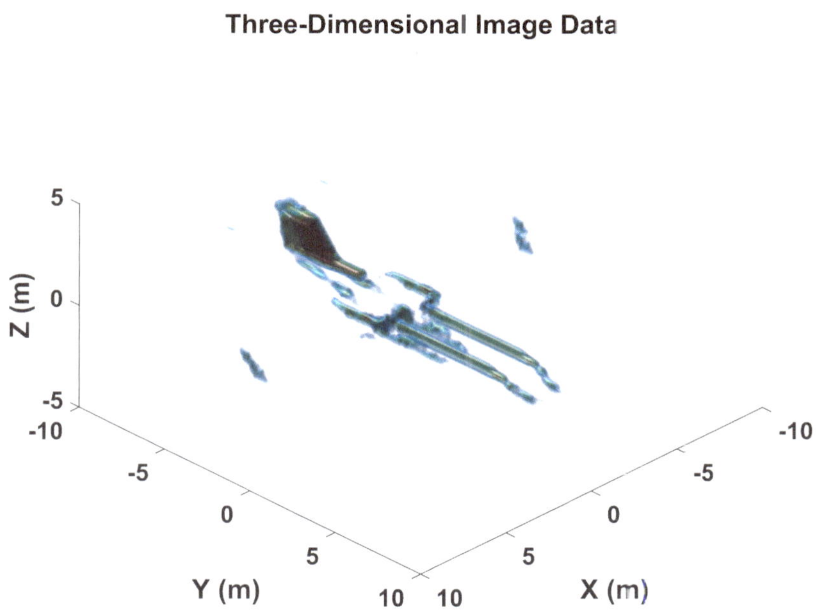

Figure 4.11 Three-dimensional imagery of a Learjet generated with the backprojection algorithm (VV polarization).

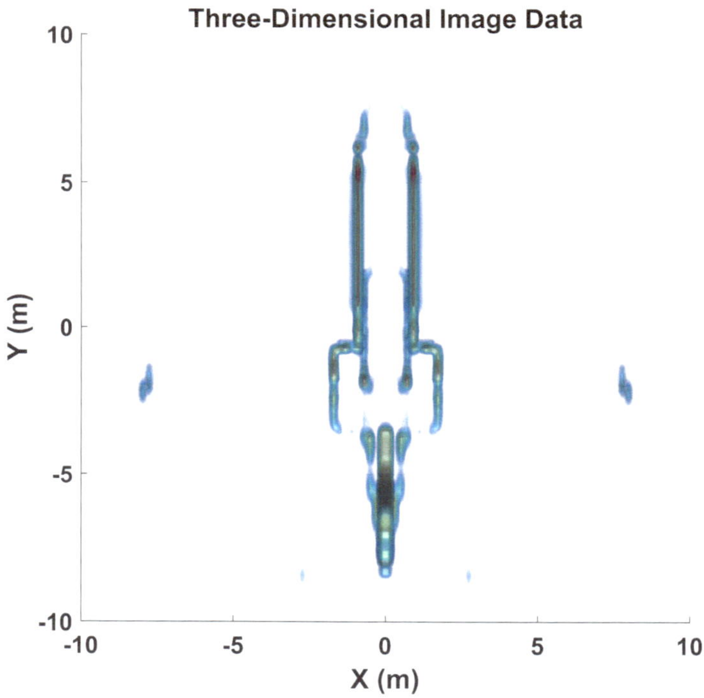

Figure 4.12 Three-dimensional imagery of a Learjet generated with the backprojection algorithm as viewed in the x–y plane (VV polarization).

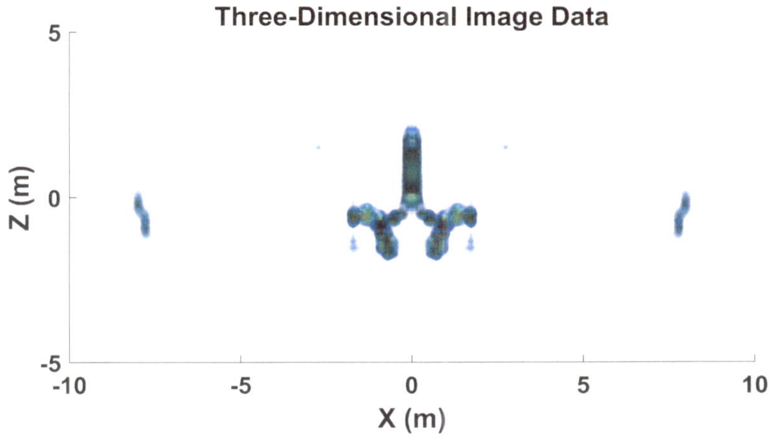

Figure 4.13 Three-dimensional imagery of a Learjet generated with the backprojection algorithm as viewed in the x–z plane (VV polarization).

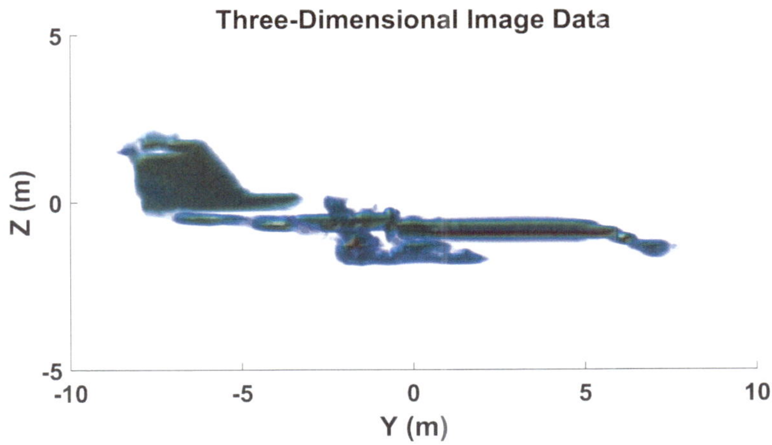

Figure 4.14 Three-dimensional imagery of a Learjet generated with the backprojection algorithm as viewed in the y–z plane (VV polarization).

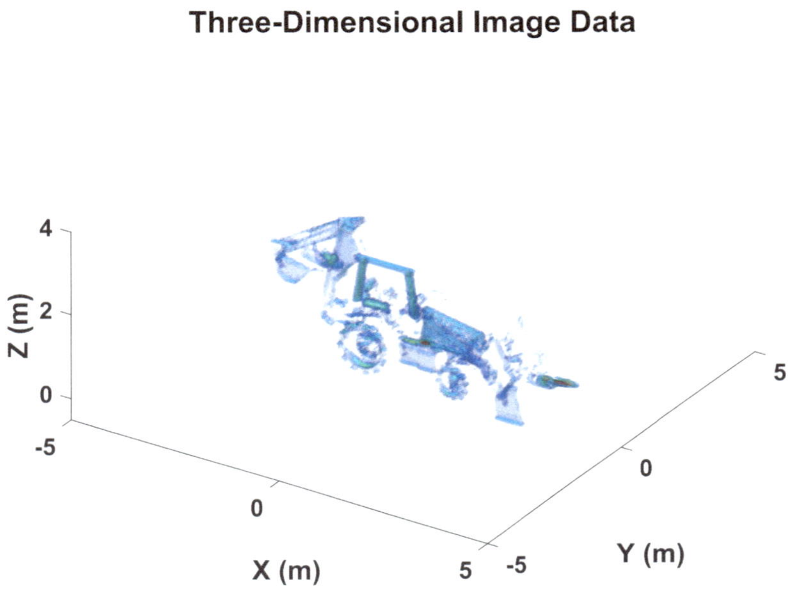

Figure 4.15 Three-dimensional imagery of a backhoe generated with the backprojection algorithm (VV polarization).

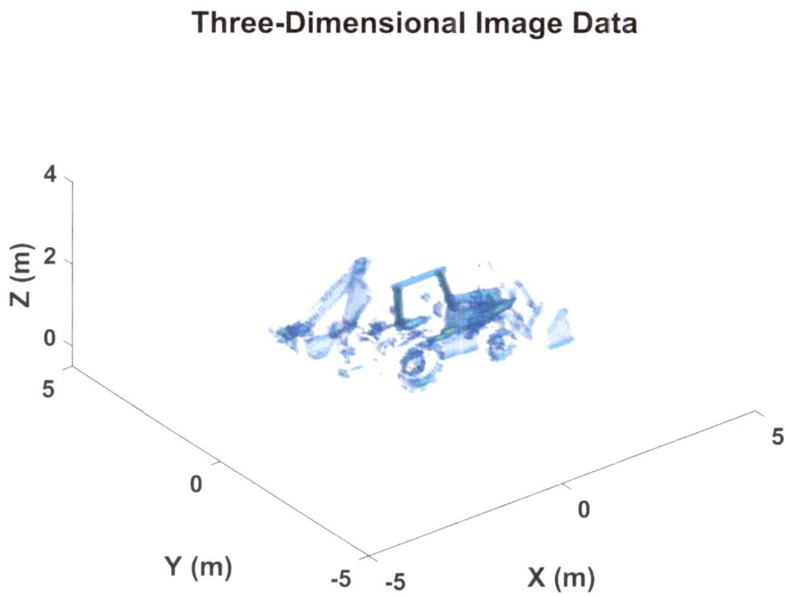

Figure 4.16 Three-dimensional imagery of a backhoe generated with the backprojection algorithm (VV polarization).

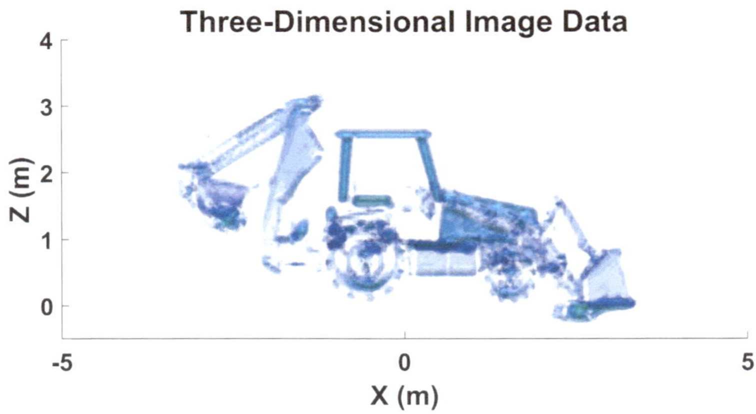

Figure 4.17 Three-dimensional imagery of a backhoe generated with the backprojection algorithm as viewed in the x–z plane (VV polarization).

PROBLEMS

4.1 What are the major advantages and disadvantages of three-dimensional SAR imaging in terms of system complexity, hardware, software, cost, and data products?

4.2 How does three-dimensional SAR imaging improve classification and discrimination of targets as compared to conventional two-dimensional imaging?

4.3 Describe the difference between InSAR and TomoSAR in terms of image formation methods and the resulting imagery.

4.4 Propose alternative data collection geometries that result in synthetic apertures suitable for three-dimensional SAR imaging.

4.5 What are the major differences between ultrawideband and traditional linear frequency modulated waveforms and how is each used in three-dimensional SAR imaging?

4.6 What are the major limitations of the polar format algorithm for three-dimensional SAR imaging?

4.7 Describe the advantages and disadvantages of using the backprojection algorithm for three-dimensional SAR imaging.

4.8 Characterize the differences in the creation of the synthetic aperture for three-dimensional SAR imaging among airborne, spaceborne, through-wall, and ISAR systems.

4.9 What is the distinction between the planar slice and linear trace versions of the Fourier slice theorem in three dimensions? Give example applications for each version.

4.10 Using the Python or MATLAB code associated with Example 4.5.2, create separate images for nose-on (azimuth = 0°) and broadside (azimuth = 90°) viewing. Describe the differences in the imagery and how those affect target classification.

References

[1] A. Reigber, F. Lombardini, F. Viviani, M. Nannini, and A. Martinez del Hoyo. The independence of the continuum hypothesis. *Proceedings of the National Academy of Sciences*. 50(6):1143–1148, 1963.

[2] Z. She, D. Gray, R. Bogner, J. Homer, and I. Longstagg. Three-dimensional space-borne synthetic aperture radar (SAR) imaging with multiple pass processing. *International Journal of Remote Sensing*, 23:4357–4382, 2002.

[3] S. DeGraaf, C. Twigg, and L. Phillips. Mono- and multistatic polarimetric sparse aperture 3D SAR imaging. *Proceedings of SPIE 6970 Algorithms for Synthetic Aperture Radar Imagery XV*, 2008.

[4] Y. Yang. *Development of a Real-time Ultra-wideband See Through Wall Imaging Radary System*. Ph.D. Thesis, University of Tennessee, 2008.

[5] Camero-Tech. Camero Xaver 800. http://www.camero-tech.com/xaver-products/xaver-800/. Accessed: 2021-02-12.

[6] S. Palm, R. Sommer, D. Janssen, A. Tessmann, and U. Stilla. Airborne circular W-band SAR for multiple aspect urban site monitoring. *IEEE Transactions on Geoscience and Remote Sensing*, 57(9):6996–7016, 2019.

[7] L. Graham. Synthetic interferometer radar for topographic mapping. *Proceedings of the IEEE*, 62(6):763–768, 1974.

[8] G. Laprade. An analytical and experimental study of stereo for radar. *Photogrammetric Engineering*, 29(2):294–300, 1963.

[9] F. Leberl, J. Raggam, and M. Kobrick. On stereo viewing of SAR images. *IEEE Transactions on Geoscience and Remote Sensing*, 23(2):110–117, 1985.

[10] F. Leberl. *Radargrammetric Image Processing*. Artech House, Norwood, 1990.

[11] H. Zebeker. Topographic mapping from interferometry synthetic aperture radar observation. *Journal of Geophysical Research*, 91:4993–4999, 1986.

[12] A. Babriel, R. Goldstein, and A. Gabriel. Crossed orbit interferometry: theory and experimental results from SIR-B. *International Journal of Remote Sensing*, 9:857–872, 1988.

[13] R. Goldstein, H. Zebker, and C. Wernet. Satellite radar interferometry: two-dimensional phase unwrapping. *Radio Science*, 32:713–720, 1988.

[14] C. Jakowatz, D. Wahl, and P. Thompson. Three-dimensional SAR imaging using cross-track coherent stereo collections. In *Conference Record of the Thirty-First Asilomar Conference on Signals, Systems and Computers*. Pacific Grove, 1997.

[15] D. Chibiao, Q. Xiaolan, X. Feng, L. Xingdong, J. Zekun, and Z. Fubo. Synthetic aperture radar three-dimensional imaging-from TomoSAR and Array InSAR to Microwave Vision. *Journal of Radars*, 8(6):693–709, 2019.

[16] DLR. TanDEM-X Science Service System. `https://tandemx-science.dlr.de/`. Accessed: 2021-02-12.

[17] K. Knaell. Three-dimensional SAR from practical apertures. *Proceedings of SPIE 2562, Radar/Ladar Processing and Applications*, 1995.

[18] S. Zhou, Y. Li, F. Zhang, L. Chen, and X. Bu. Automatic regularization of TomoSAR point clouds for buildings using neural networks. *Sensors*, 19(17):3748, 2019.

[19] M. Schmitt and X. Zhu. Demonstration of single-pass millimeterwave SAR tomography for forest volumes. *IEEE Geoscience and Remote Sensing Letters*, 13(2):202–206, 2016.

[20] X. Zhu and R. Bamler. Tomographic SAR inversion by L1 norm regularization-the compressive sensing approach. *IEEE Transactions on Geoscience and Remote Sensing*, 48(10):3839–3846, 2010.

[21] P. Pasquali, C. Prati, and F. Rocca. A 3-D SAR experiment with EMSL data. In *Proceedings of International Geoscience and Remote Sensing Symposium*. Firenze, 1995.

[22] A. Reigber and A. Moreira. First demonstration of airborne SAR tomography using multibaseline L-band data. *IEEE Transactions on Geoscience and Remote Sensing*, 5:38, 2000.

[23] G. Fornaro, F. Lombardini, and F. Serafino. Three-dimensional multipass SAR focusing: Experiments with long-term spaceborne data. *IEEE Transactions on Geoscience and Remote Sensing*, 43(4):702–714, 2005.

[24] B. Mahafza and M. Sajjadi. Three-simensional SAR imaging using linear array in transverse motion. *IEEE Transactions on Aerospace and Electronic Systems*, 32(1):499–510, 1996.

[25] F. Heliere, A. Carbone, J. Fonseca, N. Ayllon, A. Barnes, and M. Fehringer. Biomass P-band SAR. In *European Conference on Synthetic Aperture Radar*. Hamburg, 2016.

[26] F. Heliere, F. Fois, M. Arcioni, P. Bensi, M. Fehringer, and K. Scipal. Key note on BIOMASS: a P-band SAR polar-interferometric mission selected as 7th Earth Explorer mission. In *Proceedings of the Advanced RF Sensors and Remote Sensing Instruments Ka-band Earth Observation Radar Missions*. Noordwijk, 2014.

[27] W. Baek, H. Jung, and S. Chae. Precise three-dimensional mapping of the 2016 Kumamoto earthquake through the integration of SAR interferometry and offset tracking. In *IEEE International Geoscience and Remote Sensing Symposium (IGARSS)*. Brussels, 2017.

[28] D. Li, Q. Zhang, L. Li, and Y. Xi. Sparsity analysis of SAR signal and three-dimensional imaging of sparse array SAR. In *IEEE International Geoscience and Remote Sensing Symposium - IGARSS*. Melbourne, 2013.

[29] Z. Wang, R. Zhang, B. Huang, C. He, H. Kuang, and Y. Huang. A three-dimensional imaging method based on peak detection for multi-angle SAR. In *IEEE International Conference on Signal, Information and Data Processing (ICSIDP)*. Chongqing, 2019.

[30] W. Kuklinski and A. Kraay. 3D SAR imaging using a hybrid decompostion superresolution technique. *Proceedings SPIE 5427, Algorithms for Synthetic Aperture Radar Imagery XI*, 2004.

[31] M. Castelloe and D. Munson. 3-D SAR Imaging via High-Resolution Spectral Estimation Methods: Experiments. In *IEEE Proceedings of the International Conference on Image Processing*. Santa Barbara, 1997.

[32] A. Brito, S. Chan, and S. Cabrera. SAR image superresolution via 2-D adaptive extrapolation. *Multidimensional Systems and Signal Processing*, 14(1):83–104, 2003.

[33] A. Ameti, R. Fontana, E. Knight, and E. Richley. Ultra wideband technology for aircraft wireless intercommunications systems (AWICS) design. Reston, 2003.

[34] R. Fontana. Recent system applications of short-pulse ultra-wideband (UWB) technology. *IEEE Transactions on Microwave Theory and Techniques*, 52(9):2087–2104, 2004.

[35] Office of the OSD/DARPA. Assessment of ultra-wideband (UWB) technology. Technical report, Ultra-Wideband Radar Review Panel R-6280, Washington, 1990.

[36] R. Fontana. *A Brief History of UWB Communications*. Multispectral Solutions, Inc.

[37] G. Ross. *Early Motivations and History of Ultra Wideband*. Anro Engineering, Inc., Lexington.

[38] T. Barrett. *History of Ultra Wideband (UWB) Radar*. UCI, Vienna.

[39] J. Cai, M. Martorella, Q. Liu, Z. Ding, and E. Giusti. 3D ISAR imaging: The alignment problem. In *International Radar Conference (RADAR)*. Toulon, 2019.

[40] J. Fortuny. An efficient 3D near field ISAR processor. *IEEE Transactions on Aerospace and Electronics Systems*, 34(4):1261–1270, 1998.

[41] T. Liu, N. Li, R. Wang, and Y. Deng. Achieving high-quality three-dimensional InISAR imageries of maneuvering target via super-resolution ISAR imaging by exploiting sparseness. *IEEE Geoscience and Remote Sensing Letters*, 11(4):828–832, 2014.

[42] M. Pieraccini, L. Miccinesi, and N. Rojhani. RCS measurements and ISAR images of small UAVs. *IEEE Aerospace and Electronic Systems Magazine*, 32(9):28–32, 2017.

[43] X. Xu and M. Narayanan. Three-dimensional interferometric ISAR imaging for target scattering diagnosis and modeling. *IEEE Transactions on Image Processing*, 10(7):1094–1102, 2001.

[44] G. Wang, X. Xia, and V. Chen. Three-dimensional ISAR imaging of maneuvering targets using three receivers. *IEEE Transactions on Image Processing*, 10(3):436–447, 2001.

[45] G. Bergland. A guided tour of the fast Fourier transform. *IEEE Spectrum*, pages 41–52, 1969.

[46] A. Harrison. *Introduction to Radar Using Python and MATLAB*. Artech House, Norwood, 2020.

[47] A. Doerry. Basics of polar-format algorithm for processing synthetic aperture radar images. Technical Report SAND2012-3369, Sandia National Laboratories, Albuquerque, New Mexico 87185, 2012.

[48] X. Peng, W. Tan, W. Hong, C. Jiang, Q. Bao, and Y. Wang. Airborne DLSLA 3-D SAR image reconstruction by combination of polar formatting and L1 regularization. *IEEE Transactions on Geoscience and Remote Sensing*, 54(1):213–226, 2016.

[49] J. Radon and P. Parks (translator). On the determination of functions from their integral values along certain manifolds. *IEEE Transactions on Medical Imaging*, 5(4):170–176, 1986.

[50] R. Penrose. Twistor algebra. *Journal of Mathematical Physics*, 1(8):345–366, 1967.

[51] R. Penrose. Twistor quantisation and curved space-time. *International Journal of Theoretical Physics*, 1:61–99, 1968.

[52] R. Penrose. Solutions of the zero-rest-mass equations. *Journal of Mathematical Physics*, 10(1):38–39, 1969.

[53] G. Zeng. *3D Image Reconstruction*, chapter 5, pages 87–23. Springer, Berlin, 2010.

[54] A. Aouali, S. Chevalier, A. Sommier, E. Abisset-Chavanne, J. Batsale, and C. Pradere. 3D infrared thermospectroscopic imaging. *Scientific Reports*, 10(1), 2020.

[55] E. Tsyganov, P. Antich, R. Mason, R. Parkey, S. Seliounine, N. Slavine, and A. Zinchenko. *3D Reconstruction Using Optical Images*. Dallas, Texas, 2004.

[56] Carl Zeiss Industries. *Make the Invisible Visible*, 2021.

[57] GE Healthcare. Computed Tomography — GE Healthcare (United States). https://www.gehealthcare.com/products/computed-tomography.

[58] G. Hounsfield. Computerized transverse axial scanning (tomography): Part 1. *British Journal of Radiology*, 46:1016–1022, 1973.

[59] A. Cormack. Representation of a function by its line integrals with some radiological implications I. *Journal of Applied Physics*, 34:2722–2727, 1963.

[60] A. Cormack. Representation of a function by its line integrals with some radiological implications II. *Journal of Applied Physics*, 35:2908–2913, 1964.

[61] Tripoint Industries. lucernhammer : Electromagnetic signature / radar cross section prediction. http://lucernhammer.tripointindustries.com/. Accessed: 2021-12-05.

Chapter 5

Autofocus

To produce focused SAR imagery requires coherent phase-history data along the synthetic aperture. In previous chapters, perfect knowledge of platform motion, imaging geometry, and electromagnetic propagation was assumed. In real-world imaging scenarios, these parameters may not be known with a high degree of accuracy. For example, early SAR systems only used basic motion compensation to generate focused imagery, and as a result, these images experienced streaking in the along-track direction due to the presence of residual phase errors. These residual phase errors were often caused by nonlinear motion errors in the sensor platform, radar transmitter and receiver timing errors, and atmospheric inhomogeneity, which are difficult to measure. Therefore, a procedure that uses the phase-history data to determine and correct for these residual phase errors is desired. Such procedures are referred to as *autofocus*, and over the years many algorithms have been proposed, ranging from quantitative evaluation to qualitative visual inspection [1]. SAR autofocus algorithms fall into two basic categories: *model-based* and *nonparametric*, which are discussed in subsequent sections. The chapter begins with background information on SAR autofocus, followed by the development of an error model that is used throughout the chapter. Next, model-based and nonparametric autofocus techniques are covered. Map-drift and inverse filtering autofocus techniques are relatively simple methods and are presented first for illustration of autofocus concepts. This is followed by phase-gradient and minimum-entropy autofocus techniques. A few specialized autofocus algorithms are then discussed along with inherent benefits when applied to certain SAR systems and imaging modes. The chapter concludes with Python and MATLAB examples to further illustrate the concepts of autofocus algorithms.

5.1 BACKGROUND

As stated earlier, the formation of quality SAR imagery requires a set of coherent phase-history data along the synthetic aperture. Referring to Figure 5.1, the distance from the sensor platform to the center of the area to be imaged is R_0^i, for the *i*th pulse. The relative error in the measurement of R_0^i from pulse to pulse must be less than a fraction of a wavelength for image degradation to be negligible [2–4]. The absolute error in the measurement of R_0^i is not critical, as a constant offset or bias in the measurements does not affect the quality of the final image. The estimate of R_0^i is often based on inertial navigation systems (INS), GPS, and Global Navigation Satellite System (GNSS) [5–9]. Early airborne SAR systems relied solely on these types of motion measurements to form focused imagery. However, the drift errors of the measurement systems resulted in imagery that still suffered from image degradation. Under ideal conditions, state of art GNSS/INS/GPS systems are capable of determining position with accuracy in the cm range [10]. For imaging scenarios lasting only a few seconds, or in low-resolution applications, corrections to the phase-history data based only on GPS/INS/GNSS measurements may be adequate. Note, using only the motion measurements to form imagery is commonly referred to as *motion compensation* in the literature [2–5]. As shown in Chapter 3, the formation of high-resolution SAR imagery requires large bandwidth waveforms and long synthetic apertures. These parameters place more demanding requirements on the accuracy and drift rates of GNSS/INS/GPS systems. Also, these motion measurement systems cannot account for additional phase errors caused by factors such as medium inhomogeneity (e.g., dust, clouds, air turbulence) and timing errors in the radar transmitter and receiver. These errors produce artifacts such as streaking, blurring, and loss of contrast in the final imagery, as shown in Figure 5.2 [2–4, 11, 12]. The SAR image of a scene before the removal of phase errors is shown in Figure 5.2(a). With this level of image corruption little information about the scene can be extracted from the image. On the other hand, Figure 5.2(b) shows the SAR imagery of the same scene after the phase errors have been removed. It is clear that much more scene information, including spatial information related to buildings, roads, and vegetation, may be obtained from the imagery. Therefore, a systematic process that uses the phase-history data for determining and correcting for the phase error is desired. A block diagram of a generic autofocus procedure is given in Figure 5.3. Under most circumstances, the range-compressed phase-history data is used to begin the autofocus procedure. An initial image is formed, and the phase errors are estimated from this initial image data. The estimated phase errors are then used to correct the phase-history data and form a well-focused image. If the final image does not meet the requirements for image quality, then the procedure in Figure 5.3 may be repeated in an iterative fashion until the desired results are obtained or no further improvement is possible.

5.2 ERROR MODEL

To facilitate the discussion of autofocus methods and to test these algorithms, an error model that can be applied to SAR imagery in a consistent fashion is needed. Recalling (3.38), a perfectly formed image may be written in terms of two separable one-dimensional inverse discrete Fourier transforms as

$$s(x, y) = \mathcal{F}_n^{-1}\Big\{\mathcal{F}_m^{-1}\Big\{S\big(k_x^m, k_y^n\big)\Big\}\Big\}, \tag{5.1}$$

where $s(x, y)$ is the perfectly formed image and $S\big(k_x^m, k_y^n\big)$ is the phase-history data in wavenumber space. The index m is for the range dimension and n corresponds to the cross-range dimension. Applying the inverse discrete Fourier transform in the range dimension gives

$$s(x, y) = \mathcal{F}_n^{-1}\Big\{S\big(x, k_y^n\big)\Big\}, \tag{5.2}$$

where $S\big(x, k_y^n\big)$ is the range-compressed phase-history data.

In some applications, the phase errors are space variant (i.e., position dependent) and difficult to compensate for as the error varies over different regions of the image. A typical solution involves breaking the image into smaller images and applying conventional autofocus techniques to these subimages. These focused

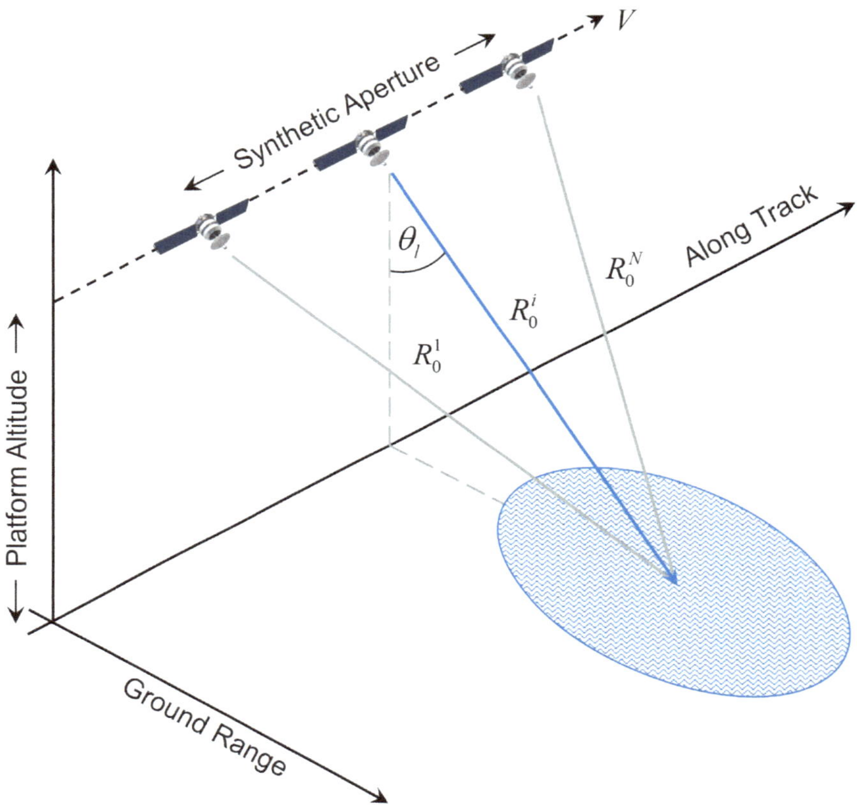

Figure 5.1 Distance R_0^i from the platform to the center of the scene.

(a)

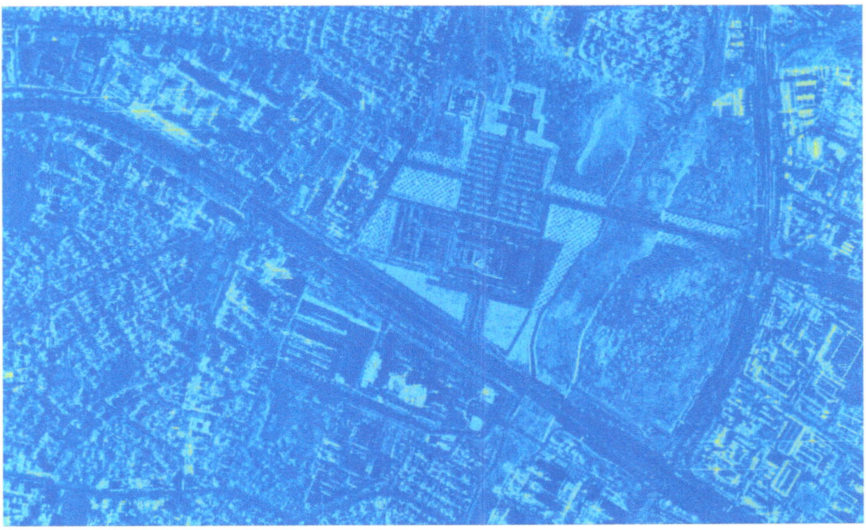

(b)

Figure 5.2 Comparison of SAR imagery (a) before and (b) after the removal of phase errors. Courtesy of ICEYE.

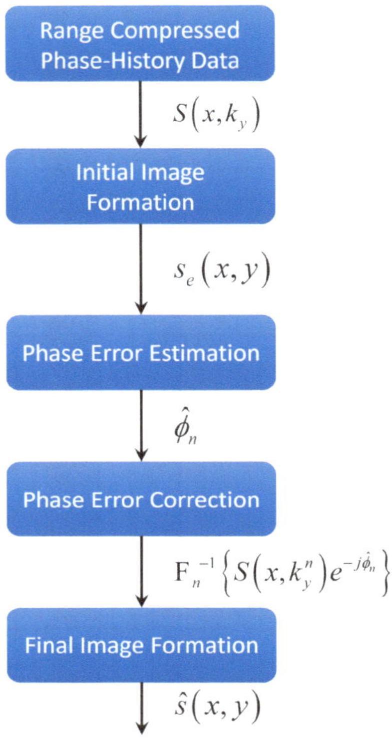

Figure 5.3 Block diagram of a generic SAR autofocus procedure.

subimages are then assembled in a patchwork fashion to form an entire focused image [1].

While there are phase errors that are position-dependent, this book treats the phase errors due to timing errors, uncompensated motion, and propagation uncertainties as errors in the demodulation time in the radar receiver. Therefore, the phase error is modeled as multiplicative and constant for a single pulse, and varying across the span of the synthetic aperture. This type of phase error is shown in Figure 5.4 and is sometimes referred to as a ribbon pattern [2, 4]. Applying a constant, multiplicative phase error to each pulse in the range-compressed phase-history data gives

$$S_e(x, k_y^n) = S(x, k_y^n)e^{j\phi_n}, \tag{5.3}$$

where ϕ_n is the phase error associated with the nth sample location, or pulse, along the synthetic aperture. The degraded image is then written as

$$s_e(x, y) = \mathcal{F}_n^{-1}\Big\{S_e(x, k_y^n)\Big\} = \mathcal{F}_n^{-1}\Big\{S(x, k_y^n)e^{j\phi_n}\Big\}. \tag{5.4}$$

The aim of autofocus algorithms is to produce an estimate of ϕ_n that is multiplied by the measured phase-history data to produce a focused image as

$$\hat{s}(x, y) = \mathcal{F}_n^{-1}\Big\{\hat{S}(x, k_y^n)e^{-j\hat{\phi}_n}\Big\}, \tag{5.5}$$

where $\hat{s}(x, y)$ is the image resulting from the autofocus algorithm, $\hat{S}(x, k_y^n)$ is the range-compressed phase-history data collected by the sensor, and $\hat{\phi}_n$ is the estimated phase error.

In general, propagation induced phase errors have a higher frequency content, (i.e., vary more rapidly across the aperture) than those generated by platform motion uncertainties. Higher frequency phase errors result in higher sidelobe levels and loss of contrast, while lower frequency phase errors result in a loss of resolution. Figure 5.5 illustrates the broadening of the response of a point scatterer when a quadratic phase error is included. Figure 5.6 illustrates the increase in sidelobe levels when a high-frequency phase error is included. Ideally, a robust autofocus technique capable of handling various types of phase errors is desired.

5.3 MODEL-BASED METHODS

Model-based autofocus techniques determine the coefficients of a polynomial expansion used to represent the phase error. Some of the more basic methods may only be able to determine quadratic type phase errors, illustrated in Figure 5.7(a), while

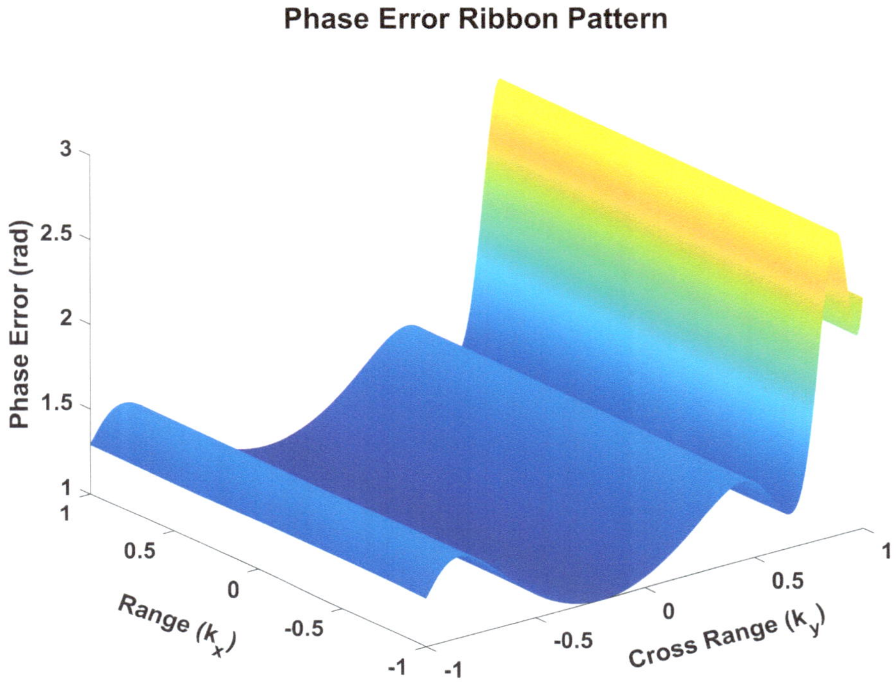

Figure 5.4 Illustration of multiplicative phase errors that are constant with range.

more sophisticated techniques are able to compensate for higher order polynomial phase errors shown in Figure 5.7(b) [1]. Model-based techniques are simple to implement and computationally efficient. However, quality imagery is only obtained when the actual phase error is accurately represented by the model. For example, modeling the phase error as quadratic when the true error is a fourth order polynomial will result in degraded imagery. As the complexity of the phase error increases, such as high-frequency or wideband errors, model-based methods typically do not perform well due to the underlying assumption that the phase errors are accurately described by a finite order polynomial [2–4].

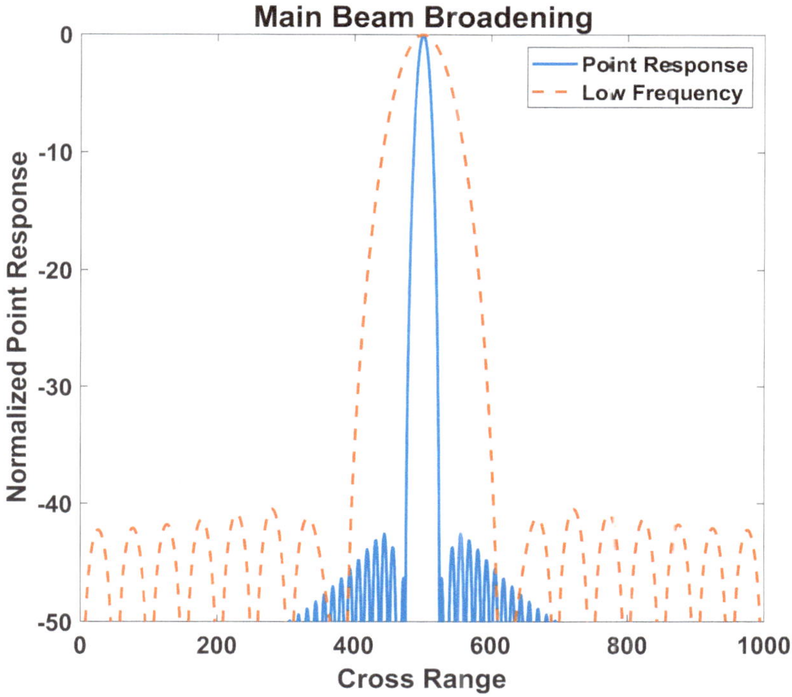

Figure 5.5 Illustration of the loss of resolution for a point scatterer with low-frequency phase errors applied.

5.3.1 Map Drift

Map-drift autofocus (MD) is a subaperture based method used to estimate quadratic phase errors present in SAR imagery [13–16]. MD divides the synthetic aperture data into two subapertures, usually of equal length, and SAR imagery is formed for each subaperture, as illustrated in Figure 5.8. MD relies on the translation property of the Fourier transform, as a linear phase shift between the two images results in a linear offset, or *drift*, between the two images, shown as ΔR in Figure 5.8. If the two images $s_1(x, y)$ and $s_2(x, y)$ differ by a linear offset, (x_0, y_0), then

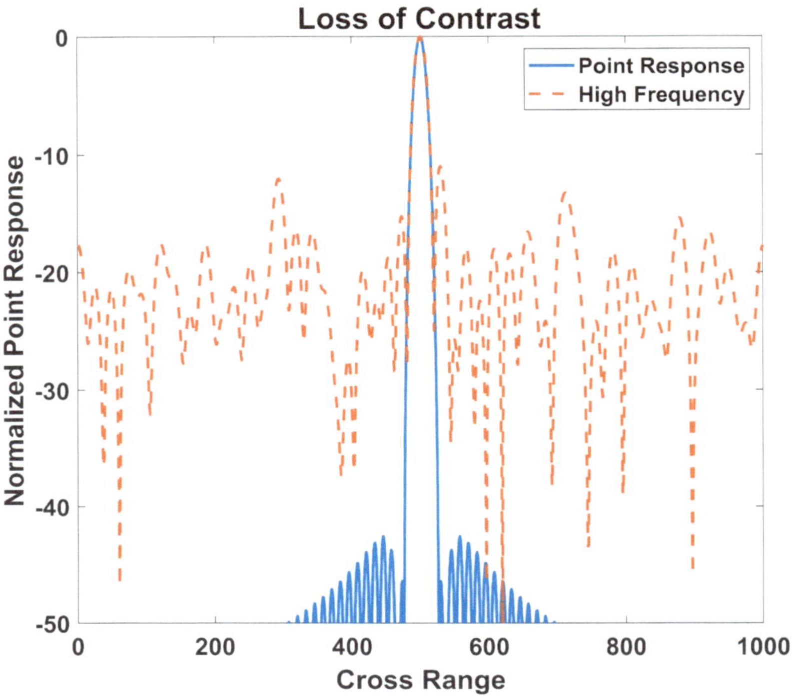

Figure 5.6 Illustration of the loss of contrast for a point scatterer with high-frequency phase errors applied.

$$s_2(x, y) = s_1(x - x_0, y - y_0). \tag{5.6}$$

Taking the Fourier transform of both images gives

$$S_2(k_x, k_y) = S_1(k_x, k_y)\, e^{-j2\pi(k_x\, x_0 + k_y\, y_0)}. \tag{5.7}$$

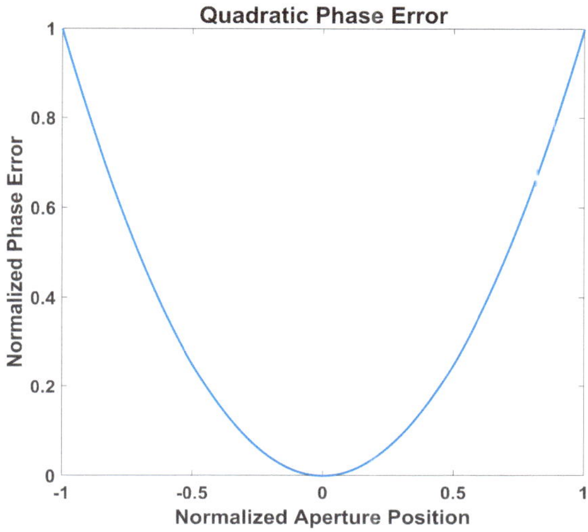

(a)

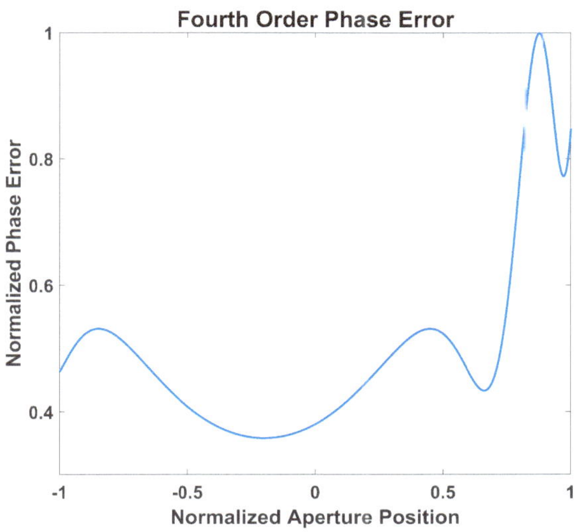

(b)

Figure 5.7 Examples of (a) quadratic and (b) fourth order phase errors.

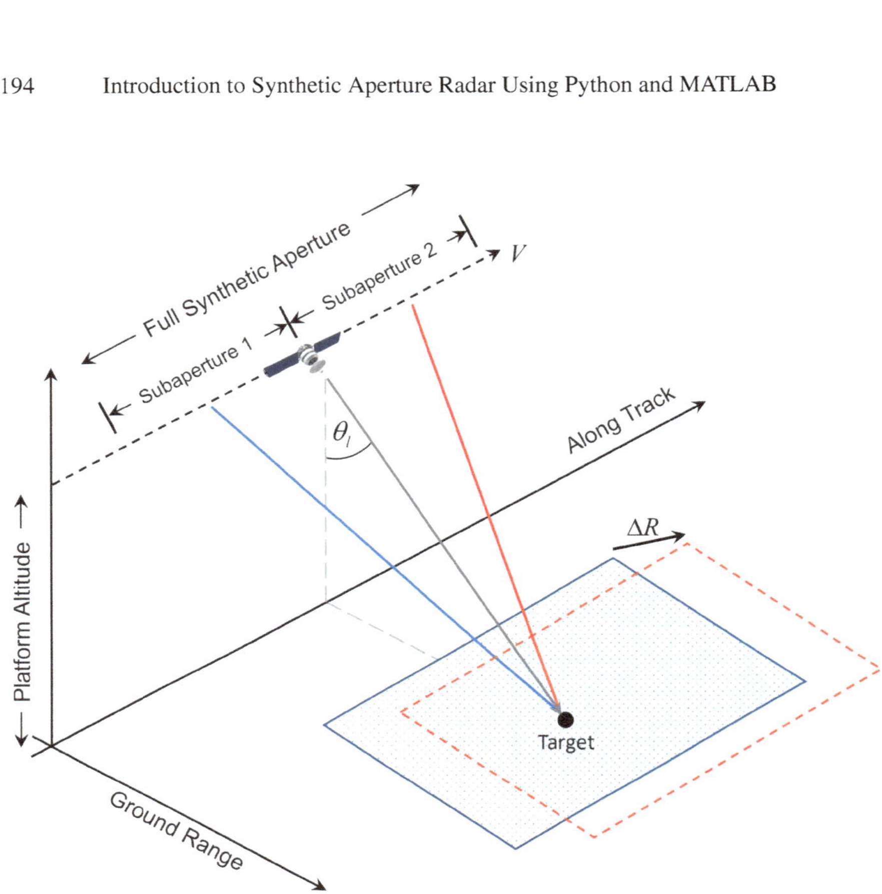

Figure 5.8 Formation of two images from two subapertures for basic map drift autofocus.

Taking the cross-power spectrum of the two images gives

$$P(S_1, S_2) = \frac{S_1(k_x, k_y)\, S_2^*(k_x, k_y)}{\left|S_1(k_x, k_y)\, S_2(k_x, k_y)\right|} = e^{-j2\pi(k_x\, x_0 + k_y\, y_0)}. \tag{5.8}$$

Finally, taking the inverse Fourier transform of (5.8) results in a *sinc* function centered at (x_0, y_0). For quadratic phase errors, the two images have a relative shift of 180 degrees per pixel, resulting in a factor of $n_0\,\pi$, where n_0 is the index of the peak of the sinc function [2, 4]. The quadratic phase error is then estimated by

$$\hat{\phi}_n = n_0\,\pi\,y^2 \qquad \text{(rad)}, \tag{5.9}$$

where $y \in [-1, 1]$ is the normalized aperture position. The processing steps required for focusing SAR imagery with MD is given in Figure 5.9.

While MD with two subapertures estimates a quadratic phase error, more complex errors can be determined by modeling the error as a polynomial expansion.

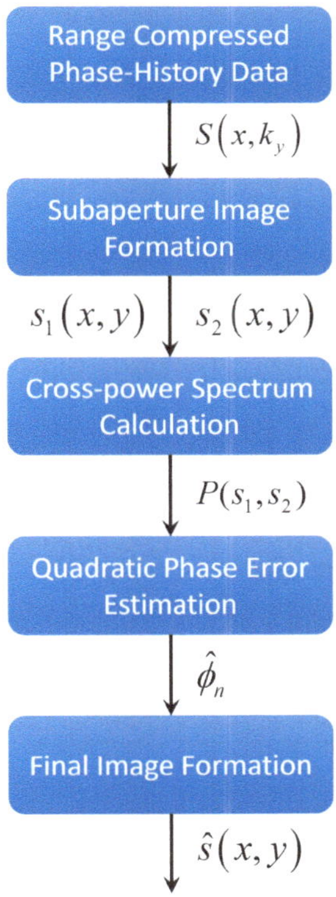

Figure 5.9 Generic map drift autofocus block diagram.

The coefficients of the polynomial are then found from the relative phase shifts in a sequence of more than two subapertures. This technique is referred to as multiple-aperture map-drift autofocus [2, 4, 13, 17]. Beginning with the quadratic term, the polynomial expansion for the phase error is expressed as

$$P(a) = a_{n-1}y^n + \cdots + a_2y^3 + a_1y^2, \tag{5.10}$$

and requires n subapertures to solve for the coefficients, a_i. Therefore, the order of the phase error to be modeled is limited by the number of subapertures which can be used to form images.

Various other approaches using MD have been proposed. These include modeling the phase errors with Fourier series, local quadratic modeling of phase errors, and two-dimensional spatially varying models [13, 15, 17–20].

5.4 NONPARAMETRIC METHODS

Nonparametric autofocus techniques do not model the phase errors as a finite order polynomial expansion and therefore do not require knowledge of the behavior of the phase errors. The most common of these methods, phase-gradient autofocus (PGA), shows excellent results in removing higher-order phase errors over many different scenes [21, 22]. Basic PGA has been extended by several different approaches including eigenvector methods, weighted least squares, quality PGA, and generalized PGA. The reader is referred to the following sources for more information on those methods [23–26]. Among other nonparametric methods are algorithms that optimize image metrics such as image sharpness, image entropy, and Fisher information [1].

5.4.1 Inverse Filtering

As an introduction to nonparametric autofocus algorithms, inverse filtering is presented as it is intuitive and relatively simple to implement. Inverse filtering depends on a single isolated point scatterer to estimate phase errors. The information found in a point scatterer of a corrupted image provides the basis for the estimates of the phase error. Depending on the type of scene being imaged, it may be difficult to identify an isolated point scatterer that produces good results. For example, urban scenes or scenes with man-made objects often contain points that are appropriate and yield good results, such as the Muscat International Airport in Oman, shown in Figure 5.10. However, landscapes with large areas of nearly homogeneous media are far less likely to contain a single isolated point scatterer capable of producing good phase-error estimates, such as Port Hedland Australia, shown in Figure 5.11. Note, the point scatterer is selected from the SAR imagery containing residual phase errors, such as that illustrated in Figure 5.2(a), which can make selection difficult.

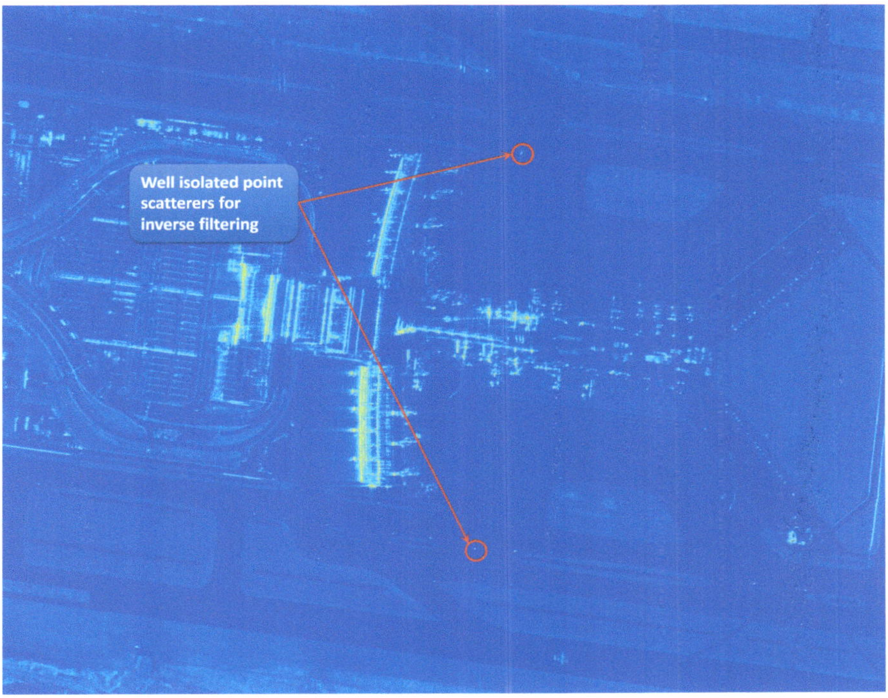

Figure 5.10 SAR image of Muscat International Airport in Oman with well isolated point scatterers for inverse filtering. Courtesy of ICEYE.

Figure 5.11 SAR image of Port Hedland Australia with few useful point scatterers for inverse filtering. Courtesy of ICEYE.

To demonstrate inverse filtering, begin with the error model in (5.4). The corrupted image may be written as the convolution of an error function with the focused image as [27]

$$s_e(x, y) = \mathcal{F}_n^{-1}\Big\{e^{j\phi_n}\Big\} * s(x, y), \tag{5.11}$$

where $(*)$ represents the convolution operator. Let the scene consist of a single point scatterer located at (x_0, y_0). The image is expressed as

$$s(x,y) = A\,\delta(x - x_0, y - y_0), \tag{5.12}$$

where A represents the reflectively of the point scatterer, and $\delta(x,y)$ is the delta function defined as

$$\delta(x,y) = \begin{cases} 1 & \text{if } \ x = 0, y = 0 \\ 0 & \text{otherwise} \end{cases} . \tag{5.13}$$

Substituting (5.12) into (5.11), the image degraded by the phase error ϕ_n may be written as

$$s_e(x,y) = \mathcal{F}_n^{-1}\Big\{e^{j\phi_n}\Big\} * A\,\delta(x - x_0, y - y_0). \tag{5.14}$$

Performing the convolution with the delta function yields

$$s_e(x,y) = A\,\mathcal{F}_n^{-1}\Big\{e^{j\phi_n}\Big\}. \tag{5.15}$$

Taking the Fourier transform of both sides results in

$$\mathcal{F}\big\{s_e(x,y)\big\} = A\,e^{j\phi_n}. \tag{5.16}$$

Therefore, measuring the phase of the Fourier transform of the image data $s_e(x,y)$ gives an estimate of the phase error within a constant as

$$\hat{\phi}_n = \angle\Big[\mathcal{F}\big\{s_e(x,y)\big\}\Big] = \angle\Big[A\,e^{j\phi_n}\Big] = \angle A + \phi_n \qquad \text{(rad)}. \tag{5.17}$$

The derivation given above is based on a single, isolated point scatterer. Imaging areas of interest are vastly more complicated than simple point scatterers. However, the phase estimate in (5.17) is approximated by identifying a reasonably well separated point-like scatterer in the scene. Data around the point scatterer corresponding to the estimated support band (see Section 2.2) is kept and the remaining image data is zeroed out. A one-dimensional discrete Fourier transform is employed to convert the image data to range-compressed phase-history data as

$$S\big(x, k_y^n\big) = \mathcal{F}_y\big\{s(x, y)\big\}. \tag{5.18}$$

The data in (5.18) is then used with (5.17) to estimate the phase error. Recall, the phase error is modeled as multiplicative and constant for a single pulse, and varies across the span of the synthetic aperture. Therefore, the final step in the inverse-filtering autofocus method is to multiply each range line in the range-compressed phase-history data by the estimate of the phase error as expressed in (5.5). The processing steps required for focusing SAR imagery with inverse filtering are shown in Figure 5.12.

5.4.2 Phase Gradient

Inverse filtering autofocus, covered in the previous section, has significant limitations. Depending on the type of scene being imaged, there may be no suitable point scatterer to use as the basis for the estimate of the phase error. However, if the information from several scatterers could be combined in some fashion, then a better estimate of the phase error should be achievable. Recalling the error model of (5.4), scatterers at different locations are effectively altered by the same phase errors. Phase gradient autofocus (PGA) is a nonparametric method developed from maximum likelihood estimation that takes advantage of this redundancy of the phase error across several scatterers [21, 28, 29].

While various methods exist to select several individual scatterers to be used in the estimation of the phase errors, the method used here is to select the brightest scatterer on each range line [2, 4, 23, 25]. The scatterers are then isolated by performing a circular shift in the cross-range direction, thereby centering the scatterers. The circular shift is then followed by a windowing operation. Figure 5.13 illustrates a degraded image along with the result of the circular shift operation. As with the inverse filtering method, an estimate of the support band of the phase errors is found and data outside of this support band is set to zero. The windowing operation preserves the information about the phase errors while rejecting information contained

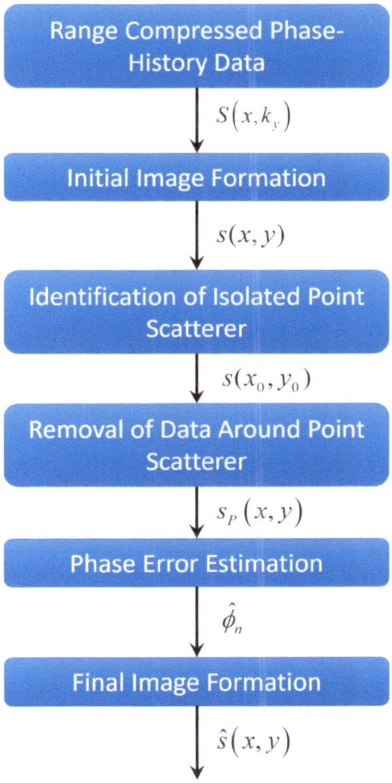

Figure 5.12 Inverse filtering autofocus block diagram.

in noise and clutter. Various methods may be used to estimate the support band of the phase errors [2–4, 22]. For this book, a noncoherent averaging scheme is employed, where the average intensity over all range lines for each cross-range position is found. This is expressed as

$$P_n = \frac{1}{M} \sum_{m=1}^{M} \left| s_c(m, n) \right|^2, \tag{5.19}$$

(a) Initial image created from phase-history data.

(b) Circularly shifted image.

Figure 5.13 Result of circularly shifting the brightest scatterer for each range line.

where P_n is the noncoherent average of the image intensity and $s_c(m,n)$ is the circularly shifted imagery. For low frequency phase errors, a simple threshold placed on P_n works well for determining the support band. This is due to low-frequency phase errors broadening the response of a point scatterer. On the other hand, high-frequency phase errors tend to raise the sidelobes of a point scatterer response. In that case, a nominal threshold may be chosen on the first iteration and then reduced for subsequent iterations. Figure 5.14 shows the selection of the window width, along with the application of the window to the circularly shifted data.

Once the SAR imagery has been circularly shifted and windowed, an estimate of the gradient of the phase error is computed. Following [21], a linear unbiased minimum variance estimate of the phase-error gradient may be written as

$$\hat{\phi}_n = \frac{\sum\limits_{m=1}^{M} \Im\left\{ s_w^*(m,n)\dot{s}_w(m,n) \right\}}{\sum\limits_{m=1}^{M} \left| s_w(m,n) \right|^2}, \tag{5.20}$$

where $s_w(m,n)$ is the circularly shifted and windowed data shown in Figure 5.14. The processing steps required for focusing SAR imagery with PGA is summarized in Figure 5.15.

5.4.3 Minimum Entropy

Minimum-entropy autofocus methods (ME) use image entropy as a quality metric for gauging image focus. In general, entropy is a measure of the smoothness of a distribution function, with smoother distributions having larger values of entropy [30]. Applying this idea to SAR imagery, lower values of entropy are an indication of sharper images, which has been shown in several works [11, 12, 30]. Thus, the goal is to find a method for determining the phase correction that minimizes the image entropy.

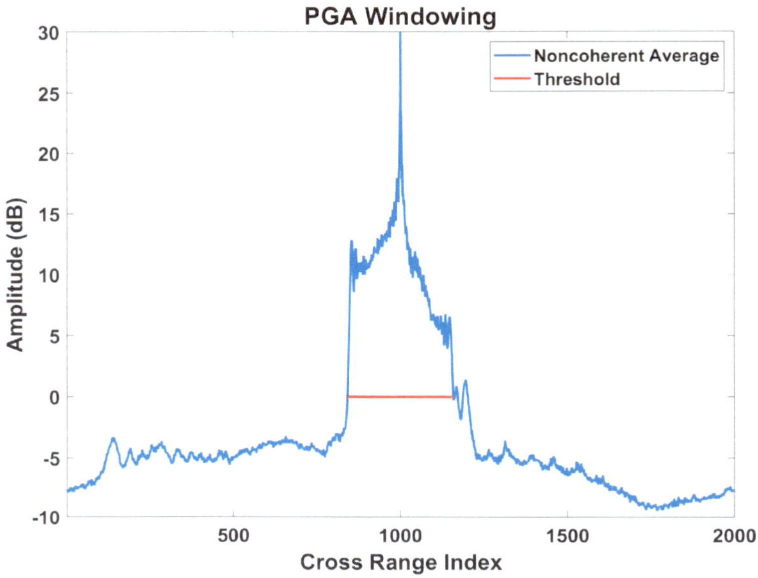

(a) Noncoherent average in cross range.

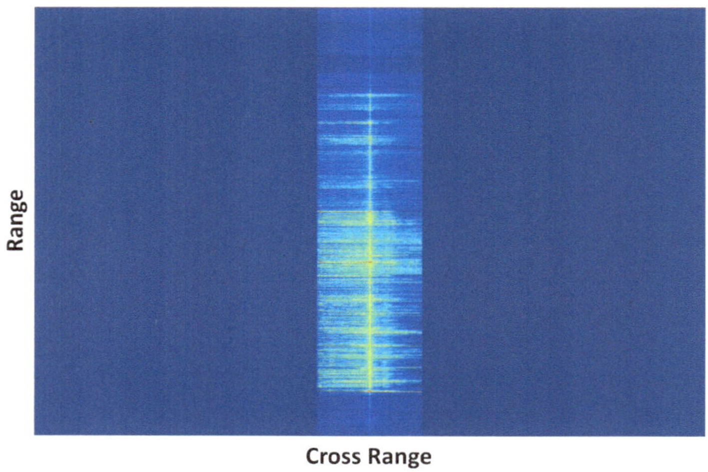

(b) Circularly shifted and windowed image.

Figure 5.14 Windowing the circularly shifted data using the noncoherent average.

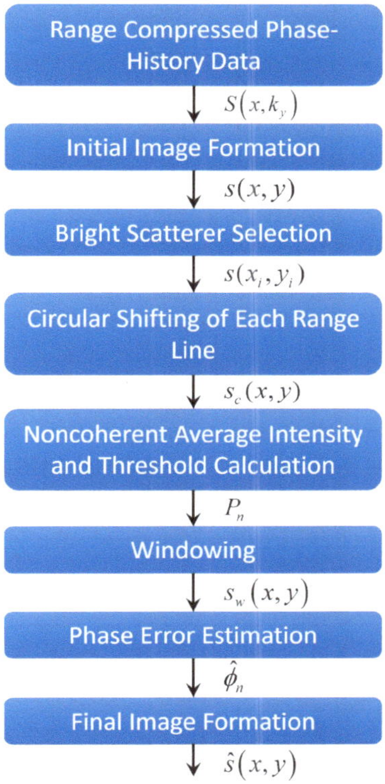

Figure 5.15 Phase gradient autofocus block diagram.

For SAR imagery, entropy is defined as [30]

$$\varepsilon = -\sum_{m=1}^{M}\sum_{n=1}^{N}|\hat{s}(m,n)|^2 \ln|\hat{s}(m,n)|^2, \tag{5.21}$$

where $\hat{s}(m,n)$ is the phase-adjusted complex-valued image given by

$$\hat{s}(m,n) = \mathcal{F}_n^{-1}\left\{\hat{S}(m,n)e^{-j\hat{\phi}_n}\right\}. \tag{5.22}$$

There is no closed-form solution for the phase correction that minimizes (5.21). Therefore, a numerical minimization technique is required. Various approaches have been taken, including gradient-based, stage-by-stage approaching, global optimization, and optimization transfer techniques [12, 31–33]. For the purpose of illustration, a relatively simple fixed-point iterative algorithm is used in this section.

The phase correction, $\hat{\phi}_n$, that minimizes the image entropy satisfies

$$\frac{\partial \varepsilon}{\partial \hat{\phi}_n} = 0. \tag{5.23}$$

Using (5.21), the partial derivative in (5.23) is written as

$$\frac{\partial \varepsilon}{\partial \hat{\phi}_n} = -\sum_{m=1}^{M}\sum_{n=1}^{N}\left[1 + \ln|\hat{s}(m,n)|^2\right]\frac{\partial |\hat{s}(m,n)|^2}{\partial \hat{\phi}_n}. \tag{5.24}$$

Using $|\hat{s}(m,n)|^2 = \hat{s}(m,n)\hat{s}^*(m,n)$, the partial derivative of the phase-adjusted image is expressed as

$$\frac{\partial |\hat{s}(m,n)|^2}{\partial \hat{\phi}_n} = 2\Re\left\{\hat{s}^*(m,n)\frac{\partial \hat{s}(m,n)}{\partial \hat{\phi}_n}\right\}. \tag{5.25}$$

Substituting (5.25) into (5.24) results in

$$\frac{\partial \varepsilon}{\partial \hat{\phi}_n} = -2\Re\left\{\sum_{m=1}^{M}\sum_{n=1}^{N}\left[1 + \ln|\hat{s}(m,n)|^2\right]\hat{s}^*(m,n)\frac{\partial \hat{s}(m,n)}{\partial \hat{\phi}_n}\right\}. \tag{5.26}$$

Using (5.22) allows (5.26) to be written as

$$\frac{\partial \varepsilon}{\partial \hat{\phi}_n} = -2\Im\left\{A(n)e^{j\hat{\phi}_n}\right\}, \tag{5.27}$$

where

$$A(n) = \sum_{m=1}^{M} \hat{S}^*(m,n)\mathcal{F}_n\Big\{\big[1 + \ln|\hat{s}(m,n)|^2\big]\hat{s}(m,n)\Big\}. \tag{5.28}$$

Finally, the phase correction is found from

$$\hat{\phi}_n = \angle A(n). \tag{5.29}$$

Note, $\hat{\phi}_n$ contains an arbitrary constant, which is an integer multiple of π. As stated earlier, a constant phase shift does not affect the quality of the final image. The complex-valued image is updated with the phase correction given in (5.29) and the process is repeated until the phase correction satisfies a convergence criteria of the form

$$\max\left\{\left|\exp\left(\hat{\phi}_n^i\right) - \exp\left(\hat{\phi}_n^{i-1}\right)\right|\right\} \leq \epsilon, \tag{5.30}$$

where ϵ is a threshold specifying the accuracy. The superscript i indicates the current iteration and i-1 is the previous iteration. The processing steps required for focusing SAR imagery with ME is given in Figure 5.16.

5.5 ADVANCED AUTOFOCUS TECHNIQUES

As described in Section 5.2, the autofocus techniques covered in this chapter are based on the assumption the phase errors are well modeled as one-dimensional errors, which is valid for many imaging scenarios. For very large scenes or imaging scenarios where the far-field approximation breaks down, the phase errors are not well modeled as one-dimensional. Much work has been performed in estimating target-dependent and spatially-varying types of phase errors. In [3, 34, 35], a Fourier transform method is used to compensate for both one-dimensional and two-dimensional phase errors. This approach has been validated on both simulated and

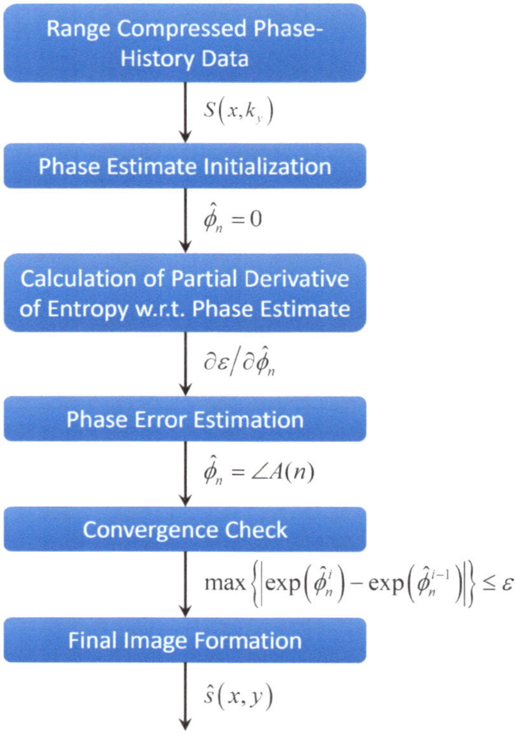

Figure 5.16 Generic minimum entropy autofocus block diagram.

measured SAR data. Two-dimensional phase errors have also been estimated with fast elliptic partial differential equation solvers [36, 37].

While PGA of Section 5.4.2 has been used with much success within the SAR community, it is often limited to imaging geometries where polar format or range-Doppler image formation techniques are applicable. A generalized phase gradient autofocus algorithm has been proposed which is also suitable for use with backprojection imaging algorithms [25]. This allows for a phase gradient based autofocus method to be used in a wider range of imaging scenarios, while simultaneously overcoming the drawback of using only a single scatterer per range line for phase-error estimation.

Nonparametric particle swarm optimization has also been proposed for use in SAR autofocus and is capable of estimating both low-frequency and high-frequency phase errors. Particle swarm optimization is a population-based stochastic optimization method based on the movement of fish schooling or birds flocking. Particle swarm methods have been highly successful over a wide range of optimization problems including natural sciences, mathematics, computer science, finance and economics, and biological sciences, to name a few [38]. Results have been successful on both simulated and measured data from RADARSAT-1 [32].

Employing a circular flight path provides a means for creating high resolution two and three-dimensional SAR imagery. However, this can lead to problems with the estimation of the phase errors. In [39], a minimum entropy-based method is used to accurately estimate the phase errors. This approach begins by applying the backprojection algorithm to form an imaging matrix. Next the power-normalized image is formed. Finally, the phase errors are estimated through minimum entropy in an iterative manner.

UAVs extend the application of SAR scenarios and imaging methods. However, uncertainties in the platform's position and orientation lead to greatly degraded SAR imagery requiring more sophisticated and robust autofocus algorithms. Minimum entropy methods based on a variational formulation with limited apertures have shown success in three-dimensional imaging from UAV SAR platforms [40]. In [41], a minimum entropy-based method is used to coherently align two-dimensional SAR imagery to form three-dimensional images.

The U.S. Army Combat Capabilities Development Command (DEVCOM) ARL has been investigating the capability of using a three-dimensional millimeter-wave SAR platform to provide navigation for aircraft through terrain and obstacles in a degraded visual environment [42]. These types of systems require very high position measurement accuracy. Therefore, the development of an effective and practical autofocus algorithm is necessary. A metric based autofocus algorithm that models the phase errors as the rotation and translation of the radar antenna frame in three-dimensional space was studied.

5.6 EXAMPLES

The sections that follow illustrate the concepts of this chapter with a few Python/MATLAB examples. In these examples, a fully focused image is used as the starting point and the error model of Section 5.2 is used to create the corrupted image. The selected autofocus algorithm is then used to estimate the phase errors and focus the corrupted image. The Python examples for this chapter are in the directory *software/python/Chapter5* and the matching MATLAB examples are in the directory *software/matlab/Chapter5*. The reader should consult Chapter 1 for information on how to execute the Python and MATLAB code associated with this book.

5.6.1 Map Drift

As a first example of SAR autofocus, MD is used to estimate and correct for a quadratic phase error. The original image for this example is the ICEYE SAR image of the city of Muscat in Oman, and is shown in Figure 5.17(a). The corrupted image is shown in Figure 5.17(b) and is created by applying the phase error given in Figure 5.18 to the original image of Figure 5.17(a). Referring to Figure 5.9, the first step in MD is to create separate images from subapertures. In this example, two subapertures are used to estimate the quadratic phase error. The cross-power spectrum is then calculated from which an estimate of the offset is found. The offset estimate is used to determine the quadratic factor and finally the estimated phase error. A comparison between the applied phase error and the estimated phase error is shown in Figure 5.18, and it is seen that MD with two subapertures accurately estimates the quadratic phase error. Finally, the estimated phase error is used to correct the corrupted image in Figure 5.17(b), resulting in the focused image given in Figure 5.19. Comparing the original image in Figure 5.17(a) with the corrected image in Figure 5.19 shows the performance of MD with two subapertures for estimating and compensating for quadratic type phase errors. As stated earlier, more subapertures may be used to estimate higher order phase errors. This example is given in the Python notebook *map_drift_example.ipynb* and the MATLAB live script *map_drift_example.mlx*.

(a) Original SAR imagery.

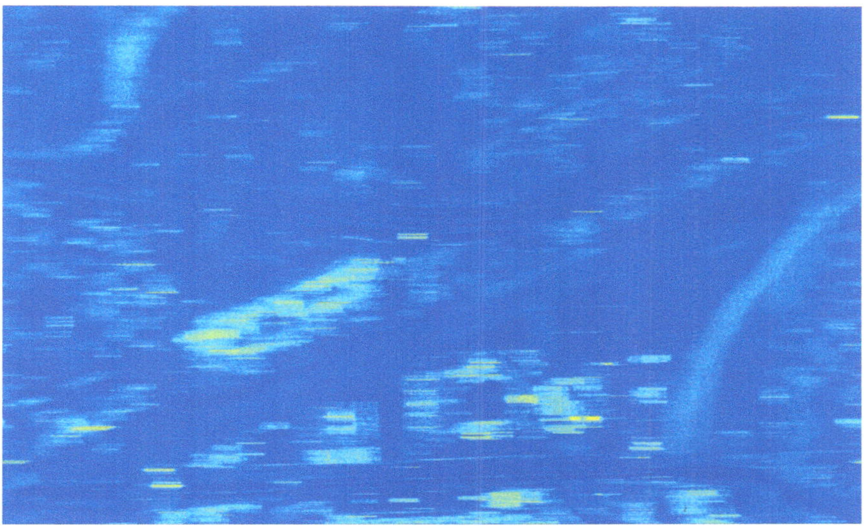

(b) SAR imagery corrupted with a quadratic phase error.

Figure 5.17 SAR imagery of Muscat, Oman corrupted with a quadratic phase error. Courtesy of ICEYE.

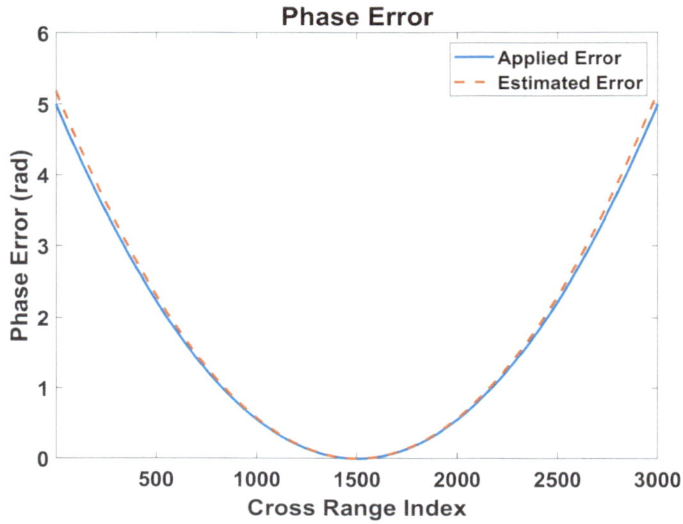

Figure 5.18 Comparison of applied phase error and phase error estimated by MD.

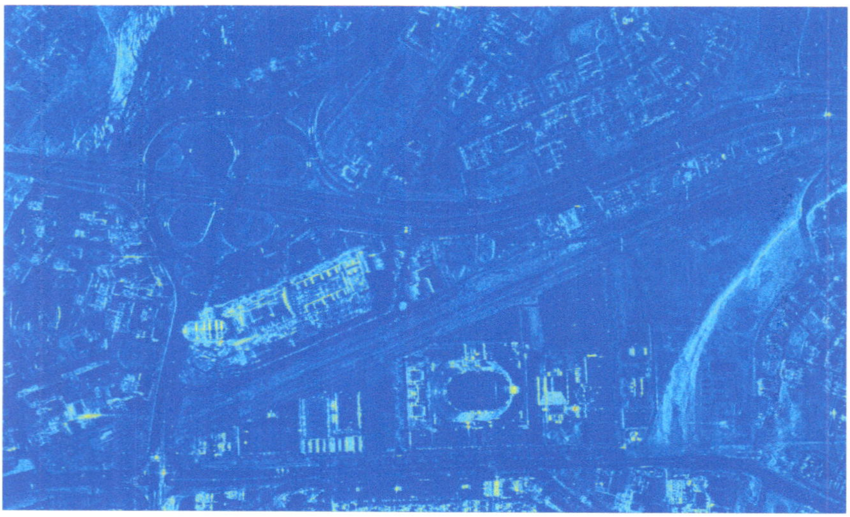

Figure 5.19 MD focused SAR image of Muscat, Oman. Courtesy of ICEYE.

5.6.2 Inverse Filtering

As a simple example of a nonparametric method, inverse filtering autofocus is used to estimate and correct for a phase error of the form $A\,x^3 + B\,x^2$, where A and B are constants, $x \in [-L_{sa}/2, L_{sa}/2]$, and L_{sa} is the synthetic aperture length. The original image used in this example is the ICEYE SAR image of the Muscat International Airport in Oman, and is shown in Figure 5.20(a) [43]. Figure 5.20(b) shows the corrupted image generated by applying the phase error shown in Figure 5.21 to the original image of Figure 5.20(a). Finally, the estimated phase error is used to correct the corrupted image in Figure 5.20(b), resulting in the focused image given in Figure 5.22. Comparing the original image in Figure 5.20(a) to the corrected image in Figure 5.22, it is seen that inverse filtering does an adequate job of estimating and correcting for the applied phase error, assuming that an isolated point scatterer can be found and used in the image. Note, the data in Figure 5.20(b) is used to select the point scatterer as this is the image created from the raw phase-history data collected by the sensor. Without knowledge of the true scatterers in the scene, it may be difficult to select an appropriate scatterer to use in the inverse filtering algorithm. This example is given in the Python notebook *inverse_filtering_example.ipynb* and the MATLAB live script *inverse_filtering_example.mlx*.

5.6.3 Phase Gradient

This example illustrates the use of PGA to estimate and correct for a fourth order phase error of the form $A\,x^4 + B\,x^2/2$, where A and B are constants, $x \in [-L_{sa}/2, L_{sa}/2]$, and L_{sa} is the synthetic aperture length. The original image used in this example is the ICEYE SAR image of Suvarnabhumi Airport in Thailand and is shown in Figure 5.23(a) [43]. The original image is then corrupted by the phase error shown in Figure 5.24, resulting in the image shown in Figure 5.23(b).

The phase error for this example was estimated with five iterations of PGA. For comparison, the estimated phase error and applied phase error are shown in Figure 5.24. The estimated phase error matches well over a large region and differs at the ends of the aperture. This difference may be reduced with more iterations of PGA. The convergence criteria may be met before an exact match to the phase error is achieved. Also, the true phase error in measured phase-history data is not

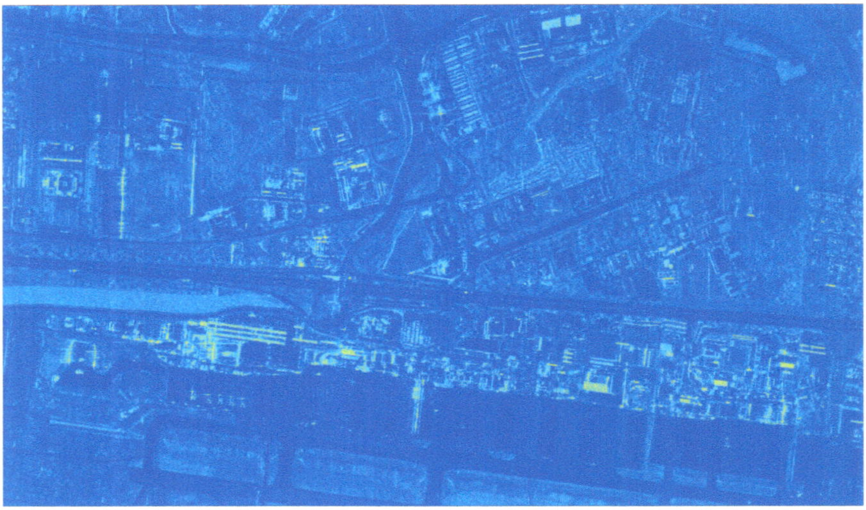

(a) Original SAR imagery.

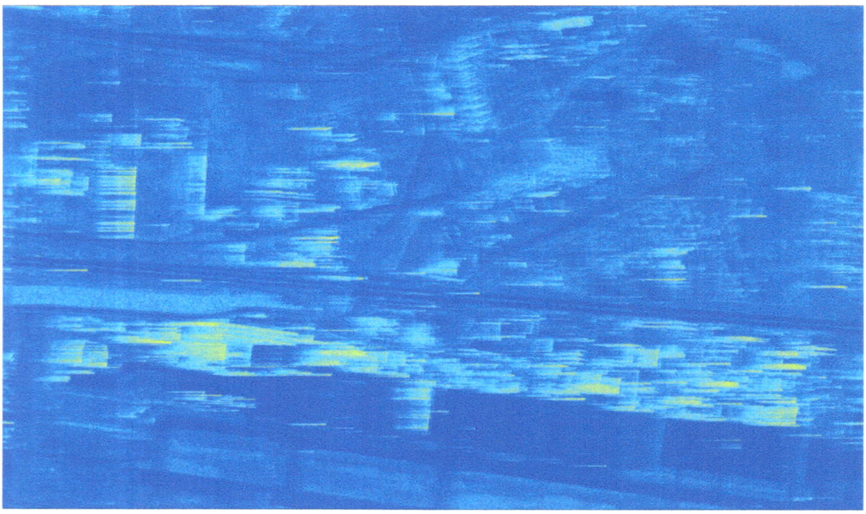

(b) SAR imagery corrupted with a quadratic phase error.

Figure 5.20 SAR imagery of Muscat International Airport corrupted with a cubic phase error. Courtesy of ICEYE.

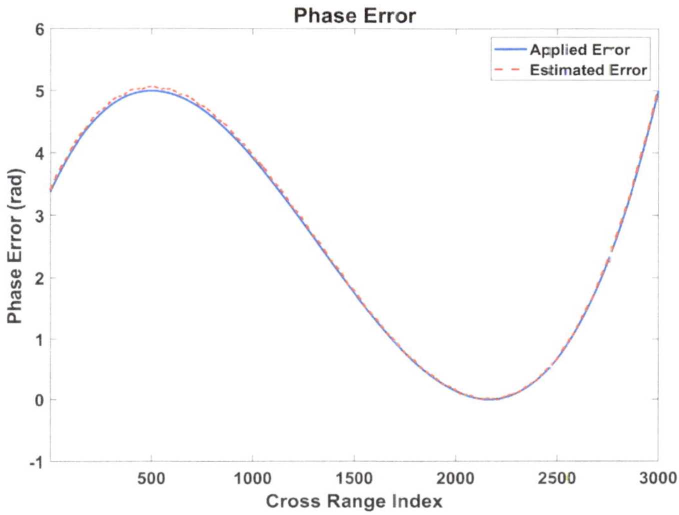

Figure 5.21 Comparison of applied phase error and phase error estimated with inverse filtering.

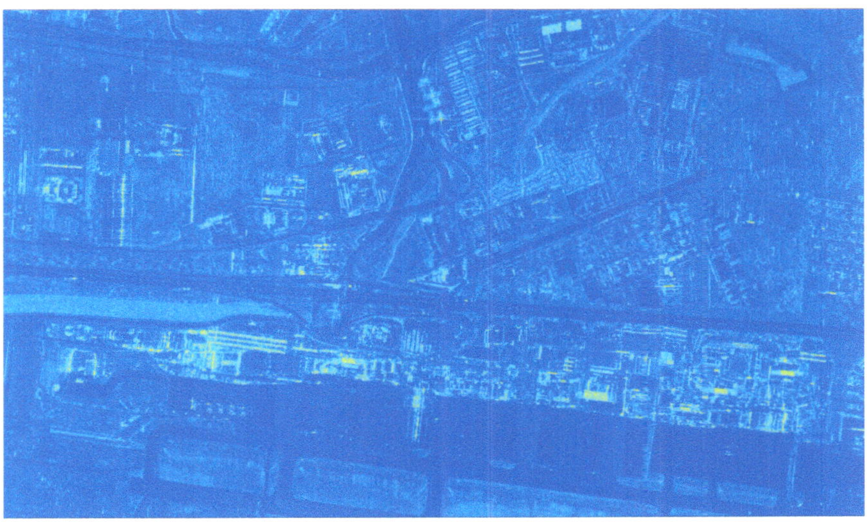

Figure 5.22 SAR image of Muscat International Airport focused with inverse filtering. Courtesy of ICEYE.

known exactly. The estimated phase error is used to correct the corrupted image in Figure 5.23(b) and the focused image is shown in Figure 5.25. This illustrates the quality of the focused imagery even with small errors in the estimated phase error. This example is given in the Python notebook *phase_gradient_example.ipynb* and the MATLAB live script *phase_gradient_example.mlx*.

5.6.4 Minimum Entropy

For the final example of SAR autofocus, ME is used to estimate and correct for a phase error of the form $A\,x^2 + B\,\cos^2(x^4)$, where A and B are constants, $x \in [-L_{sa}/2, L_{sa}/2]$, and L_{sa} is the synthetic aperture length. The original image used in this example is the ICEYE SAR image of Kuala Lumpur International Airport in Malaysia, and is shown in Figure 5.26(a) [43]. The phase error of Figure 5.27 is applied and results in the image given in Figure 5.26(b). The minimum entropy algorithm is then used to estimate the phase error using (5.28) and (5.29). The phase error for this example was estimated with only a single iteration of ME. For comparison, the estimated phase error and applied phase error are shown in Figure 5.27. This illustrates the ability of ME to estimate somewhat arbitrary phase errors. The estimated phase error is then used to correct the corrupted image in Figure 5.26(b) and the focused image is shown in Figure 5.28. This example is given in the Python notebook *minimum_entropy_example.ipynb* and the MATLAB live script *minimum_entropy_example.mlx*.

PROBLEMS

5.1 Characterize SAR imagery with and without the use of an autofocus algorithm.

5.2 Describe the differences between traditional motion compensation and autofocus for SAR image generation.

5.3 What are the limiting factors and major problems with using traditional motion compensation only?

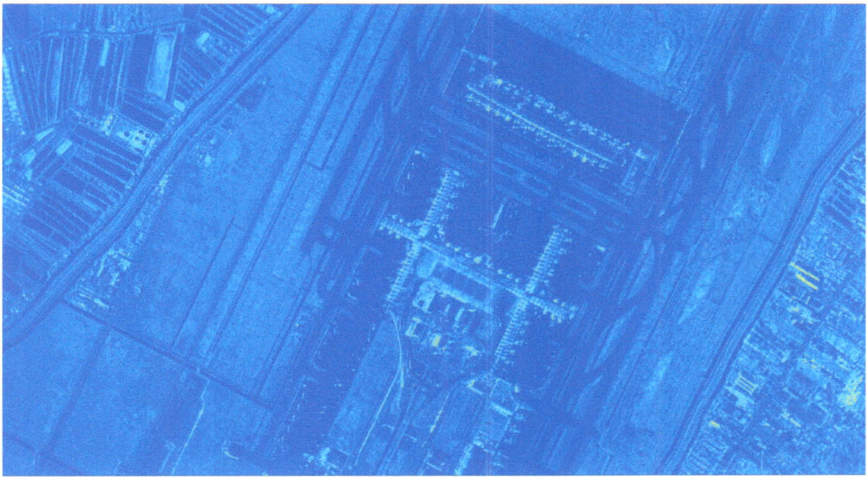

(a) Original SAR imagery.

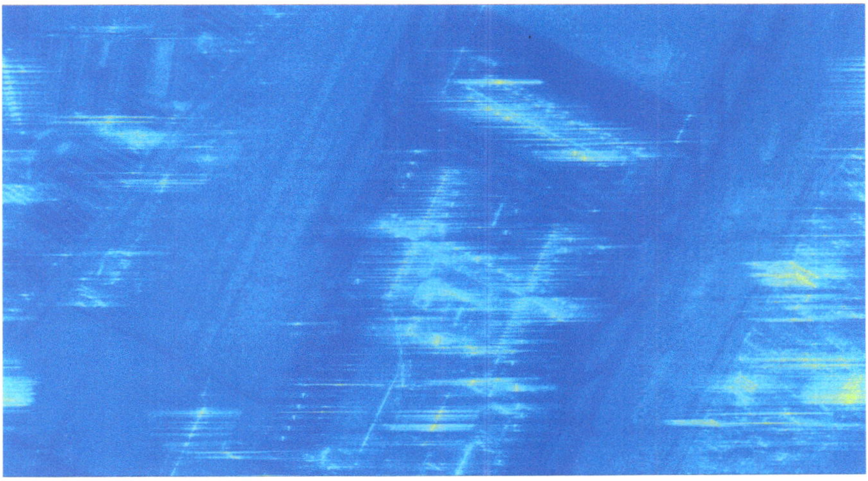

(b) SAR imagery corrupted with a fourth-order phase error.

Figure 5.23 SAR imagery of Suvarnabhumi Airport in Thailand corrupted with a fourth-order phase error. Courtesy of ICEYE.

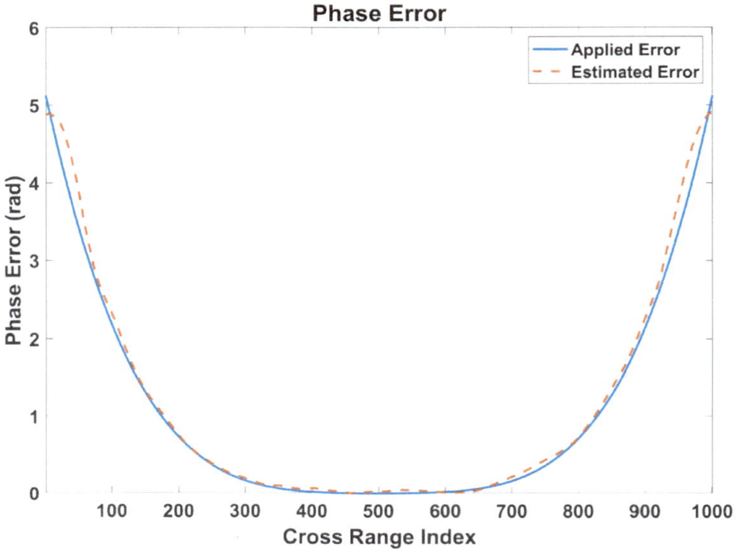

Figure 5.24 Comparison of applied phase error and phase error estimated with PGA.

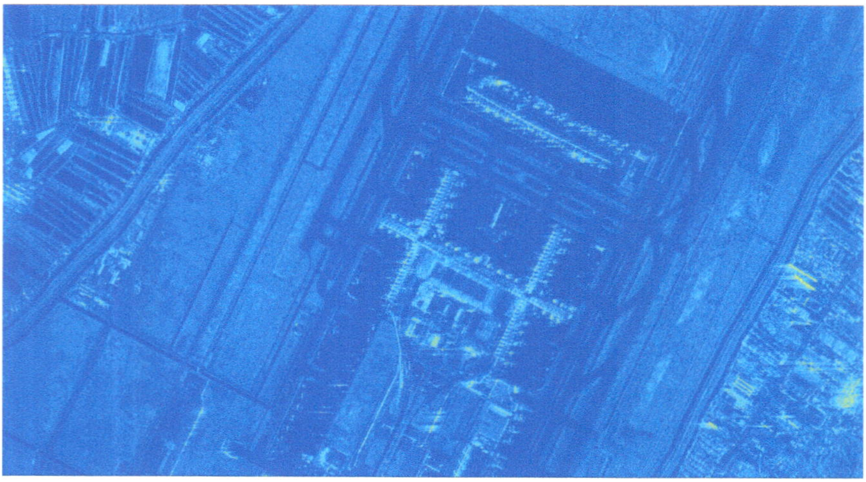

Figure 5.25 PGA focused SAR image of Suvarnabhumi Airport. Courtesy of ICEYE.

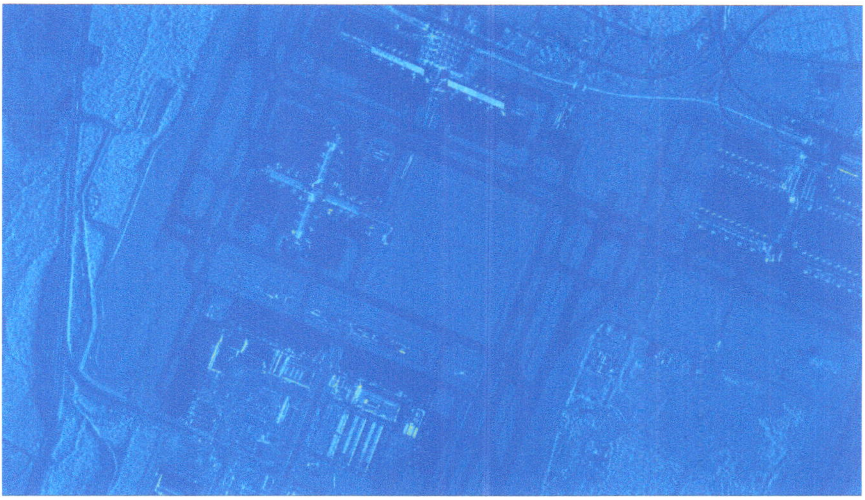

(a) Original SAR imagery.

(b) SAR imagery corrupted by phase errors.

Figure 5.26 SAR imagery of Kuala Lumpur International Airport in Malaysia corrupted by phase errors. Courtesy of ICEYE.

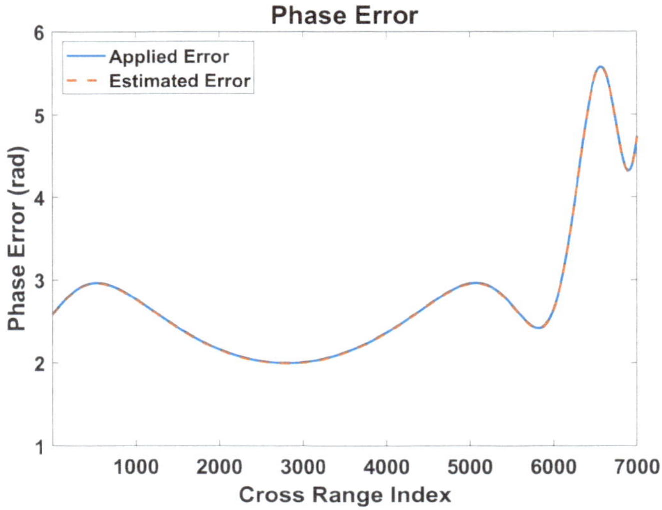

Figure 5.27 Comparison of the applied phase error and ME estimated phase error.

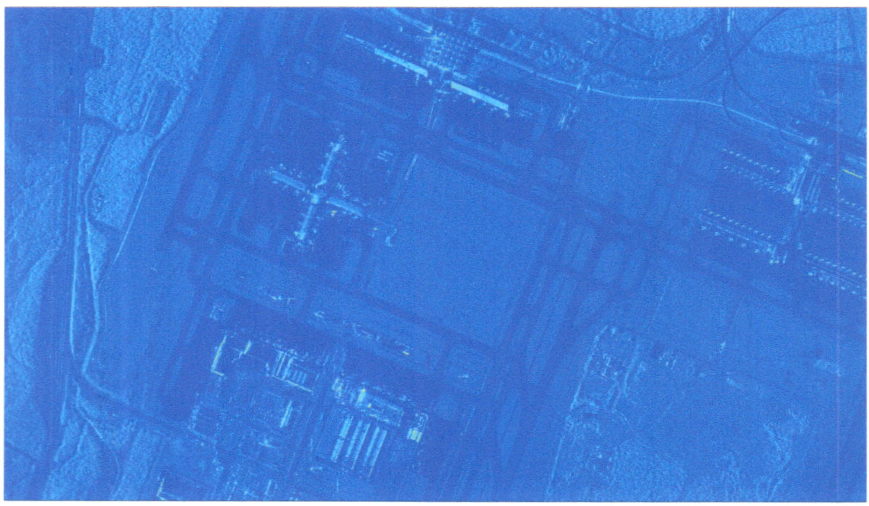

Figure 5.28 ME focused image of of Kuala Lumpur Airport. Courtesy of ICEYE.

5.4 Other than motion errors, what are other mechanisms that contribute to phase errors?

5.5 Give state-of-the-art measurement accuracies for GNSS/INS/GPS systems.

5.6 What is the major assumption used in the development of the error model of Section 5.2?

5.7 How would the error model of Section 5.2 be modified to account for spatially varying phase errors?

5.8 Describe the major differences between model-based and nonparametric autofocus techniques.

5.9 Illustrate an imaging scenario where it would be advantageous to use a model-based method rather than a nonparametric method.

5.10 How can map-drift autofocus be extended to estimate fifth order phase errors?

5.11 What is the major limiting factor when extending map-drift autofocus to higher order polynomial type phase errors?

5.12 What are the major advantages of using phase gradient autofocus versus inverse filtering?

5.13 Give examples of imaging scenarios where inverse filtering would not produce good results.

5.14 What is the fundamental difference between minimum entropy autofocus and phase gradient autofocus?

5.15 Describe imaging scenarios where minimum entropy autofocus would excel compared to phase gradient autofocus.

5.16 List some other image metrics that are used in nonparametric autofocus algorithms.

References

[1] V. Koo, T. Lim, and H. Chuah. A comparison of autofocus algorithms for SAR imagery. In *Progress In Electromagnetics Research Symposium*. Hangzhou, 2005.

[2] C. Jakowatz, D. Wahl, P. Eichel, D. Ghiglia, and P. Thompson. *Spotlight-Mode Synthetic Aperture Radar: A Signal Processing Approach*. Springer, New York, 1996.

[3] D. Hawkins. *Synthetic Aperture Imaging Algorithms: with application to wide bandwidth sonar*. PhD thesis, University of Canterbury, Christchurch, New Zealand, October 1996.

[4] W. Carrara, R. Majewski, and R. Goodman. *Spotlight Synthetic Aperture Radar: Signal Processing Algorithms*. Artech House, Norwood, 1995.

[5] J. Kirk. A discussion of digital processing in synthetic aperture radar. *IEEE Transactions on Aerospace and Electronic Systems*, 11(3):326–337, May 1975.

[6] J. Kirk. Motion compensation for synthetic aperture radar. *IEEE Transactions on Aerospace and Electronic Systems*, 11(3):338–348, May 1975.

[7] J. Mims and J. Farrell. Synthetic aperture imaging with maneuvers. *IEEE Transactions on Aerospace and Electronic Systems*, 8(4):410–418, July 1972.

[8] U.S. Spaceforce. GPS: The Global Positioning System. https://www.gps.gov. Accessed: 2021-10-11.

[9] NovAtel. *An Introduction to GNSS*. NovAtel Inc., Calgary, 2nd edition, 2015.

[10] SBG Systems. Ellipse series - miniature inertial navigation sensors. https://www.sbg-systems.com/products/ellipse-series. Accessed: 2021-10-11.

[11] A. Yegulalp. Minimum entropy SAR autofocus. In *Proceedings of the Adaptive Sensor Array Processing (ASAP) Workshop*, Lexington, 1999.

[12] T. Kragh. Monotonic iterative algorithm for minimum-entropy autofocus. In *Proceedings of the Adaptive Sensor Array Processing (ASAP) Workshop*, Lexington, 2006.

[13] T. Calloway and G. Donohoe. Subaperture autofocus for synthetic aperture radar. *Transactions on Aerospace and Electronic Systems*, 30(2):617–621, April 1994.

[14] T. Calloway, C. Jakowatz, P. Thompson, and P. Eichel. Comparison of synthetic-aperture radar autofocus techniques: Phase gradient versus subaperture. In *Proceedings of the Advanced Signal Processing Algorithms, Architectures, and Implementations II. International Society for Optics and Photonics*, pages 353–364, San Diego, July 1991.

[15] Z. Liu, S. Yang, Z. Feng, Q. Gao, and M. Wang. Fast SAR autofocus based on ensemble convolutional extreme learning machine. *Journal of Remote Sensing*, 13(14), 2021.

[16] G. Bendor and T. Gedra. Single-pass fine-resolution SAR autofocus. In *Proceedings of IEEE National Aerospace and Electronics Conference NAE-CON*, pages 482–488, Dayton, May 1983.

[17] O. Bezvesilniy, I. Gorovyi, and D. Vavriv. Estimation of phase errors in SAR data by local-quadratic map-drift autofocus. In *Proceedings of the 19th International Radar Symposium*, 2012.

[18] H. Cantalloube and C. Nahum. Motion compensation and autofocus of range/Doppler or two-dimensional processing for airborne synthetic aperture radar. *Aerospace Science and Technology*, 2(4):251–263, May 1998.

[19] H. Cantalloube and C. Nahum. Multiscale local map-drift driven multilateration SAR autofocus using fast polar format algorithm image synthesis. *IEEE Transactions on Geoscience and Remote Sensing*, 49(10):3730–3736, May 2010.

[20] G. Wang, M. Zhang, Y. Huang, L. Zhang, and R. Wang. Robust two-dimensional spatial-variant map-drift algorithm for UAV sar autofocusing. *Journal of Remote Sensing*, 11(3), 2019.

[21] P. Eichel and C. Jakowatz. Phase gradient algorithm as an optimal estimator of the phase derivative. *Optics Letters*, 14(20):1101–1103, 1989.

[22] D. Wahl, P. Eichel, D. Ghiglia, and C. Jakowatz. Phase gradient autofocus - a robust tool for high resolution SAR phase correction. *IEEE Transactions on Aerospace and Electronic System*, 30(3):827–835, 1994.

[23] A. Azouz and Z. Li. Improved phase gradient autofocus algorithm based on segments of variable lengths and minimum entropy phase correction. In *IEEE China Summit and International Conference on Signal and Information Processing*, pages 194–198, 2014.

[24] H. Tang, H. Shi, and C. Qi. Study on improvement of phase gradient autofocus algorithm. *First International Workshop on Education Technology and Computer Science*, pages 58–61, 2009.

[25] A. Evers and J. Jackson. Generalized phase gradient autofocus using semidefinite relaxation phase estimation. *IEEE Transactions on Computational Imaging*, 6:291–303, 2020.

[26] D. Thompson, J. Bates, and D. Arnold. Extending the phase gradient autofocus algorithm for low-altitude stripmap mode SAR. In *Proceedings of the 1999 IEEE Radar Conference. Radar into the Next Millennium (Cat. No.99CH36249)*, pages 36–40, 1999.

[27] A. Harrison. *Introduction to Radar Using Python and MATLAB*. Artech House, Norwood, 2020.

[28] P. Eichel, D. Ghiglia, and C. Jakowatz. Speckle processing method for synthetic-aperture radar phase correction. *Optics Letters*, 14(1):1–3, January 1989.

[29] P. Eichel, D. Ghiglia, and C. Jakowatz. Phase correction system for automatic focusing of synthetic aperture radar. *U.S. Patent*, 4:229, May 1990.

[30] J. Wang, D. Kasilingam, X. Liu, and Z. Zhou. ISAR minimum-entropy phase adjustment. In *Proceedings of the 2004 IEEE Radar Conference*, pages 197–200, 2004.

[31] T. Lim, V. Koo, H. Ewe, and H. Chuah. High frequency phase error reduction in SAR using particle swarm optimization. *Journal of Electromagnetic Waves and Applications*, 21(6):795–810, 2007.

[32] T. Lim, H. Ewe V. Koo, and H. Chuah. A SAR autofocus algorithm based on particle swarm optimization. *Progress in Electromagnetics Research B*, (1):159–176, 2008.

[33] C. Jakowatz and D. Wahl. Eigenvector method for maximum-likelihood estimation of phase errors in synthetic-aperture radar imagery. *Journal of the Optical Society of America A*, 10(12):2539–2546, December 1993.

[34] T. Isernia, V. Pascazio, R. Pierri, and G. Schirinzi. Image reconstruction from fourier transform magnitude with applications to synthetic aperture radar imaging. *Journal of the Optical Society of America*, 13(5):922–934, May 1996.

[35] T. Isernia, V. Pascazio, R. Pierri, and G. Schirinzi. Synthetic aperture radar imaging from phase corrupted data. *IEE Proceedings on Radar, Sonar, and Navigation*, 143(4):268–274, August 1996.

[36] D. Ghiglia and L. Romero. Direct phase estimation from phase differences using fast elliptic partial differential equation solvers. *Optics Letters*, 14(20):1107–1109, October 1989.

[37] D. Ghiglia and G. Mastin. Two-dimensional phase correction of synthetic-aperture-radar imagery. *Optics Letters*, 14(20):1104–1106, October 1989.

[38] B. de Almeida and V. Leite. *Particle Swarm Optimization: A Powerful Technique for Solving Engineering Problems*. Intech Open Book Series, 2019.

[39] B. Zhang, X. Zhang, and S. Wei. A circular SAR image autofocus algorithm based on minimum entropy. In *2015 IEEE 5th Asia-Pacific Conference on Synthetic Aperture Radar (APSAR)*, pages 152–155, 2015.

[40] I. Fletcher, C. Watts, E. Miller, and D. Rabinkin. Minimum entropy autofocus for 3D sar images from a uav platform. *IEEE*, pages 1–5, 2016.

[41] T. Kragh. Minimum-entropy autofocus for three-dimensional sar imaging. Technical report, BAE Systems - Society of Photo-Optical Instrumentation Engineers, 2009.

[42] L. Nguyen. 3D image autofocus for millimeter-wave forward-looking SAR. In *Passive and Active Millimeter-Wave Imaging XXIV*, volume 11745, pages 13–21. International Society for Optics and Photonics, SPIE, 2021.

[43] ICEYE. Example SAR data from ICEYE. https://www.iceye.com/downloads/datasets. Accessed: 2021-09-21.

Chapter 6

Image Registration

SAR image registration is the process of aligning multiple complex images of a common scene captured by the same or different sensors from different viewpoints, at different times, and possibly different operational modes. Image registration is key to a variety of remote sensing applications including change detection, image fusion, multitemporal analysis, three-dimensional reconstruction, interferometry, and tomographic imaging [1, 2]. Image registration is a challenging task, as the images have significant intensity variations and geometric differences including effects such as translation, rotation, scale, shear, and illumination These lead to imaging warping, which must be estimated and removed to achieve the subpixel accuracy required for advanced image processing techniques [3–5]. Several diverse approaches to the image registration problem have been proposed. These include geometrical image registration, interferometric registration using spectral diversity, multistep least-squares matching, and analytic search methods [2, 6–12]. With improved resolution and shorter revisit times of modern sensor systems, the size of SAR data is growing rapidly and the registration of multiple SAR images results in a very large computational burden [13]. High performance computing methods, including parallelization of algorithms, have been studied and implemented with success [5, 14–17]. The chapter begins with an overview of some of the more popular image registration methods. This is followed by coverage of phase correlation techniques. Next, feature based methods including Harris corner, scale-invariant feature transform (SIFT), and speeded-up robust features (SURF) are covered. Methods for warp function extraction, including random sample consensus (RANSAC), fast least trimmed squares (Fast-LTS), and extended fast least trimmed squares (EF-LTS), are studied. The chapter concludes with a few Python and MATLAB examples to further illustrate the concepts of SAR image registration.

6.1 METHODS

SAR image registration methods are typically separated into two categories, intensity-based methods and feature-based methods [18]. Intensity-based methods typically use some form of cross correlation or mutual information techniques [19–21]. Cross correlation methods operate on the pixel values directly, and mutual information techniques often operate in the frequency domain. These methods have a high computational cost and suffer from limitations when dealing with images with large geometric differences [22]. Feature-based methods identify and extract image features such as corners, lines, and contours, and then employ methods to match the extracted features among the differing images. These methods can be impacted by speckle, which is caused by random interference from multiple scattering that occurs in each resolution cell [23]. In general, feature-based methods have better performance over a wide range of image registration scenarios and are computationally more efficient than intensity-based methods [20, 24, 25]. The Harris corner detector is used as a point-feature method and has a response that is related to the local neighboring gradient distribution of a point [26–28]. Harris corner does not have a feature descriptor as many of the other featured-based image registration methods do. As such, Harris corner is only invariant to translation and rotation and is successful for simple image warping. One of the most commonly used feature-based methods is SIFT, which extracts features based on automatic scale selection [24, 29, 30]. The main drawback of SIFT is long execution times [31]. This problem is further complicated by speckle and limits its application to general image registration problems. Therefore, many variations of SIFT have been proposed to deal with these limitations [32]. SURF is another popular feature-based image registration method commonly used for SAR imagery [33–36]. SURF feature detection and description have been shown to outperform SIFT in terms of speed, repeatability, robustness, and distinctiveness [25]. It has also been used in many other areas including satellite image registration and seabed recognition based on sonar imagery [37, 38].

6.2 PHASE CORRELATION

Phase correlation methods for image registration use the image data in the spatial-frequency domain to register images that are translated, rotated, and scaled with respect to one another [39]. Since translation, rotation, and scale manifest in a particular manner in the spatial-frequency domain, the parameters required to register the images are readily found from properties of the Fourier transform. Phase correlation methods typically have low computational burden compared to other image registration methods and are usually not affected by noise [2].

6.2.1 Translation

For translation, consider two images $s_1(x, y)$ and $s_2(x, y)$ with Fourier transforms $S_1(k_x, k_y)$ and $S_2(k_x, k_y)$ such that the images only differ by a translation, as illustrated in Figure 6.1. This translation is expressed as

$$s_1(x', y') = s_2(x - x_0, y - y_0), \tag{6.1}$$

where the translation is (x_0, y_0). The Fourier shift theorem states that a delay in the time domain corresponds to a linear phase term in the frequency domain [40]. Therefore

$$s_2(x - x_0, y - y_0) \longleftrightarrow S_2(k_x, k_y)\, e^{-j2\pi(k_x\, x_0 + k_y\, y_0)}. \tag{6.2}$$

Taking the cross-power spectrum of the two images gives [39]

$$P(S_1, S_2) = \frac{S_1(k_x, k_y)\, S_2^*(k_x, k_y)}{\left|S_1(k_x, k_y)\, S_2(k_x, k_y)\right|} = e^{-j2\pi(k_x\, x_0 + k_y\, y_0)}. \tag{6.3}$$

Finally, taking the inverse Fourier transform of (6.3) results in a *sinc* function centered at the translation (x_0, y_0) required to register the two images.

6.2.2 Rotation

To demonstrate rotation, consider $s_1(x, y)$ to be a rotated version of $s_2(x, y)$ such that

$$s_1(x', y') = s_2\big(x \cos\theta_0 - y \sin\theta_0, x \sin\theta_0 + y \cos\theta_0\big), \tag{6.4}$$

where θ_0 is the rotation between the two images, as shown in Figures 6.1(a) and 6.2(a). A new rotated coordinate system is defined as

$$\begin{bmatrix} x' \\ y' \end{bmatrix} = \begin{bmatrix} \cos\theta_0 & -\sin\theta_0 \\ \sin\theta_0 & \cos\theta_0 \end{bmatrix} \begin{bmatrix} x \\ y \end{bmatrix}. \tag{6.5}$$

The Fourier rotation theorem states that the rotation of a function by an angle θ_0 results in the rotation of the Fourier transform by the same angle [40]. This is expressed as

$$s_2(x', y') \longleftrightarrow S_2\big(k_x \cos\theta_0 - k_y \sin\theta_0, k_x \sin\theta_0 + k_y \cos\theta_0\big). \tag{6.6}$$

If the rotation is represented as a translational shift with polar coordinates as

$$\big|S_1(r, \theta)\big| = \big|S_2(r, \theta - \theta_0)\big|, \tag{6.7}$$

using the techniques in Section 6.2.1, the rotation angle θ_0 may then be found. For illustration, consider $S_2(r, \theta)$ partitioned into two sections $S_2^a(r, \theta)$ and $S_2^b(r, \theta)$ as shown in Figure 6.3. For simplicity, allow $S_2^b(r, \theta)$ subtends the angle θ_0. A rotation by θ_0 does not affect the r coordinate. The effect on the θ coordinate is shown in Figure 6.3. The polar transformation of $S_2(r, \theta)$ shows that $S_2^a(r, \theta)$ extends from 0 to $2\pi - \theta_0$ before the rotation and θ_0 to 2π after the rotation. Similarly, $S_2^b(r, \theta)$ extends from $2\pi - \theta_0$ before rotation and 0 to θ_0 after rotation. Therefore, the

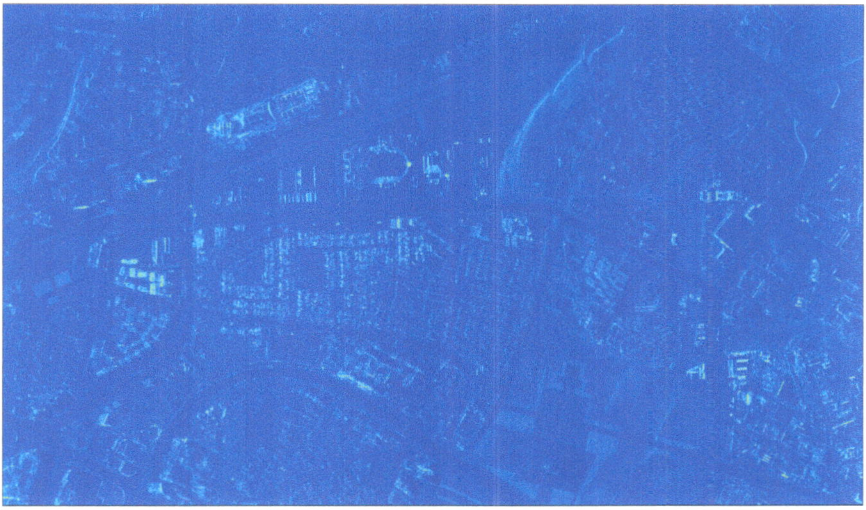

(a) SAR imagery used as the reference.

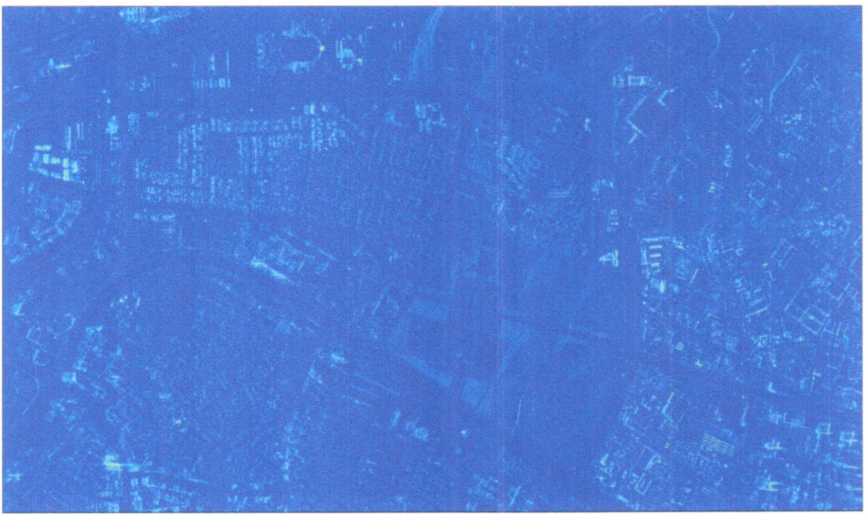

(b) SAR imagery with translation.

Figure 6.1 Comparison of SAR imagery of Muscat, Oman with a translational difference. Courtesy of ICEYE.

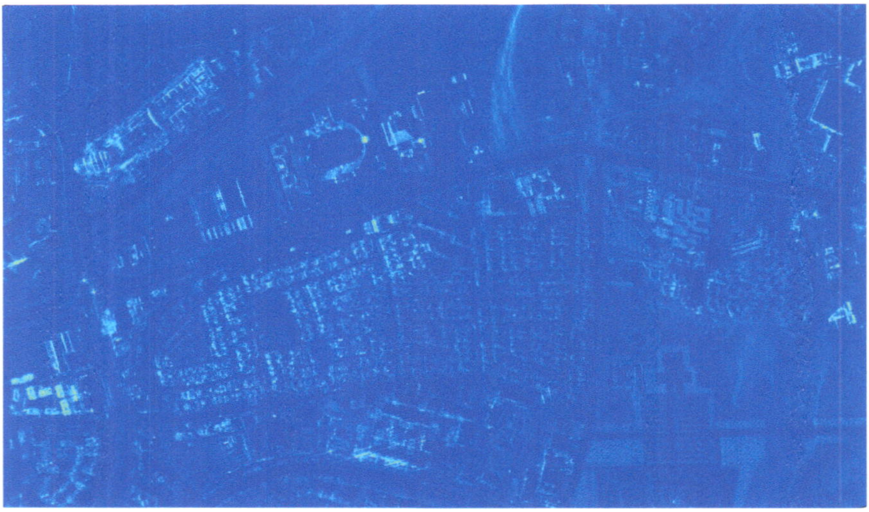

(a) SAR imagery with a rotation of 16°.

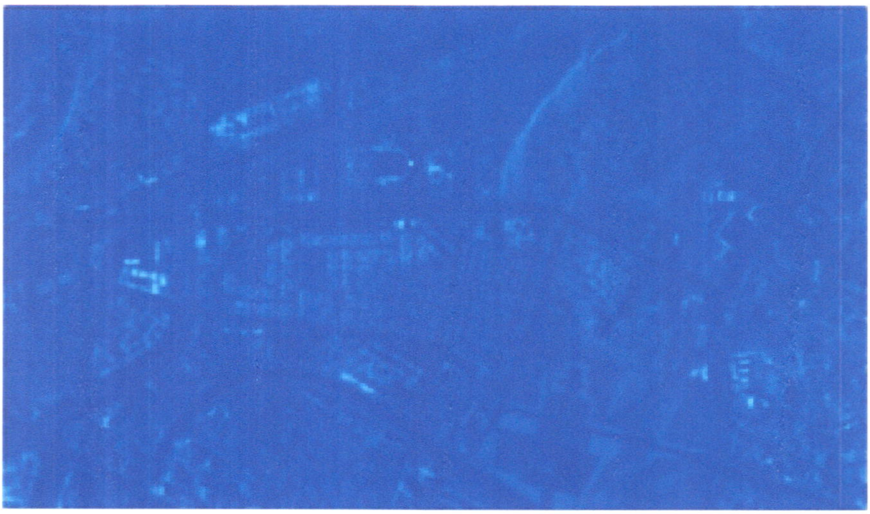

(b) SAR imagery with a Gaussian scale factor of 4.

Figure 6.2 Comparison of SAR imagery of Muscat, Oman with (a) rotation and (b) scale change. Courtesy of ICEYE.

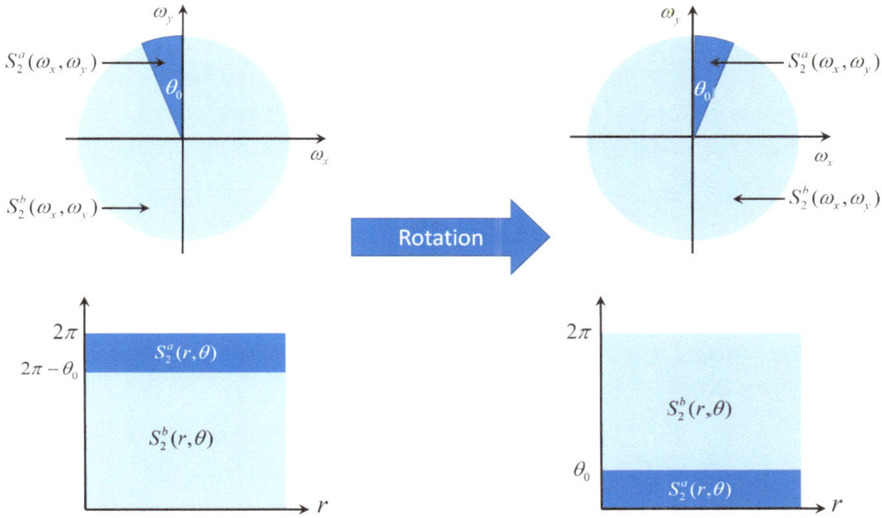

Figure 6.3 Effect of image rotation on the spatial-frequency domain image data.

rotation corresponds to a translation after the polar transform and may be found using the techniques of Section 6.2.1.

6.2.3 Scale Change

To illustrate scale change, consider the situation where $s_1(x, y)$ is a scaled version of $s_2(x, y)$, as illustrated in Figure 6.2(b). A scale change between the two images may be written as

$$s_1(x', y') = s_2(\alpha\, x, \beta\, y). \tag{6.8}$$

The new scaled coordinate system is defined as

$$\begin{bmatrix} x' \\ y' \end{bmatrix} = \begin{bmatrix} \alpha & 0 \\ 0 & \beta \end{bmatrix} \begin{bmatrix} x \\ y \end{bmatrix}. \tag{6.9}$$

Note that α is the scale change in the x direction and β is the scale change in the y direction. The Fourier scale theorem, which is sometimes referred to as the similarity theorem, states that a compression or dilation by factors of (α, β) in the time domain results in a compression or dilation by factors of $(1/\alpha, 1/\beta)$ in the frequency domain. This allows (6.8) to be expressed in the spatial-frequency domain as

$$S_1(k_x, k_y) = \frac{1}{|\alpha\,\beta|}\, S_2\left(\frac{k_x}{\alpha}, \frac{k_y}{\beta}\right). \tag{6.10}$$

The scale change may now be transformed to a translation by converting the wavenumber axes to a logarithmic scale as [41]

$$S_1(\omega_x, \omega_y) = S_2(\omega_x - \ln\alpha, \omega_y - \ln\beta), \tag{6.11}$$

where $\omega_x = \ln k_x$ and $\omega_y = \ln k_y$. Recalling (6.2) and (6.3), the scale factors (α, β) are found from the peak of the sinc function, which is located at $(\ln\alpha, \ln\beta)$. The translation, rotation, and scale are given by the weak affine transform expressed as [42]

$$\begin{bmatrix} x' \\ y' \\ 1 \end{bmatrix} = \begin{bmatrix} \alpha\cos\theta_0 & -\beta\sin\theta_0 & -x_0 \\ \alpha\sin\theta_0 & \beta\cos\theta_0 & -y_0 \\ 0 & 0 & 1 \end{bmatrix} \begin{bmatrix} x \\ y \\ 1 \end{bmatrix}. \tag{6.12}$$

Warp functions are often more complicated than the transform in (6.12) and require more robust image registration methods.

6.3 HARRIS CORNER

The Harris corner detector is a point-feature method and has a response that is the weighted sum of the determinant and squared trace of the first-order moment matrix which describes the local neighboring gradient distribution of a point [26–28]. The Harris corner detector is often used in computer vision to extract corners and develop features, and is an improvement upon the Moravec corner detector [43]. Corners are local points at the junction of two edges of dominant and different directions, where edges are determined by sudden changes in image brightness. Figure 6.4 illustrates constant, edge, and corner features. The points where two edges meet to create corners are invariant to translation, rotation, and illumination. These attributes are very useful when creating features based on corners. Many variations of corner detectors have been studied, with Harris being one of the simplest, most efficient, and most reliable [44]. Even though the Harris corner detector has some inherent limitations, it remains an important method for many applications.

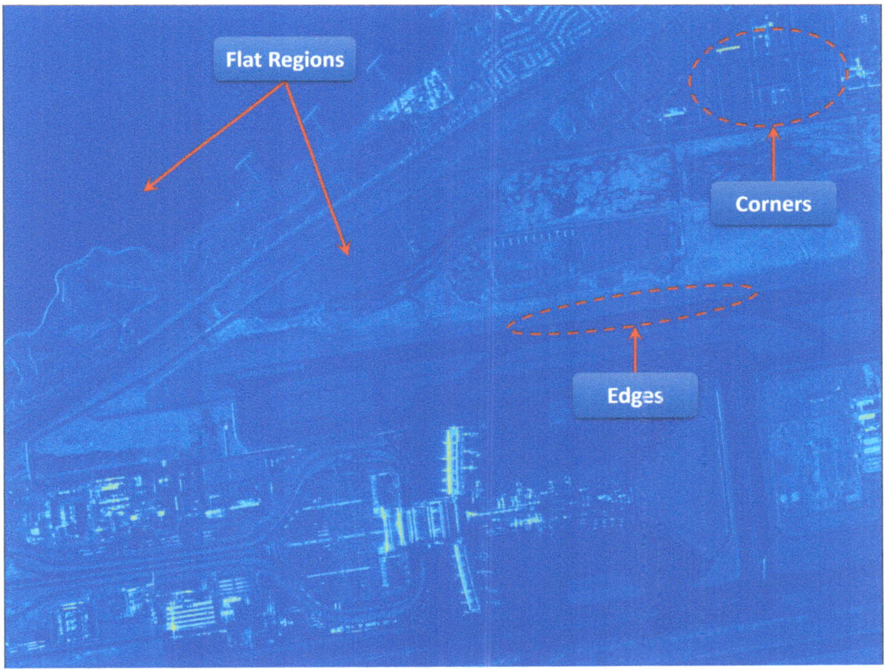

Figure 6.4 Typical corners, edges, and flat regions found in imagery. SAR imagery of Muscat International Airport in Oman. Courtesy of ICEYE.

6.3.1 Corner Detection

To detect corners, a window around an individual pixel is formed and is referred to as the reference window. The uniqueness of the reference window is found by creating a test window of the same size as the reference window and shifting this test window in various directions and calculating the similarity between the reference window and the test window. To illustrate corner detection, first consider an image where the pixels are fairly constant, as shown in Figure 6.5. Selecting the reference window around a certain pixel and moving the test window to surrounding locations shows similarity to all test window locations. Next, examine the image given in Figure 6.6, where there are bright areas adjacent to darker areas. Following the same procedure of selecting a reference window around a given pixel and then moving the test window to different locations shows similarity at other locations as highlighted in Figure 6.6. This case is more unique than the first case in that it matches at fewer test window locations, but still has some ambiguity. Finally, study the image illustrated in Figure 6.7 with several bright pixels in a distinct pattern. Taking a reference window around the highlighted region and shifting the test window results in only a single location that is highly similar. This makes the pixel an interest point because it is locally unique and therefore potentially useful in matching with other images.

Watch this animation of corner detection.

A measure of similarity between the reference window and the test window may be found from the sum squared difference of the two. This is expressed as [45]

$$s(x, y, \Delta x, \Delta y) = \sum_{(i,j)\in W} \Big[I(x+\Delta x+i, y+\Delta y+j) - I(x+i, y+j) \Big]^2, \tag{6.13}$$

where

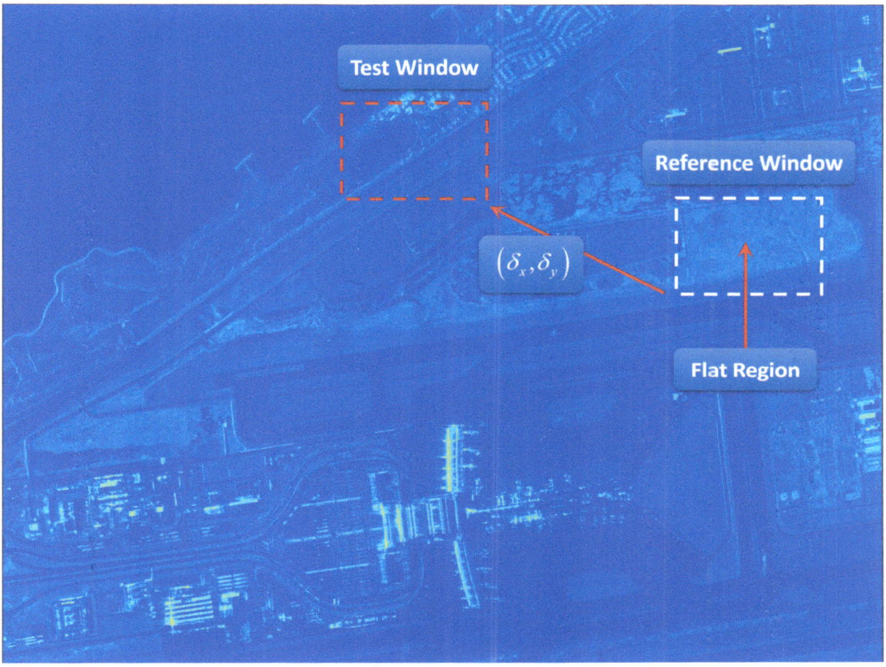

Figure 6.5 Reference and test windows in a flat region. SAR imagery of Muscat International Airport in Oman. Courtesy of ICEYE.

$$
\begin{aligned}
I(x,y) &= \text{image pixels,}\\
(\Delta x,\Delta y) &= \text{shift from reference window,}\\
(i,j)\in W &= \text{window pixels,}\\
s(x,y,\Delta x,\Delta y) &= \text{measure of similarity between reference}\\
&\ \text{and shifted windows.}
\end{aligned}
$$

The first term inside the summation represents the test window, while the second term is the reference window. If the two regions are similar, the expression in (6.13) goes to zero. If the regions are not similar, this becomes a large positive value. To generalize this approach, a Gaussian window is added to the similarity measure in (6.13), and is written as

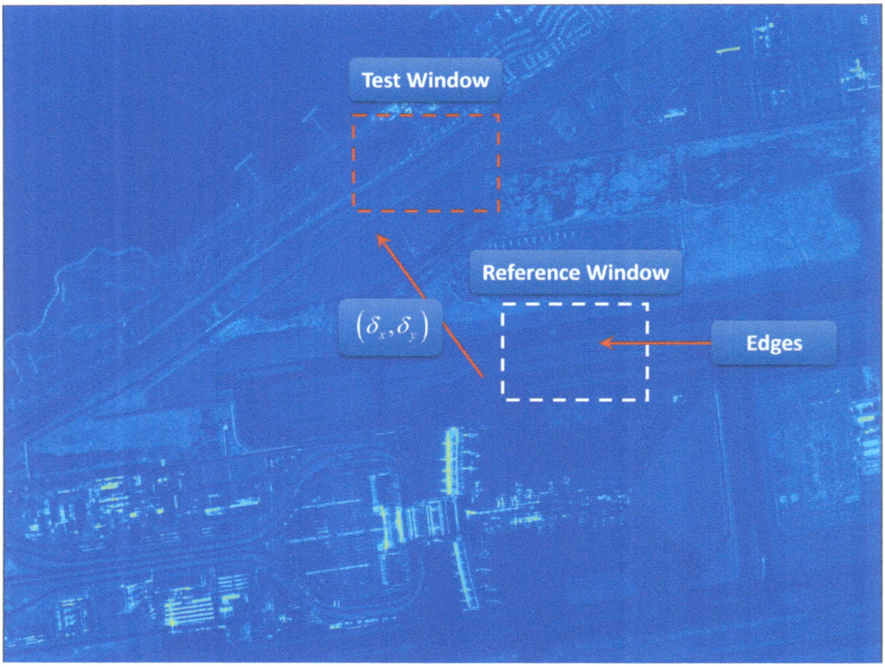

Figure 6.6 Reference and test windows in an edge region. SAR imagery of Muscat International Airport in Oman. Courtesy of ICEYE.

$$s(x, y, \Delta x, \Delta y) = \sum_{(i,j)\in w} \mathbf{W}(i, j)\times$$

$$\Big[I(x + \Delta x + i, y + \Delta y + j) - I(x + i, y + j)\Big]^2, \quad (6.14)$$

where $\mathbf{W}(i, j)$ is the Gaussian weighting matrix. The Gaussian weighting gives more emphasis to test windows close to the reference window than those further away. Using a Taylor series expansion, the expression in (6.14) may be rewritten as

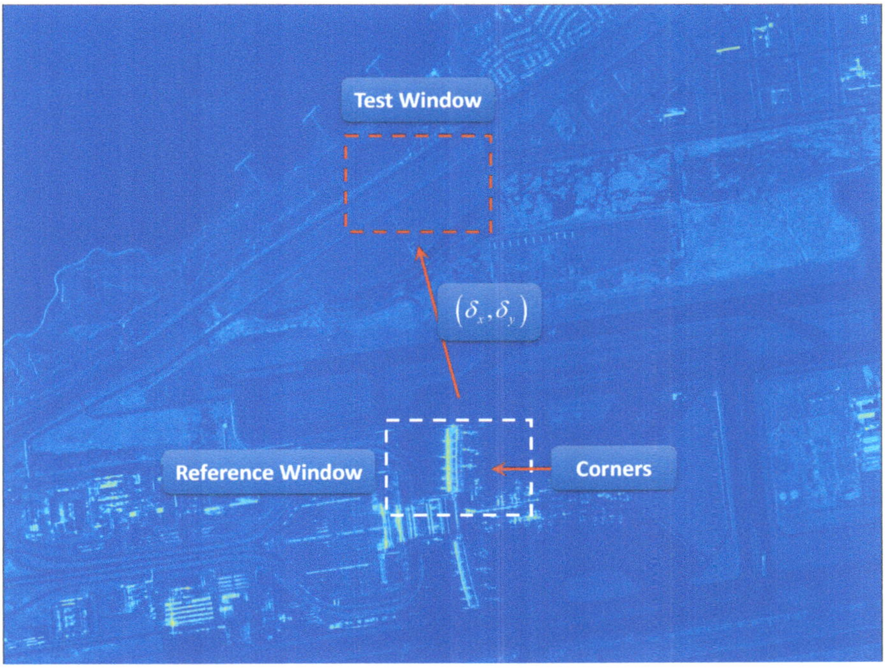

Figure 6.7 Reference and test windows for distinct pixel locations. SAR imagery of Muscat International Airport in Oman. Courtesy of ICEYE.

$$s(x, y, \Delta x, \Delta y) = \begin{bmatrix} \Delta x & \Delta y \end{bmatrix} \mathbf{H} \begin{bmatrix} \Delta x \\ \Delta y \end{bmatrix}, \tag{6.15}$$

where the matrix **H** is a structure tensor and often referred to as the Harris matrix. The Harris matrix describes the neighboring gradient distribution of the pixel and is expressed as [45]

$$\mathbf{H} = \begin{bmatrix} G(\sigma) * I_x^2 & G(\sigma) * I_x I_y \\ G(\sigma) * I_x I_y & G(\sigma) * I_y^2 \end{bmatrix}, \tag{6.16}$$

where I_x and I_y are the first order partial derivatives in the x and y directions, respectively. $G(\sigma)$ is a Gaussian smoothing kernel and $*$ denotes convolution [46]. Since the Harris matrix is based on image gradients, changes in illumination are eliminated by the gradient operator. The Harris matrix is 2×2, which gives rise to two eigenvalues, which provide useful information about the image around the current pixel. From Table 6.1, it is seen that if both eigenvalues are small this represents a flat or constant portion of the image, such as that of a homogeneous region. If either of the eigenvalues are large and the other small, this represents an edge in that part of the image. If both eigenvalues are large, this is considered a point of interest and corresponds to a local extremum. In other words, it is a distinctive point that may be used for matching in other images. The Harris corner detector finds these interest points by effectively finding pixels with two large eigenvalues. It does so by taking the scaled difference between the determinant and the trace of the matrix **H**. This is often referred to as the Harris *response* of a pixel and is written as [27, 42, 43, 45]

$$R = |\mathbf{H}| - \kappa(tr(\mathbf{H}))^2, \tag{6.17}$$

where κ is taken to be in the range [0.04–0.06]. If the response R exceeds a specified threshold, then the pixel is taken to be a corner. The Harris matrix is semi-definite Hermitian and the eigenvalues are positive. Therefore, (6.17) can be written as [47]

Table 6.1

Image Region Associated with Eigenvalues of Harris Matrix

Eigenvalue — λ_1	*Eigenvalue — λ_2*	*Image Region*
Small	Small	Flat (Constant)
Small	Large	Edge
Large	Small	Edge
Large	Large	Peak

$$R = \lambda_1\lambda_2 - \kappa(\lambda_1 + \lambda_2)^2, \tag{6.18}$$

where λ_1 and λ_2 are the two eigenvalues of the Harris matrix. The response R is determined by the eigenvalues of $\mathbf{H}$ and any unitary transformation of $\mathbf{H}$ does not affect the detection of a corner. The Harris corner detector is therefore invariant to a Euclidean transformation given by [42]

$$\begin{bmatrix} x' \\ y' \\ 1 \end{bmatrix} = \begin{bmatrix} \cos\theta & -\sin\theta & -x_0 \\ \sin\theta & \cos\theta & -y_0 \\ 0 & 0 & 1 \end{bmatrix} = \begin{bmatrix} x \\ y \\ 1 \end{bmatrix}, \tag{6.19}$$

where x_0 and y_0 are translations in the x and y directions and θ is the rotation.

6.3.2 Corner Matching

Rather than trying to match individual corners between two images, an $N \times N$ window of pixels around each corner is formed and a similarity metric is calculated. The window approach is used as individual pixels are not unique. Consider two images A and B, and extracted corners given by

$$\mathbf{C}_a = \{c_1^a, c_2^a, \cdots, c_K^a\} \tag{6.20}$$

$$\mathbf{C}_b = \{c_1^b, c_2^b, \cdots, c_L^b\}. \tag{6.21}$$

Each corner in $\mathbf{C}_a$ is compared against every corner in $\mathbf{C}_b$, which results in a $K \times L$ search problem, where K is the number of extracted corners in image A and L is the number of extracted corners in image B. Commonly used metrics include the sum of absolute differences (SAD), sum of squared differences (SSD), and zero-normalized cross correlation (ZNCC). These metrics are defined as [45, 48]

$$SAD = \sum_{(x,y)\in W} \left| I_1(x,y) - I_2(x,y) \right|, \tag{6.22}$$

$$SSD = \sum_{(x,y) \in W} \left[I_1(x,y) - I_2(x,y) \right]^2, \tag{6.23}$$

and

$$ZNCC = \frac{\sum_{(x,y) \in W} I_1(x,y) \cdot I_2(x,y)}{\sum_{(x,y) \in I} I_1(x,y) \cdot \sum_{(x,y) \in W} I_2(x,y)}. \tag{6.24}$$

SAD is the simplest metric and has the lowest computational burden. However, SAD is more sensitive to outliers than SSD or ZNCC. ZNCC has the highest computational cost, but is the most precise of the three metrics, and is the least sensitive to changes in intensity [45, 48]. A match is considered when the metric is below a specified threshold. While these are the most popular in image processing, there are many more metrics including Chebyshev, Minkowski, and Canberra [48].

6.4 SCALE INVARIANT FEATURE TRANSFORM

SIFT is a powerful feature detection method in the field of computer vision that can be used for various applications such as image registration, object recognition, gesture recognition, video tracking, and wildlife identification [24, 49, 50]. SIFT finds local features in an image, referred to as *keypoints*, and creates associated descriptors that are invariant to translation, rotation, and scale change as given in (6.12). Since the descriptors are robust, matching the descriptors between two images allows SIFT to be a very effective method for image registration. Since the introduction of SIFT, there have been numerous variations and competitive methods developed and the reader is referred to the following references for more information on these methods [29, 32, 33, 42]. The SIFT algorithm consists of the following four processes [24]

1. Construction of a Gaussian scale space of all possible scale changes.
2. Identification of local features to be used as keypoints.
3. Assignment of one or more orientations to each keypoint.

4. Generation of the descriptor for each keypoint.

6.4.1 Gaussian Scale-Space Construction

The SIFT algorithm makes use of a multiscale image representation to handle imagery at different scales. This is accomplished by representing the imagery as a collection of smoothed images defined by a single parameter. This smoothing reduces noise in the imagery, simulates blurring that would be present in images taken at different ranges, and allows for stable features to be identified [24]. Under a number of valid assumptions, a Gaussian function is a unique scale-space kernel based on causality [51, 52]. Using a variable-scale Gaussian kernel, the imagery in scale space is written as [24]

$$L(x, y; t) = G(x, y; t) * s(x, y), \tag{6.25}$$

where

$G(x, y; t)$	=	variable-scale Gaussian function,
$s(x, y)$	=	input image,
t	=	scale parameter,
$L(x, y; t)$	=	scale space of image $s(x, y)$.

The Gaussian kernel is defined by

$$G(x, y; t) = \frac{1}{2\pi t^2} \exp\left(\frac{-(x^2 + y^2)}{2t^2}\right). \tag{6.26}$$

Figure 6.8 highlights the result of applying the Gaussian smoothing function to the original image. The loss of detail is visible as the Gaussian smoothing simulates the image being zoomed out. To create the multiscale representation, the original image is scaled by octaves and subsequent smoothed images are created for each octave. This idea is illustrated in Figure 6.9 where each octave scale space is convolved with Gaussians resulting in the set of scale space images. To enhance features for more efficient detection of keypoints, a difference of Gaussian (DoG) technique is employed [24, 42]. This results in another set of images created by subtracting every smoothed image from the previous image in the same octave, which is written as

$$D(x, y; t) = L(x, y; kt) - L(x, y; t), \tag{6.27}$$

where k is the separation between any two nearby smoothed images. Figure 6.10 illustrates the DoG approach for the first octave only and the results are compelling.

6.4.2 Keypoint Selection

Keypoint selection may be thought of as two separate steps. The first step is to find the local extrema, followed by the removal of unstable extrema. To determine a local minimum or maximum, each pixel in the DoG representation is compared to its eight adjacent pixels in the same scale and the nine adjacent pixels in the scale above and in the scale below, as illustrated in Figure 6.11. This location is selected as a potential keypoint if the pixel value is larger or smaller than all the other 26 pixel locations. Unstable keypoints are often associated with low contrast or edge responses. The details of determining and removing unstable extrema is beyond the scope of this book. However, the reader is referred to the following references for more details [24, 27, 49].

6.4.3 Orientation Assignment

An orientation is assigned to each of the keypoints identified in the previous section. The assignment is based on local image properties and is made such that the keypoint descriptors are invariant to rotation. To ensure the descriptors are rotation invariant, a region of pixels around the keypoint are used to create a normalized reference orientation. The scale, t, of the current keypoint is used for all calculations. The gradients around the keypoint located at a pixel location (i_x, i_y) are calculated by

$$\nabla_x = L(i_x + 1, i_y) - L(i_x - 1, i_y), \tag{6.28}$$

$$\nabla_y = L(i_x, i_y + 1) - L(i_x, i_y - 1). \tag{6.29}$$

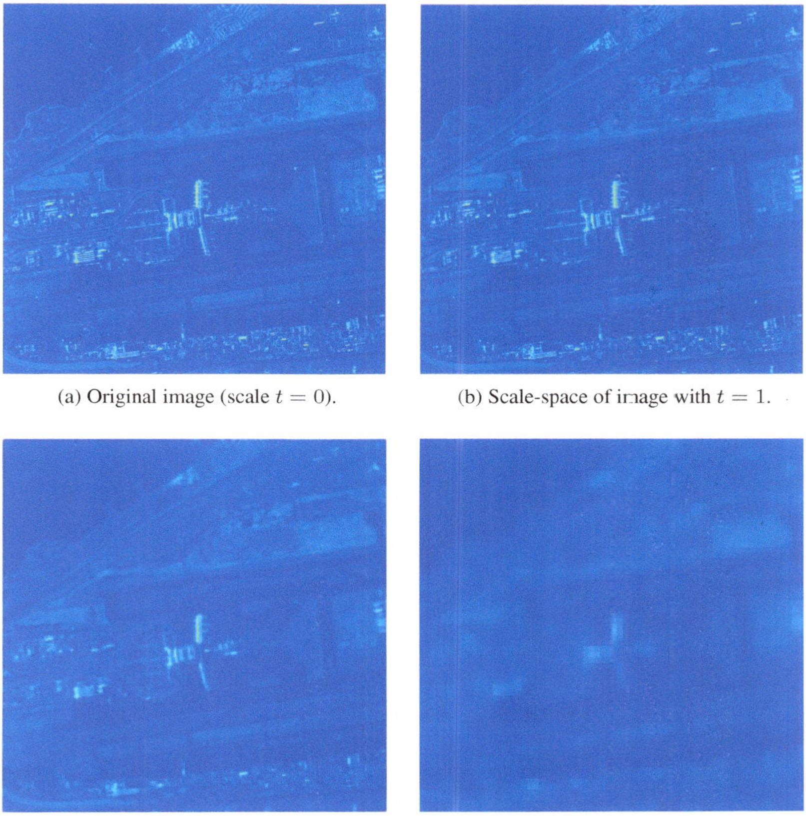

(a) Original image (scale $t = 0$).

(b) Scale-space of image with $t = 1$.

(c) Scale-space of image with $t = 4$.

(d) Scale-space of image with $t = 16$.

Figure 6.8 Gaussian smoothing kernel applied to the original image $s(x, y)$. SAR imagery of Muscat International Airport in Oman. Courtesy of ICEYE.

The gradient magnitude and orientation are then computed by

$$A(i_x, i_y) = \sqrt{\nabla_x^2 + \nabla_y^2}, \tag{6.30}$$

$$\psi(i_x, i_y) = \tan^{-1}(\nabla_y / \nabla_x). \tag{6.31}$$

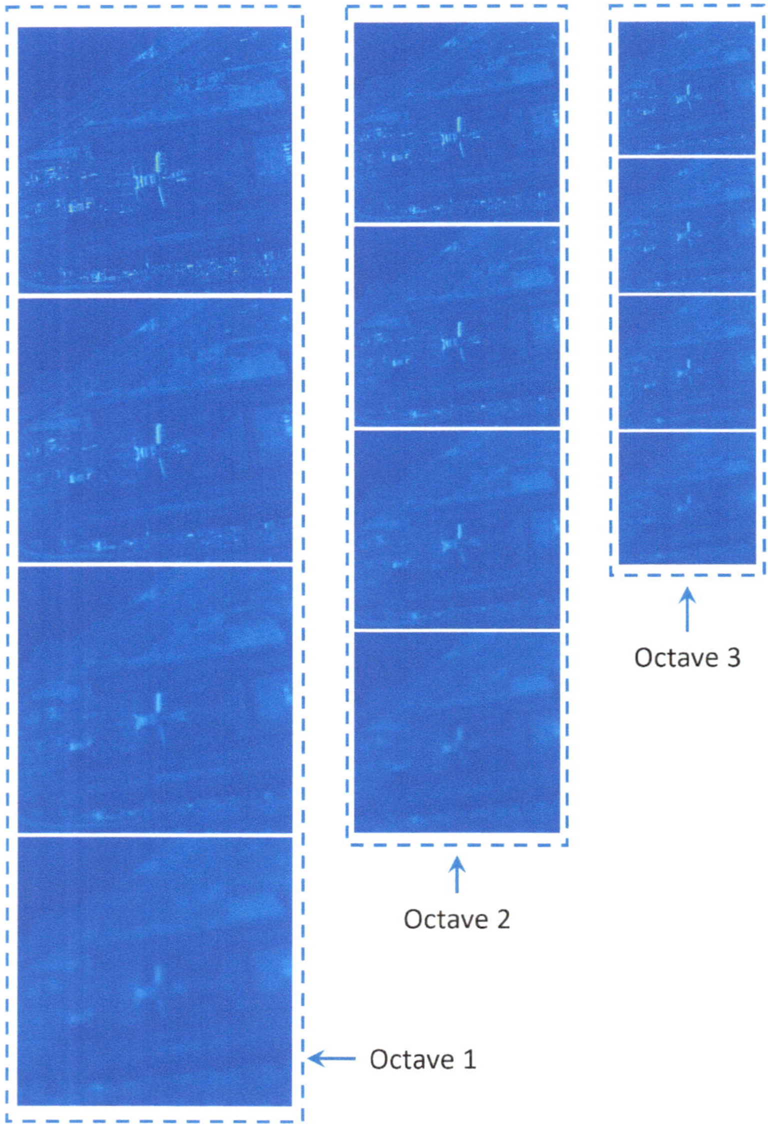

Figure 6.9 Scale-space representation $L(x, y; t)$ of original image $s(x, y)$. SAR imagery of Muscat International Airport in Oman. Courtesy of ICEYE.

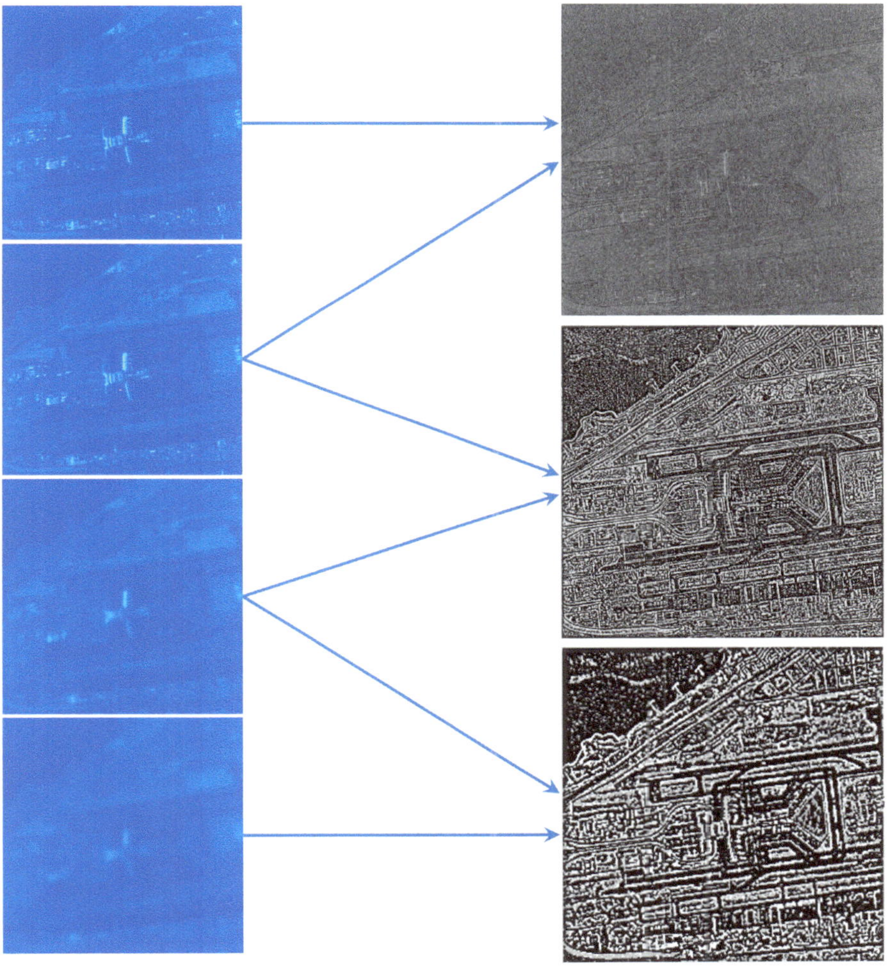

Figure 6.10 DoG representation for the first octave. SAR imagery of Muscat International Airport in Oman. Courtesy of ICEYE.

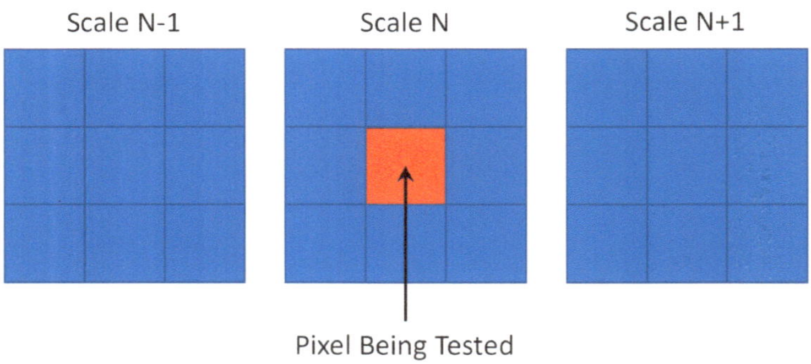

Figure 6.11 Comparison of surrounding pixels for local extrema detection.

Next, a histogram is created from the magnitudes and angles of the gradients associated with pixel locations in the region around the keypoint. The angle samples in the histogram are weighted by the associated magnitudes. The predominant orientation is indicated by a peak in the histogram, as illustrated in Figure 6.12. In addition, any other peak in the histogram that is 80% of the maximum may be used to create additional keypoints.

6.4.4 Keypoint Descriptor Creation

A keypoint descriptor is a unique encoded version of the spatial distribution of the gradient orientation in the region around the keypoint [24, 41, 42]. To create the descriptors, an $M \times M$ region of pixels around the keypoint is used. This region is then subdivided into $N \times N$ regions. For each subregion, a histogram with bins is created from the magnitude and orientation. To ensure invariance to rotation, the descriptor coordinates and gradient orientations are rotated relative to the keypoint orientation. Since the calculations are normalized over subregions, the descriptors will be somewhat invariant to illumination (i.e., if the illumination changes the brightness in a similar fashion over the subregion, the gradient values will not be affected). Finally, the descriptor is created from the vector of all the orientation histogram entries, as shown in Figure 6.13. Typically, a 4×4 region is selected and divided into 4×4 subregions and 8 bins are taken. This leads to a descriptor length

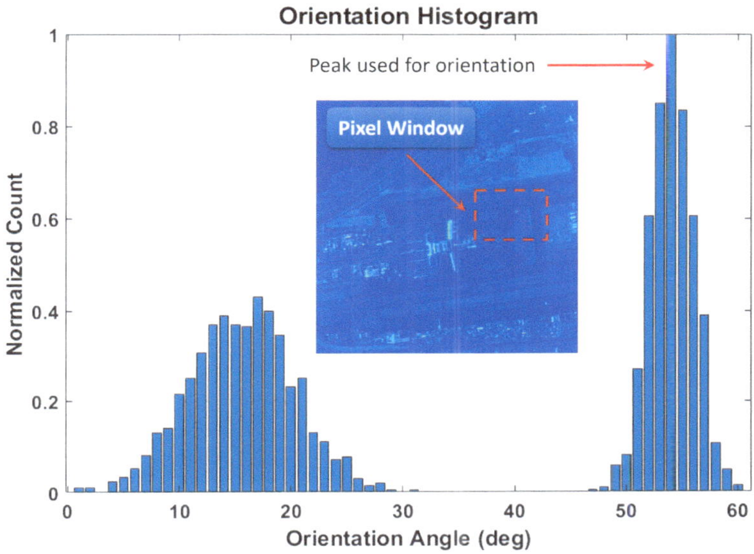

Figure 6.12 Orientation histogram for scale-invariant keypoints.

of 128. In literature, this is often referred to as the *feature vector* and is expressed as [24]

$$\mathbf{d}_p = h_k^{ij}, \tag{6.32}$$

where

$$\begin{aligned} i &= 1 \cdots N, \\ j &= 1 \cdots N, \\ k &= 1 \cdots K, \\ p &= (i-1)NK + (j-1)N + k, \\ h_k^{ij} &= k\text{th bin of the } ij\text{th histogram}, \\ \mathbf{d} &= \text{keypoint descriptor}. \end{aligned}$$

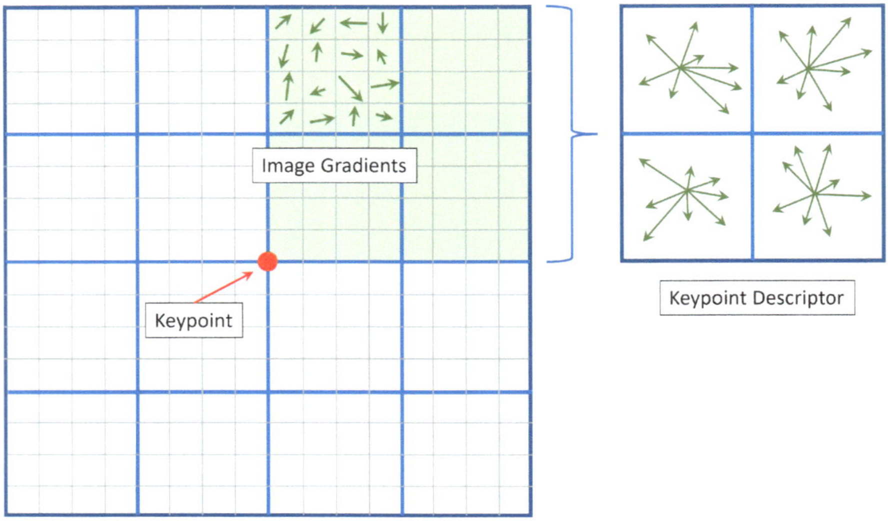

Figure 6.13 Illustration of keypoint descriptor creation.

6.4.5 Descriptor Matching

Matching the keypoint descriptors between two images, or between an image and a large database, is one of the main purposes of keypoint detection and description. Using images $s_a(x, y)$ and $s_b(x, y)$, denote the set of M_d and N_d descriptors for each image as $\mathbf{D}_a$ and $\mathbf{D}_b$, respectively. This is expressed as

$$\mathbf{D}_a = \{\mathbf{d}_1^a, \mathbf{d}_2^a, \cdots, \mathbf{d}_{M_d}^a\} \tag{6.33}$$

$$\mathbf{D}_b = \{\mathbf{d}_1^b, \mathbf{d}_2^b, \cdots, \mathbf{d}_{N_d}^b\}, \tag{6.34}$$

Matches may be found by considering each descriptor in the set $\mathbf{D}_a$ and searching for the descriptor in the set $\mathbf{D}_b$ that minimizes the Euclidean distance. A match is only considered valid if the minimum Euclidean distance is below a specified threshold as

$$\mathbf{d}_{match} = \underset{\mathbf{d} \in \mathbf{D}_b}{\operatorname{argmin}} ||\mathbf{d} - \mathbf{d}_a|| < T_{abs}, \tag{6.35}$$

where

$\mathbf{d}$	=	each descriptor from set $\mathbf{D}_b$,
$\mathbf{d}_a$	=	descriptor from set $\mathbf{D}_a$,
T_{abs}	=	absolute threshold for a valid match,
$\mathbf{d}_{match}$	=	the descriptor from set $\mathbf{D}_b$ that matches with $\mathbf{d}_a$.

While matches may be found in this manner, implementations of the SIFT algorithm often employ an adaptive threshold based on the second closest match in (6.35). Both of these approaches to matching have limitations and more robust matching techniques have been developed [49].

6.5 SPEEDED-UP ROBUST FEATURES

The SIFT algorithm of the previous section is computationally intensive and is therefore relatively slow when compared to competing methods of image registration [49]. SURF is similar to SIFT by providing a representation and comparison of images invariant to translation, rotation, and scale. In SIFT, the DoG was used to approximate the Laplacian of the Gaussian kernel. The advantage of SURF is in the fast approximation of the Laplacian of the Gaussian kernel using box filters. The convolution with a box filter is simply calculated with integral images and in parallel for varying image scales. This approach is computationally efficient enough to support real-time object recognition and tracking. SURF consists of the following basic algorithmic steps [25]:

1. Identification of interest points.
2. Assignment of one or more orientations to each interest point.
3. Generation of the descriptor for each interest point.

Note, literature discussing SURF often refers to *interest points*, which are analogous to keypoints in SIFT [25, 31, 42].

6.5.1 Interest Point Selection

SURF relies on the determinant of a Hessian-matrix approximation for selecting both the location and scale of interest points [53]. At a given pixel location in an image, described by $\mathbf{p} = (x, y)$, the Hessian matrix is written as

$$\mathcal{H}(\mathbf{p}, \sigma) = \begin{bmatrix} L_{xx}(\mathbf{p}, \sigma) & L_{xy}(\mathbf{p}, \sigma) \\ L_{xy}(\mathbf{p}, \sigma) & L_{yy}(\mathbf{p}, \sigma) \end{bmatrix}, \tag{6.36}$$

where σ is the scale factor and $L_{xx}(\mathbf{p}, \sigma)$, $L_{yy}(\mathbf{p}, \sigma)$, and $L_{xy}(\mathbf{p}, \sigma)$ are the second order partial derivatives of the Gaussian in the x-, y-, and xy-directions convolved with the image at $\mathbf{p}$ [25, 42]. Box filters, denoted by D_{xx}, D_{yy}, and D_{xy}, are used as approximations for the Gaussian second order partial derivatives. The approximation using box filters is found with the use of *integral images*, which are computationally efficient and independent of size [25, 42, 54]. Integral images are summed-area tables used to quickly and efficiently generate the sum of values in a rectangular subset of a grid (i.e., the average intensity within an image). The value of the integral image at any point (x, y) is the sum of all pixels in a rectangular region formed by the origin and (x, y), as shown in Figure 6.14 [53, 55]. This is expressed as

$$I_{\Sigma}(x, y) = \sum_{x'=0}^{x} \sum_{y'=0}^{y} I(x', y'), \tag{6.37}$$

where $I(x, y)$ is the input image and $I_{\Sigma}(x, y)$ is the integral image. The integral image is computed efficiently in a single pass over the image as [54]

$$I_{\Sigma}(x, y) = I(x, y) + I_{\Sigma}(x, y - \Delta y) + I_{\Sigma}(x - \Delta x, y) - I_{\Sigma}(x - \Delta x, y - \Delta y). \tag{6.38}$$

Once the integral image has been computed, evaluating the sum of intensities over any rectangular area only requires four array references independent of the size

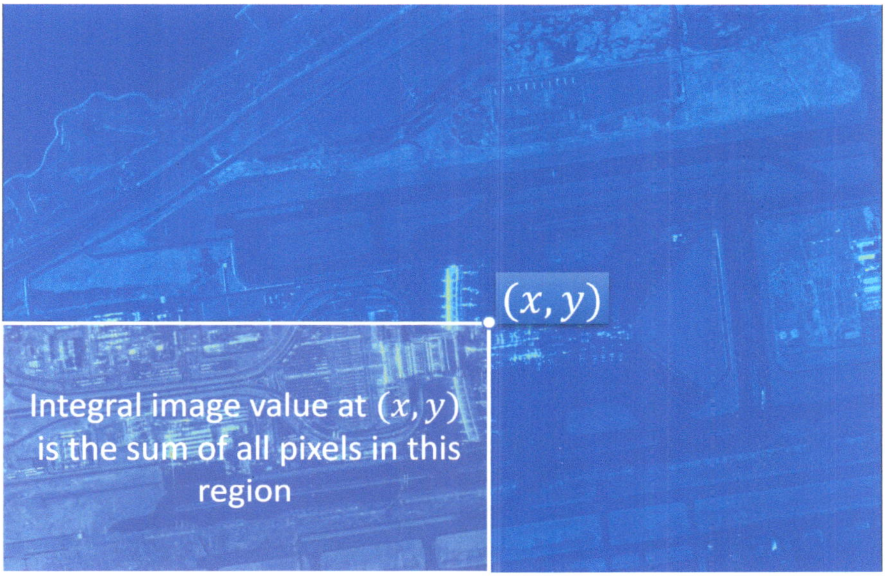

Figure 6.14 Integral image for calculation of the sum of pixels over a rectangular region.

of the area. Referring to Figure 6.14, the sum of the image intensity in the region defined by $A = (x_0, y_0)$, $B = (x_1, y_0)$, $C = (x_0, y_1)$, and $D = (x_1, y_1)$ is

$$\sum_{x=x_0}^{x_1} \sum_{y=y_0}^{y_1} I(x, y) = I_\Sigma(A) - I_\Sigma(B) - I_\Sigma(C) + I_\Sigma(D). \tag{6.39}$$

The result of using 9 × 9 box filters to approximate the Gaussian second order partial derivatives is shown in Figure 6.15. As indicated in [25], the performance of the box filter approximation is comparable to, and in some cases better than, the original Gaussians. The determinant of the approximate Hessian matrix is then written as

$$\left|\mathcal{H}_a\right| = D_{xx}D_{yy} - (wD_{xy})^2, \tag{6.40}$$

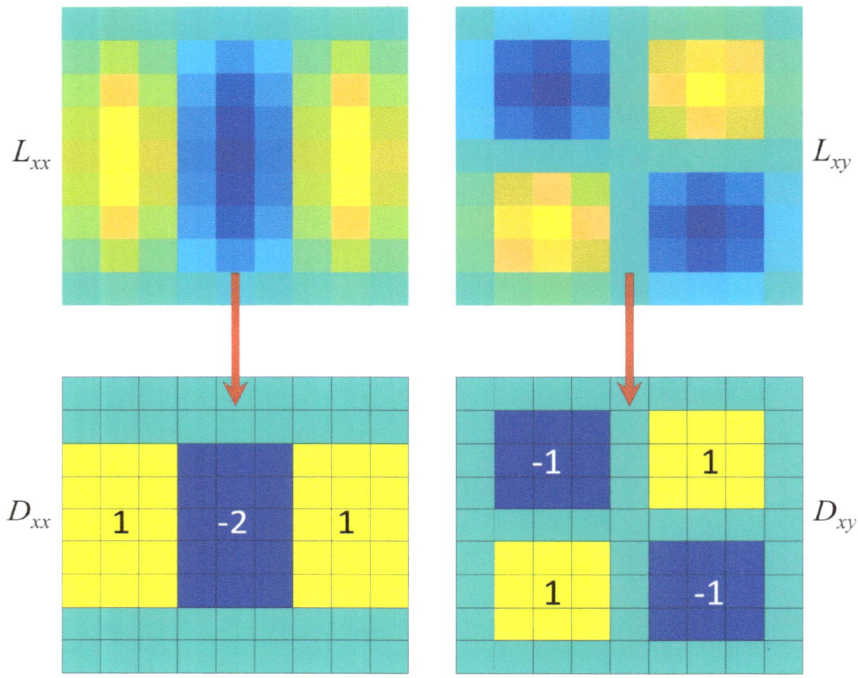

Figure 6.15 Gaussian second-order partial derivatives and corresponding box filter approximations.

where $\mathcal{H}_a$ is the approximated Hessian matrix and w is the relative weight of the filter response and is typically taken to be 0.9 [25].

The scale space is analyzed by upscaling the filter size rather than iteratively reducing image size, as illustrated in Figure 6.16. No aliasing arises in this approach as there is no downsampling of the image. As with SIFT, the scale space is divided into octaves. Each octave then gets subdivided into different scale levels. A series of filter responses is then created by convolving the same input image with a filter that is increasing in size. The location of interest points is then found by interpolating the maxima of the approximated Hessian matrix in both scale and image space [25].

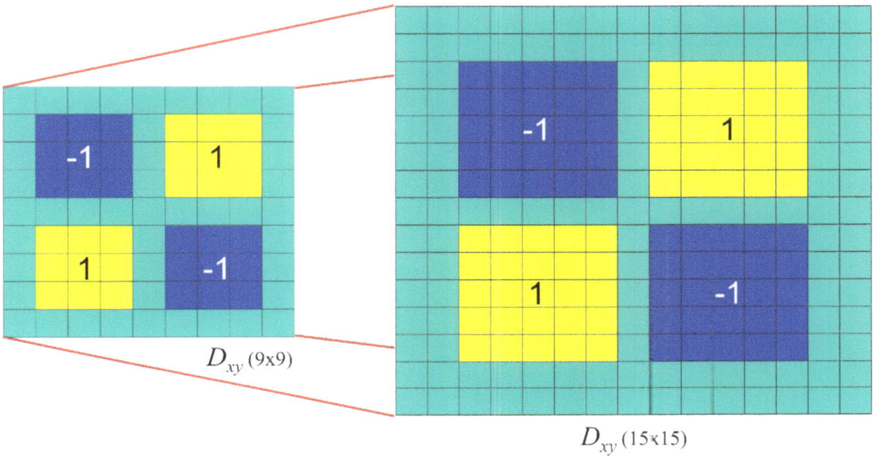

Figure 6.16 Box filter upscaling from 9 × 9 to 15 × 15 with constant cost due to the use of integral images.

6.5.2 Orientation Assignment

Similar to other methods, SURF assigns an orientation to each of the interest points found in the previous section. To begin, the Haar-wavelet responses in a circular region about the interest point is calculated in the x- and y- directions [56]. The circular region has a radius of $6s$, where s is the scale at which the interest point was found. The sampling and Haar-wavelet responses are also computed at scale s. Since the wavelets become large at larger scales, integral images are again employed to perform fast filtering. The wavelet responses are then weighted with a Gaussian centered at the interest point. The sum of the vertical and horizontal wavelet responses is calculated in a scanning area of $\pi/3$. The choice for the size of the scanning area has been studied carefully as this directly affects the accuracy of the orientation assignment [25, 31, 57]. These summed responses represent a local orientation vector, as shown in Figure 6.17. The scanning area is then rotated, and the sum of the responses recalculated. The orientation with the largest vector is chosen to be the orientation assigned to the interest point.

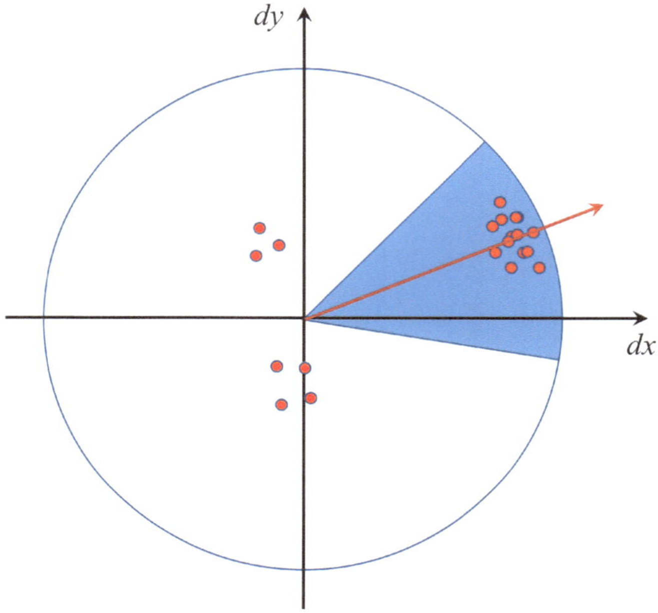

Figure 6.17 Orientation assignment from Haar-wavelet weighted responses.

6.5.3 Interest Point Descriptors

The descriptors for SURF are based on the sum of Haar-wavelet responses. The process of creating the descriptor for an interest point begins by forming a square region around the interest point under consideration. The size of the square region is taken to be $20s$, as shown in Figure 6.18. Next, the region is split into 4×4 subregions. For each of these subregions, the Haar-wavelet response is computed as illustrated in Figure 6.18, where dx and dy are the wavelet responses in the horizontal and vertical directions relative to the orientation assigned to the interest point under consideration. Each subregion has a four-dimensional descriptor for its underlying intensity structure. For the ith subregion, this is written as

$$\mathbf{v}_i = \Big[\Sigma dx,\ \Sigma dy,\ \Sigma|dx|,\ \Sigma|dy|\Big]. \tag{6.41}$$

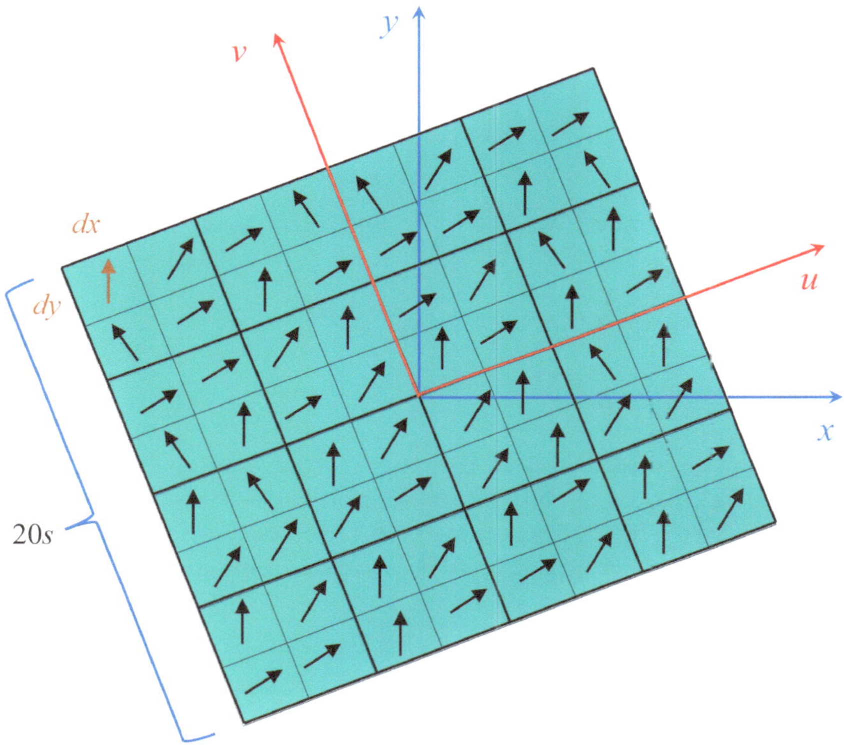

Figure 6.18 SURF interest point descriptor over a 20s region.

Therefore, the descriptor for each interest point has a length of 64, as compared to 128 for SIFT which results in fewer computations and faster descriptor matching.

Note, Figure 6.18 is for illustration purposes as the wavelet responses are computed in the original image frame and then interpolated onto the rotated frame, rather than rotating the entire image first and then computing the wavelet responses [25].

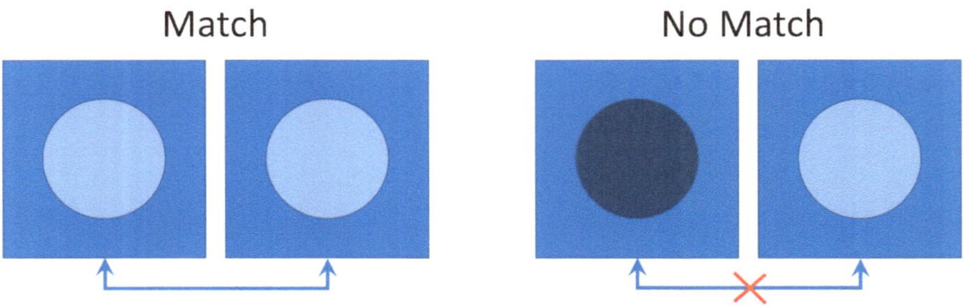

Figure 6.19 Feature contrast for fast indexing and matching.

6.5.4 Descriptor Matching

SURF uses string-based descriptors as given in (6.41) and $L1$- and $L2$-norms may be used for matching. In this book, descriptor matching methods based on nearest neighbor distance ratio (NNDR) of Section 6.4.5 are used. SURF facilitates fast indexing, which is accomplished by taking the trace of the approximate Hessian matrix. This is written as

$$\text{tr}(\mathcal{H}_a) = D_{xx} + D_{yy}. \tag{6.42}$$

This is equivalent to the sign of the Laplacian, which identifies contrast (i.e., bright features on dark backgrounds and the inverse case). Since the Laplacian is computed during the detection of interest points, as discussed in Section 6.5.1, no additional computational cost is incurred. For matching purposes, only descriptors with similar types of contrast are considered, as illustrated in Figure 6.19. This approach results in much faster descriptor matching without any impact on the robustness of the descriptors.

6.6 ORIENTED FAST AND ROTATED BRIEF

Oriented FAST and rotated BRIEF (ORB) is a viable alternative to SIFT and SURF, which are described in Sections 6.4 and 6.5 [58]. ORB was developed mainly due to SIFT and SURF being patented algorithms [58]. Note, the SIFT patent expired in 2020 and the SURF patent remains in effect [59, 60]. A very important aspect of ORB is it came from OpenCV Labs, is not patented, and is free to use [58]. ORB is built on the well-known features from accelerated segment test (FAST) keypoint detector [61] and the binary robust independent elementary features (BRIEF) descriptor [62]. ORB performs as well as SIFT and better than SURF for feature detection and is almost two orders of magnitude faster [58].

6.6.1 FAST

For a given pixel $\mathbf{p}(x, y)$, FAST compares the brightness of the surrounding 16 pixels as shown in Figure 6.20. If eight or more pixels are much lighter or darker than the pixel under consideration, a keypoint is declared. Using this selection process, FAST keypoints contain information about the location of edges in the imagery. The FAST features do not have orientation or multiscale structure. The ORB algorithm adds a multiscale representation consisting of a sequence of downsampled versions of the original source imagery. Using FAST to detect keypoints at each level make ORB partially scale invariant [58].

ORB assigns an orientation to each keypoint based on intensity gradients surrounding the point. The intensity change is found from the centroid, which is expressed by

$$C = \frac{1}{m_{00}}(m_{10}, m_{01}), \tag{6.43}$$

where the moments are given by

$$m_{pq} = \sum_{x,y} x^p y^q I(x, y). \tag{6.44}$$

A vector from the keypoint, $\mathbf{p}(x, y)$ to the centroid in (6.43) is created and the orientation is computed as

$$\theta = \tan^{-1}\left(\frac{m_{01}}{m_{10}}\right) \qquad \text{(rad)}. \tag{6.45}$$

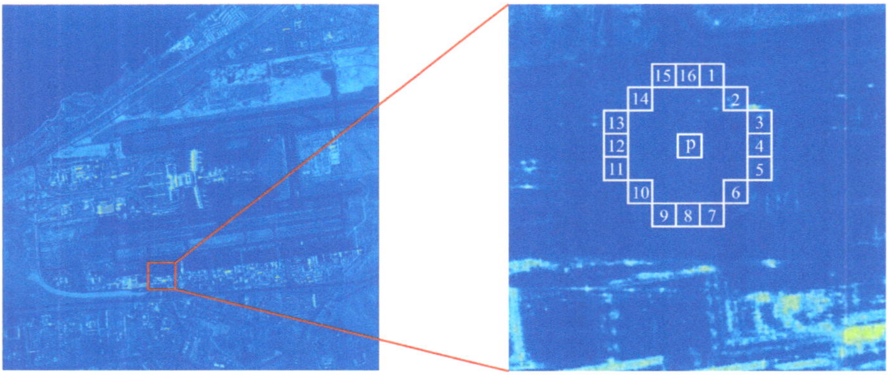

Figure 6.20 FAST keypoint selection using 16 pixels surrounding **p**.

6.6.2 BRIEF

BRIEF converts all the keypoints found by FAST into a binary vector, which is the keypoint descriptor for ORB [58]. The BRIEF algorithm begins by choosing a random pair of pixels around the keypoint. The first pixel in the pair is selected from a Gaussian distribution centered on the keypoint. The second pixel is drawn from a Gaussian distribution centered on the first pixel. If the first pixel is brighter than the second pixel, a binary 1 is assigned to the descriptor, otherwise a binary 0 is assigned. This process is typically repeated 128 times to obtain a 128-big keypoint descriptor. BRIEF is not rotation invariant and performance suffers significantly for rotations of more than a few degrees [58]. The ORB algorithm orients BRIEF based on the rotations found in the previous section.

The ORB binary descriptor for a keypoint location (x_i, y_i) is expressed by the $2 \times N$ matrix

$$S = \begin{bmatrix} x_1, \cdots, x_N \\ y_1, \ldots, y_N \end{bmatrix}. \tag{6.46}$$

The keypoint orientation is then used to create a rotated version of the binary descriptor, which is written as

$$S_\theta = R(\theta)S, \tag{6.47}$$

where $R(\theta)$ is the rotation matrix. The rotations are typically subdivided into 12° sectors and a lookup table of descriptors is computed [62]. If the orientation of the keypoints is consistent across different views, the correct set of points will be used in the creation of the ORB descriptor [58].

Table 6.2 is given to compare the main performance features of the image registration algorithms covered in the previous sections. ORB is the fastest algorithm and SIFT tends to be the most robust across a wide spectrum of image registration problems [63]. For a special case of a rotation of $\pi/2$, ORB and SURF tend to perform slightly better than SIFT. It is worth noting ORB keypoints are concentrated near the center of the image, whereas SIFT and SURF keypoints are distributed over a larger area of the image [63]. In general, SURF tends to provide the best tradeoff between speed and robustness [42]. Many variations of these image registration methods have been proposed and implemented, each with advantages, disadvantages, and caveats. Table 6.2 is meant to be a general guide and not strict for all imaging scenarios.

Table 6.2

Comparison of Image Registration Techniques

Parameter	*Harris Corner*	*SIFT*	*SURF*	*ORB*
Feature invariance	Translation	Translation, rotation, scale	Translation, rotation, scale	Translation, rotation, partial scale
Feature extraction	Fastest	Slow	Fast	Fast
Feature accuracy	Pixel	Subpixel	Subpixel	Subpixel
Descriptor matching	Slow	Fast	Fastest	Fast
Robustness	Poor	Best	Good	Good

6.7 WARP FUNCTION EXTRACTION

Once the descriptors between two images have been matched, the *geometrical warp* function needs to be found. Geometrical warp is a transformation of pixel locations in a primary image to the pixel locations in a secondary image and is caused by local distortions, sensor parameters, temporal variations, target relief, and scenario geometry [13, 42]. If there are no errors in the descriptor matching, the warp function may be found by fitting the matching descriptors with the least squares method [64]. All practical image registration applications contain errors in the descriptor matches due to system noise, spatial and temporal decorrelation, environmental interference, and limitations in descriptor creation and matching algorithms. Therefore, robust algorithms for determining the warp function, which are insensitive to these mismatches, are desired. Perhaps the most widely used method for feature-based SAR image registration is RANSAC [65]. In addition to RANSAC, Fast-LTS and EF-LTS methods have been adopted [42]. The reader is referred to the following sources for other feature-based matching and warp function retrieval algorithms [2, 19, 66].

6.7.1 Formulation

To formalize the warp function extraction process, SAR imagery is written as a continuous complex function in $\mathbb{R}^2$. Following the treatment in [13], the primary image is expressed as

$$I_p(\mathbf{x}_p): \quad \mathbf{x}_p \in \Omega_p \subset \mathbb{R}^2, \tag{6.48}$$

where $\mathbf{x}_p$ is the position vector (x, y) in the image domain Ω_p. Similarly, the secondary image is written as

$$I_s(\mathbf{x}_s): \quad \mathbf{x}_s \in \Omega_s \subset \mathbb{R}^2. \tag{6.49}$$

To determine the warp function, a two-dimensional coordinate transform, τ, needs to be found such that the transformed secondary image given by $I_s\big(\tau(\mathbf{x}_p)\big)$ is spatially aligned with $I_p(\mathbf{x}_p)$. SAR digital imagery can be thought of as sampled

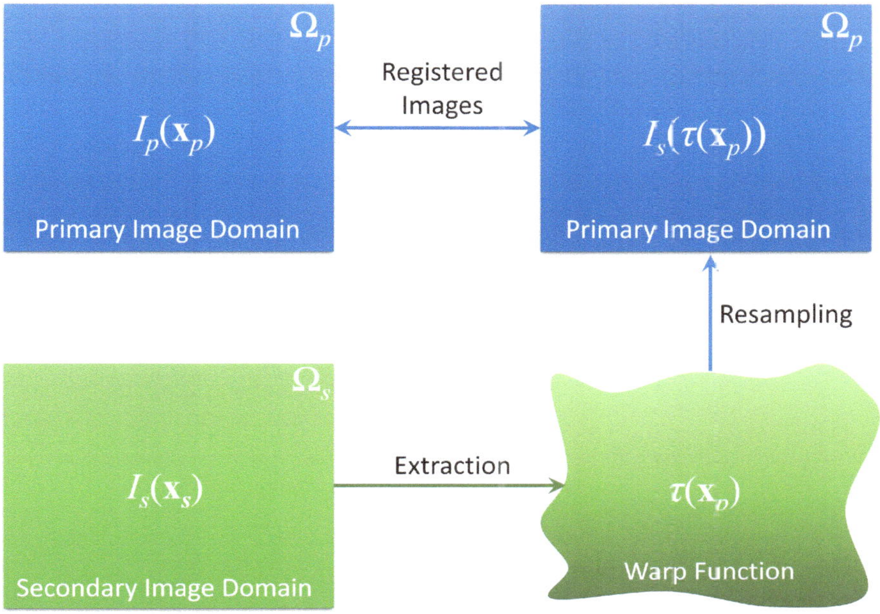

Figure 6.21 Process for registering the primary and secondary images.

versions of the continuous complex functions in (6.48) and (6.49). Once the warp function has been determined, the image pixels in the primary and secondary images are aligned, as illustrated in Figure 6.21.

SAR imagery cannot be represented as a central projection [67] and the warp model depends on the sensor parameters, scenario geometry, and target relief. However, under certain reasonable assumptions, the warp function can be approximated as a polynomial [12]. This is covered in more detail in the following sections.

6.7.2 RANSAC

RANSAC is one of the most widely used methods for extracting the geometrical warp function from a set of matched descriptors [42]. It is an iterative technique useful for datasets containing mismatches that should not influence the estimated warp function [65, 68]. In that sense, it can be thought of as a mismatch detection technique. While nondeterministic, the probability of a valid result increases with an increasing number of iterations [65].

RANSAC differs from methods that make use of the entire dataset by randomly sampling a minimal subset to estimate the warp function. The entire dataset is then checked for a consensus set of descriptor matches. This process is repeated until the largest consensus set is realized. The basic algorithm consists of the following steps:

1. Sample a random minimal subset.
2. Calculate the model parameters.
3. Determine the points within a specified tolerance.
4. Compute the percentage of inliers and compare to the threshold.

The steps given above are repeated until the percentage of inliers is above the threshold or until a maximum number of iterations is reached.

RANSAC works best for datasets containing both inliers and outliers. The inliers consist of correctly matched descriptors that accurately represent the warp function, while outliers are mismatched descriptors and should not have an impact on the determination of the warp function. Results produced with RANSAC are only trustworthy if any of the minimal subsets can accurately estimate the warp function. It is often difficult to produce similar estimates of the warp function for different subsets. This is due to factors such as noise and local distortion. Typically, SAR geometrical warping varies from pixel to pixel and a low-order polynomial model only accurately represents a global warping rather than a local distortion. However, a more stable version of RANSAC may be implemented by using more than the minimal subset of matched descriptors and applying a loss function [42]. This better estimates the support of the warp function than that obtained from the minimal subset.

6.7.3 Fast-LTS

With the limitations associated with the least squares method, more robust regression techniques have been investigated [42]. An effective approach to this problem is the least trimmed squares method (LTS) [69]. Rather than minimizing the sum of squared residual over all matched descriptors, LTS minimizes the sum of squared residuals over a subset of descriptors. The remaining samples are not used in the estimation and therefore do affect the outcome. In fact, LTS is so insensitive to errors in the descriptor matching that up to 50% of the datasets can be outliers and a reasonable estimation of the warp function still obtained [42, 70, 71].

LTS fits the linear model given by

$$y_i = x_{i1}\theta_1 + x_{i2}\,\theta_2 + \cdots + x_{ip}\,\theta_p + e_i, \quad i = 1, \cdots, n, \tag{6.50}$$

which may be written more compactly as

$$y_i = \mathbf{X}_i^T\,\boldsymbol{\Theta} + e_i, \quad i = 1, \cdots, n, \tag{6.51}$$

where

$$\begin{aligned}
\mathbf{X}_i^T = [x_{i1}, x_{i2}, \cdots, x_{ip}] &= \text{explanatory variable,}\\
\boldsymbol{\Theta} = [\theta_1, \theta_2, \cdots, \theta_p] &= \text{parameter to be estimated,}\\
e_i &= \text{error term,}\\
n &= \text{sample size,}\\
p &= \text{number of coefficients,}\\
y_i &= \text{response variable.}
\end{aligned}$$

The least squares estimate is written as [64]

$$\hat{\boldsymbol{\Theta}} = \arg\min \sum_{i=1}^{h} \left(\mathbf{r}^2\right)_i(\boldsymbol{\Theta}). \tag{6.52}$$

The term inside the sum is the loss function of the least trimmed squares estimator and is given by

$$Q = \sum_{i=1}^{h} \left(\mathbf{r}^2\right)_i(\hat{\boldsymbol{\Theta}}), \tag{6.53}$$

where

$$\mathbf{r} = [r_1, r_2, \cdots, r_n]^T, \tag{6.54}$$

and

$$r_i = y_i - \mathbf{X}_i^T \hat{\mathbf{\Theta}}. \tag{6.55}$$

Here, $(\mathbf{r}^2)_i$ is the ith element of the squared residuals in ordered form such that

$$(\mathbf{r}^2)_1 \leq (\mathbf{r}^2)_2 \leq \cdots \leq (\mathbf{r}^2)_n, \tag{6.56}$$

and h is referred to as the trimming constant. LTS performs regression by fitting the subset of length h to minimize the least squared residuals. The trimming constant determines the LTS estimator breakdown point, as $n - h$ matching descriptors with the largest residuals do not affect the estimator [69, 71]. The maximum breakdown point is then $h = (n + p + 1)/2$. This means the LTS fit remains in a bounded region when $(n - p)/2$ or fewer samples are outliers (mismatched descriptors). This is equivalent to finding the h-subset with the smallest least squares objective function. Therefore, the LTS regression estimate is the least squares fit to h points.

A major limitation of LTS is dealing with large datasets. The computational burden grows far too fast as the dataset becomes large. This limits the use of LTS in many applications. To overcome this limitation, an improvement known as Fast-LTS was developed [71]. Using any approximation of the LTS regression coefficients, a subsequent approximation can be calculated that results in a lower loss function. This is the basis of the Fast-LTS method, and is often referred to as the concentration step or C-step [42]. Fast-LTS also make use of selective iteration, nested extensions, and intercept adjustment. For small datasets, Fast-LTS usually finds the exact LTS; for large data sets it gives more accurate results than LTS and is faster by orders of magnitude [71].

6.7.3.1 Concentration Step

The concentration step is a crucial step in the Fast-LTS method Following the treatment in [71], begin with a p-variate $\mathbf{X}_i$ and response variable y_i. Then select $H_1 \subset [1, \cdots, n]$, where $|H_1| = h$. The loss factor is then given by

$$Q_1 = \left(\mathbf{r}_1^2\right)_i (\hat{\mathbf{\Theta}}_1), \tag{6.57}$$

and

$$\left(\mathbf{r}_1^2\right)_i = y_i - \mathbf{X}_i^T \hat{\mathbf{\Theta}}_1, \tag{6.58}$$

where

$$\hat{\mathbf{\Theta}}_1 = \left(\hat{\theta}_{11}, \cdots, \hat{\theta}_{p1}\right). \tag{6.59}$$

Next, construct H_2 such that

$$\left[\left|(\mathbf{r}_1)_i\right|; i \in H_2\right] = \left[\left|(\mathbf{r}_1)_1\right|, \left|(\mathbf{r}_1)_2\right|, \cdots, \left|(\mathbf{r}_1)_h\right|\right] \tag{6.60}$$

and $\left[(\mathbf{r}_1)_1 \leq (\mathbf{r}_1)_2 \leq \cdots \leq (\mathbf{r}_1)_h\right]$ are the ordered residuals. Finding the least squares fit $\hat{\mathbf{\Theta}}_2$ of h observations in H_2 results in

$$Q_2 = \sum_{i \in H_2} \left(\mathbf{r}_1^2\right)_i \leq \sum_{i \in H_2} \left(\mathbf{r}_1^2\right)_i \leq Q_1. \tag{6.61}$$

The concentration step may be expressed algorithmically by the following steps:

1. Select the h-subset H_{old}.

2. Calculate $\hat{\mathbf{\Theta}}_{old}$ for H_{old}.

3. Calculate the residuals $(\mathbf{r}_{old})_i$ for $i = 1, \cdots, n$.

4. Sort the residuals to give the permutation π such that $\left|(\mathbf{r}_{old})_{\pi(1)}\right| \leq \left|(\mathbf{r}_{old})_{\pi(2)}\right| \leq \cdots \leq \left|(\mathbf{r}_{old})_{\pi(n)}\right|$.

5. Set $H_{new} = \left[\pi(1), \pi(2), \cdots, \pi(h)\right]$.

6. Calculate $\hat{\boldsymbol{\Theta}}_{new}$ for H_{new}.

The concentration step above is performed in an iterative fashion and $Q_1 \geq Q_2 \geq \cdots \geq Q_m$ is positive and therefore converges in a finite number of iterations. Since H_{new} is computed without fully sorting the n residuals $\mathbf{r}_{old}$, the concentration step takes O(n) time complexity. The procedure above represents a simple version of the Fast-LTS algorithm, improvements in speed and performance are obtained with a more sophisticated creation of the initial subset, selective iterations, and nested extensions. These techniques are beyond the scope of this book and the reader is referred to the following excellent references on the subject [69, 71–73].

6.7.4 EF-LTS

For SAR image registration, the Fast-LTS methods need to be extended to fit a two-dimensional polynomial regression of the form [42]

$$x_{1i} = \sum_{j=0}^{m} \sum_{k=0}^{m-j} a_{jk}\, x_{2i}^{j}\, y_{2i}^{k} + \alpha_i, \tag{6.62}$$

$$y_{1i} = \sum_{j=0}^{m} \sum_{k=0}^{m-j} b_{jk}\, x_{2i}^{j}\, y_{2i}^{k} + \beta_i, \qquad i = 1, \cdots, n, \tag{6.63}$$

where

n	=	number of descriptor matches,
m	=	order of the polynomial,
a, b	=	polynomial coefficients,
α_i, β_i	=	zero-mean Gaussian error terms,
(x_{1i}, y_{1i})	=	descriptor match in the first image,
(x_{2i}, y_{2i})	=	descriptor match in the second image.

This is written more compactly as

$$x_{1i} = \mathbf{X}_i^T \, \mathbf{\Theta} + \alpha_i, \tag{6.64}$$

$$y_{1i} = \mathbf{X}_i^T \, \mathbf{\Psi} + \beta_i, \tag{6.65}$$

where $\mathbf{\Theta}$ and $\mathbf{\Psi}$ are the parameters to be estimated and are expressed as

$$\mathbf{\Theta} = [\theta_1, \theta_2, \cdots, \theta_p]^T = [a_{00}, a_{01}, \cdots, a_{N0}]^T, \tag{6.66}$$

$$\mathbf{\Psi} = [\psi_1, \psi_2, \cdots, \psi_p]^T = [b_{00}, b_{01}, \cdots, b_{N0}]^T. \tag{6.67}$$

The explanatory variable is written as

$$\mathbf{X}_i^T = \left[X_{i1}, X_{i2}, \cdots, X_{ip}\right]^T = [1, y_{2i}, \cdots, y_{2i}^N, x_{2i}, \cdots, x_{2i}^N]^T. \tag{6.68}$$

The warp function estimation problem is then transformed into two optimization problems with loss factors given by

$$Q_x = \sum_{i=1}^{h} \left(\mathbf{r}_x^2\right)_i(\hat{\mathbf{\Theta}}), \tag{6.69}$$

$$Q_y = \sum_{i=1}^{h} \left(\mathbf{r}_y^2\right)_i(\hat{\mathbf{\Theta}}), \tag{6.70}$$

where

$$\mathbf{r}_x = \left[r_{x1}, r_{x2}, \cdots, r_{xn}\right]^T, \tag{6.71}$$

$$\mathbf{r}_y = \left[r_{y1}, r_{y2}, \cdots, r_{yn}\right]^T, \tag{6.72}$$

and

$$r_{xi} = x_{2i} - \mathbf{X}_i^T \, \hat{\boldsymbol{\Theta}}, \tag{6.73}$$

$$r_{yi} = y_{2i} - \mathbf{X}_i^T \, \hat{\boldsymbol{\Psi}}. \tag{6.74}$$

Similar to the previous section, $\left(\mathbf{r}_x^2\right)_i$ and $\left(\mathbf{r}_y^2\right)_i$ are the ith elements of the squared residuals in ordered form such that

$$\left(\mathbf{r}_x^2\right)_1 \leq \left(\mathbf{r}_x^2\right)_2 \leq \cdots \leq \left(\mathbf{r}_x^2\right)_n, \tag{6.75}$$

$$\left(\mathbf{r}_y^2\right)_1 \leq \left(\mathbf{r}_y^2\right)_2 \leq \cdots \leq \left(\mathbf{r}_y^2\right)_n. \tag{6.76}$$

While a solution to (6.64) can be found by decomposing the two-dimensional problem into two independent one-dimensional problems, the matching descriptor locations are reliant on one another. EF-LTS was developed to combine the two one-dimensional regressions into a single efficient two-dimensional regression [42].

The EF-LTS algorithm begins with an initial selection of p descriptor matches and performs a least squares fit to find the initial $\hat{\boldsymbol{\Theta}}_{old}$ and $\hat{\boldsymbol{\Psi}}_{old}$. The residuals are then

$$\mathbf{r}_{x_{old}} = \left[r_{x1_{old}}, r_{x2_{old}}, \cdots, r_{xn_{old}}\right]^T, \tag{6.77}$$

$$\mathbf{r}_{y_{old}} = \left[r_{y1_{old}}, r_{y2_{old}}, \cdots, r_{yn_{old}}\right]^T, \tag{6.78}$$

where

$$r_{xi_{old}} = x_{1i} - \mathbf{X}_i^T \, \hat{\mathbf{\Theta}}_{old}, \tag{6.79}$$

$$r_{yi_{old}} = y_{1i} - \mathbf{X}_i^T \, \hat{\mathbf{\Theta}}_{old}. \tag{6.80}$$

The h-subsets are then created as

$$H_{x_{old}} = \left[\pi_x(1), \pi_x(1), \cdots, \pi_x(h)\right] \subset [1, 2, \cdots, n] \tag{6.81}$$

$$H_{y_{old}} = \left[\pi_y(1), \pi_y(1), \cdots, \pi_y(h)\right] \subset [1, 2, \cdots, n], \tag{6.82}$$

such that

$$\left(\mathbf{r}^2_{x_{old}}\right)_{\pi_x(1)} \leq \left(\mathbf{r}^2_{x_{old}}\right)_{\pi_x(2)} \leq, \cdots, \leq \left(\mathbf{r}^2_{x_{old}}\right)_{\tau_x(n)}, \tag{6.83}$$

$$\left(\mathbf{r}^2_{y_{old}}\right)_{\pi_y(1)} \leq \left(\mathbf{r}^2_{y_{old}}\right)_{\pi_y(2)} \leq, \cdots, \leq \left(\mathbf{r}^2_{y_{old}}\right)_{\pi_y(n)}. \tag{6.84}$$

The concentration step is then performed to find the new parameter estimates $\hat{\mathbf{\Theta}}_{new}$ and $\hat{\mathbf{\Psi}}_{new}$ and the new residuals

$$\mathbf{r}_{x_{new}} = \left[r_{x1_{new}}, r_{x2_{new}}, \cdots, r_{xn_{new}}\right]^T, \tag{6.85}$$

$$\mathbf{r}_{y_{new}} = \left[r_{y1_{new}}, r_{y2_{new}}, \cdots, r_{yn_{new}}\right]^T, \tag{6.86}$$

where

$$r_{xi_{new}} = x_{1i} - \mathbf{X}_i^T \, \hat{\mathbf{\Theta}}_{new}, \tag{6.87}$$

$$r_{yi_{new}} = y_{1i} - \mathbf{X}_i^T \, \hat{\mathbf{\Theta}}_{new}. \tag{6.88}$$

The estimate error scales are then found from

$$\sigma_x = A\sqrt{\frac{1}{h}\sum_{i=1}^{h}\left(\mathbf{r}^2_{x_{new}}\right)_i}, \tag{6.89}$$

$$\sigma_y = B\sqrt{\frac{1}{h}\sum_{i=1}^{h}\left(\mathbf{r}^2_{y_{new}}\right)_i}. \tag{6.90}$$

A and B are constants for Gaussian consistency [69]. Two weighting factors are computed by

$$w_{xi} = \begin{cases} 1 & \text{if} \quad \left|\frac{r_{xi}}{\sigma_x}\right| \leq 2.5 \\ 0 & \text{otherwise} \end{cases}, \tag{6.91}$$

$$w_{yi} = \begin{cases} 1 & \text{if} \quad \left|\frac{r_{yi}}{\sigma_y}\right| \leq 2.5 \\ 0 & \text{otherwise} \end{cases}. \tag{6.92}$$

The correspondence in x and y is then taken to be

$$w_i = w_{xi} \wedge w_{yi}. \tag{6.93}$$

The final estimations are then found from the weighted least squares optimization

$$\hat{\mathbf{\Theta}}_{final} = \arg\min \sum_{i=1}^{n} w_i r^2_{xi}, \tag{6.94}$$

$$\hat{\mathbf{\Psi}}_{final} = \arg\min \sum_{i=1}^{n} w_i r^2_{yi}. \tag{6.95}$$

6.8 EXAMPLES

The sections that follow illustrate the concepts of SAR image registration with a few Python/MATLAB examples. The Python examples for this chapter are in the directory *software/python/Chapter6* and the matching MATLAB examples are in the directory *software/matlab/Chapter6*. The reader should consult Chapter 1 for information on how to execute the Python and MATLAB code associated with this book.

6.8.1 Harris Corner

As a first example dealing with image registration, the Harris corner detector is used to select a number of corners based on a threshold value related to the maximum probability of the resulting corners. During corner detection, a probability is assigned to each selected corner. The list of corners is then pruned based on an optimum threshold value that varies with different imagery. The source data for this example is the SAR imagery of the Kuala Lumpur international airport in Malaysia. Figure 6.22(a) shows the result of filtering corners with a probability of less than 25% of the maximum probability. Similarly, Figures 6.22(b)–(d) show the corners with thresholds of 10%, 5%, and 1% respectively. This example is given in the Python notebook *harris_corner_example.ipynb* and the MATLAB live script *harris_corner_example.mlx*.

6.8.2 Phase Correlation

For this example, consider image registration using phase correlation techniques. The image of Muscat International Airport in Oman is the primary image and is shown in Figure 6.23(a). The secondary image is created by applying a rotation of 27°, translation of (487, 146), and scale factor of 0.7 to the primary image, and is shown in Figure 6.23(b). Using the image registration techniques of Section 6.2 results in an estimated rotation of 26.98°, translation of (487, 145), and a scale factor of 0.7009. Applying these estimated image registration parameters to the secondary image results in the image shown in Figure 6.24(b). Figures 6.24(a) and (b) show good agreement with the final image registration. This example is given in

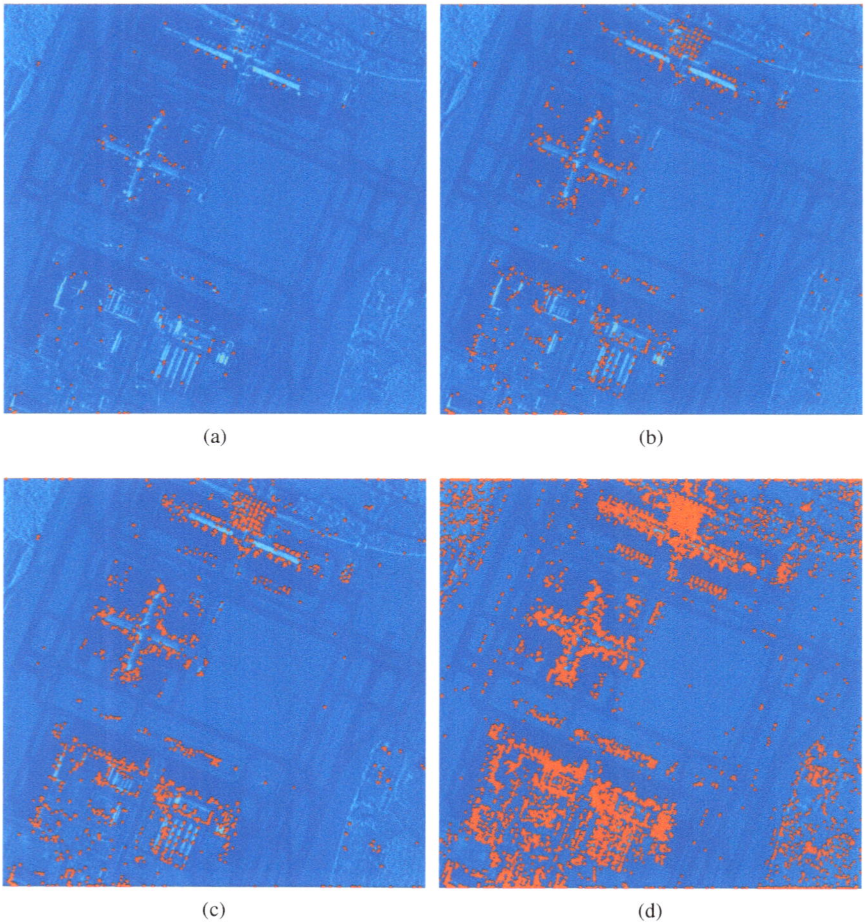

Figure 6.22 Harris corner selection with threshold values of (a) 25%, (b) 10%, (c) 5%, and (d) 1% of the maximum probability. SAR imagery of Kuala Lumpur International Airport in Malaysia. Courtesy of ICEYE.

the Python notebook *phase_correlation_example.ipynb* and the MATLAB live script *phase_correlation_example.mlx*.

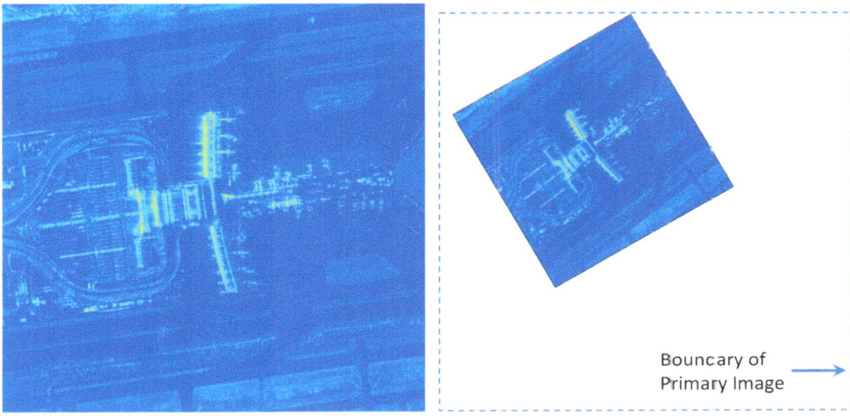

(a) (b)

Figure 6.23 Phase correlation (a) primary and (b) secondary images used for image registration. SAR imagery of Muscat International Airport in Oman. Courtesy of ICEYE.

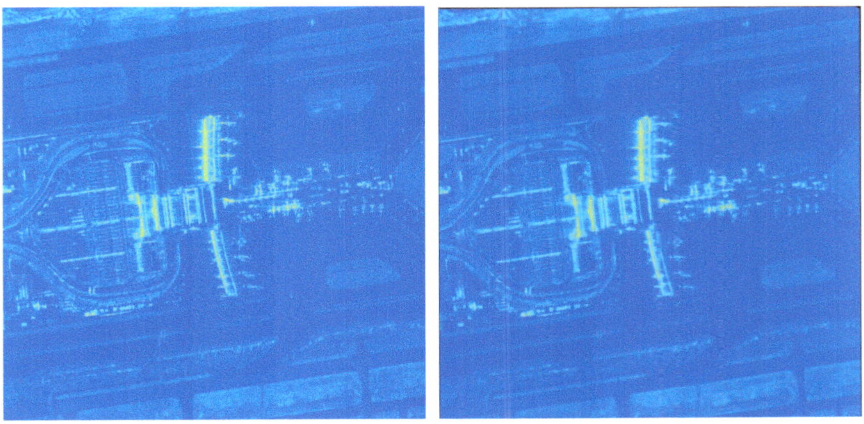

(a) (b)

Figure 6.24 Phase correlation (a) primary and (b) transformed images used for image registration. SAR imagery of Muscat International Airport in Oman. Courtesy of ICEYE.

6.8.3 SIFT

The next example of SAR image registration uses the SIFT algorithm with imagery of Kuala Lumpur International Airport in Malaysia. The primary and secondary images are given in Figures 6.25(a) and 6.25(b), respectively. The secondary image is created by applying an affine transform to the primary image. To define a unique two-dimensional affine transformation, two sets of three points are needed. The first set is the original position of the points and the second set is the position after the warping. The original positions are (0, 0), (N_x, 0), and (0, N_y), and the positions after warping are (0, $0.1N_y$), ($0.85N_x$, $0.1N_y$), and ($0.2N_x$, $0.9N_y$). This was followed by a rotation of 23° and a scaling of 0.8. Figure 6.26 shows the resulting keypoints and top 20 matches between the primary and secondary images. Using the matching keypoints, the warp function was then extracted. This warp function was applied to the secondary image to bring it into the primary image domain as outlined in Figure 6.21. Finally, the registered images are shown in Figure 6.27. Comparison of Figures 6.27(a) and 6.27(b) shows good performance by SIFT in registering SAR imagery. This example is given in the Python notebook *sift_example.ipynb* and the MATLAB live script *sift_example.mlx*.

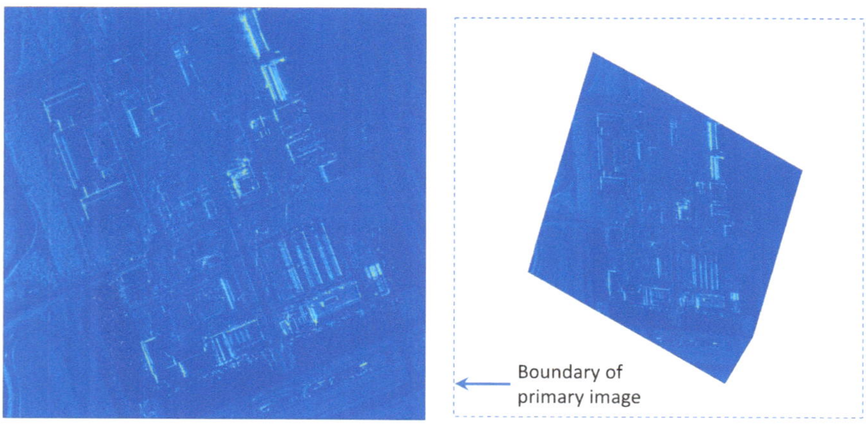

(a) (b)

Figure 6.25 SIFT (a) primary and (b) secondary images used for image registration. SAR imagery of Kuala Lumpur Airport. Courtesy of ICEYE.

Figure 6.26 Top 20 matching SIFT keypoints between the primary and secondary images. SAR imagery of Kuala Lumpur Airport. Courtesy of ICEYE.

6.8.4 ORB

In this final example of SAR image registration, the ORB algorithm is used to register imagery of Tokyo, Japan. The primary and secondary images are given in Figures 6.28(a) and 6.28(b), respectively. The secondary image is created by applying an affine transform to the primary image. The two-dimensional affine transform in this example uses an original set of three points given by (0, 0), (N_x, 0), and (0, N_y), and the positions after warping are (0, $0.1N_y$), ($0.78N_x$, $0.1N_y$), and ($0.2N_x$, $0.85N_y$). This was followed by a rotation of 23° and a scaling of 0.8. Figure 6.29 shows the resulting keypoints and top 20 matches between the primary and secondary images. Using the matching keypoints, the warp function was then extracted. This warp function was applied to the secondary image to bring it into the primary image domain as outlined in Figure 6.21.

Finally, the registered images are shown in Figure 6.30. Comparison of Figures 6.30(a) and 6.30(b) shows good performance by ORB in registering SAR imagery. This example is given in the Python notebook *orb_example.ipynb* and the MATLAB live script *orb_example.mlx*.

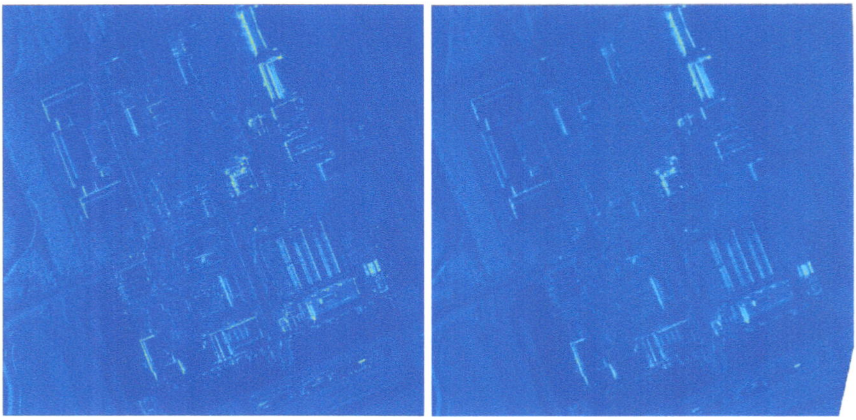

(a) (b)

Figure 6.27 Comparison of SIFT (a) primary and (b) transformed secondary images. SAR imagery of Kuala Lumpur Airport. Courtesy of ICEYE.

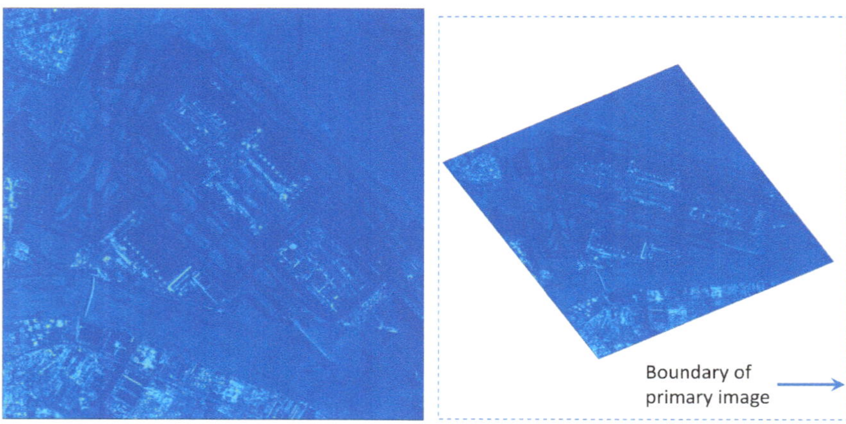

(a) (b)

Figure 6.28 ORB (a) primary and (b) secondary images used for image registration. SAR imagery of Tokyo, Japan. Courtesy of ICEYE.

Figure 6.29 Top 20 matching ORB keypoints between the primary and secondary images. SAR imagery of Tokyo, Japan. Courtesy of ICEYE.

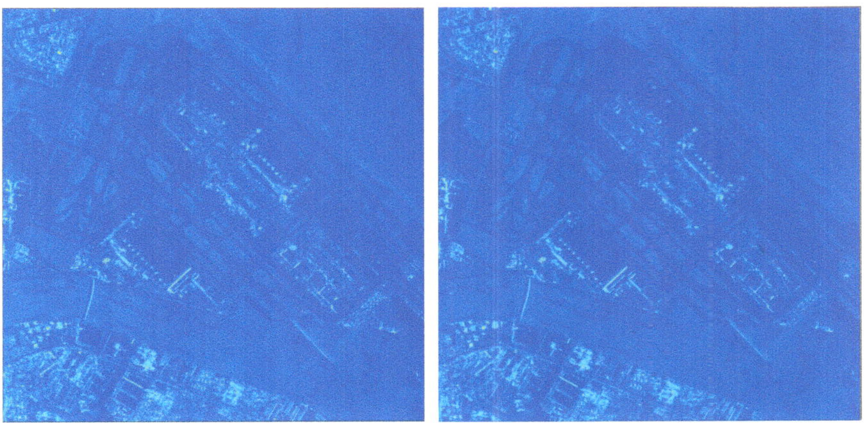

(a) (b)

Figure 6.30 Comparison of ORB (a) primary and (b) transformed secondary images. SAR imagery of Tokyo, Japan. Courtesy of ICEYE.

PROBLEMS

6.1 What are some SAR imaging modes that are highly dependent on accurate image registration?

6.2 Describe the major differences between intensity-based and feature-based SAR image registration techniques.

6.3 Why do phase correlation methods have a low computational burden compared to other image registration methods?

6.4 Compare the Moravec corner detector to the Harris corner detector in terms of robustness to decorrelation.

6.5 Describe the differences in keypoint descriptors between SIFT, SURF, and ORB.

6.6 What are the main advantages and disadvantages of using ORB compared to SIFT or SURF?

6.7 Why does ORB have a significantly lower computational burden as compared to SIFT?

6.8 Describe the importance of locating keypoints and creating descriptors invariant to translation, rotation, and scale change.

6.9 Compare the major differences in descriptor matching in SIFT and SURF.

6.10 Give the major advantages and disadvantages of using RANSAC versus LTS techniques for extracting the warp function.

References

[1] Wikipedia Contributors. Image registration. https://en.wikipedia.org/wiki/Image_registration. Accessed: 2022-01-03.

[2] L. Brown. A survey of image registration techniques. *ACM Computing Surveys*, 24(4):325–376, December 1992.

[3] G. Fornaro, G. Francescheiit, R. Lanari, E. Sansoti, and M. Tesauro. Global and local phase-unwrapping techniques: A comparison. *Journal of the Optical Society of America A*, 14:2702–2708, 1997.

[4] G. Fornaro and E. Sansosti. A two-dimensional region growing least squares phase unwrapping algorithm for interferometric SAR processing. *IEEE Transactions on Geoscience and Remote Sensing*, 37(5):2215–2226, 1999.

[5] P. Imperatore, A. Pepe, and R. Lanari. Multichannel phase unwrapping: Problem topology and dual-level parallel computational model. *IEEE Transaction on Geoscience and Remote Sensing*, 53(10):5774–5793, 2015.

[6] Q. Tian and M. Huhns. Algorithms for subpixel registration. *Computer Vision, Grpahics, and Image Processing*, 35:220–223, 1986.

[7] R. Scheiber and A. Moreira. Coregistration of interferometric SAR images using spectral diversity. *IEEE Transactions on Geoscience and Remote Sensing*, 38(5):2179–2191, 2000.

[8] M. Liao, H. Lin, and Z. Zhang. Automatic registration of InSAR data based on least-square matching and multi-step strategy. *Photogrammetric Engineering and Remote Sensing*, 70(10):1139–1144, 2004.

[9] E. Sansosti, P. Berardino, M. Manunta, F. Serafino, and G. Fornaro. Geometrical SAR image registration. *IEEE Transactions on Geoscience and Remote Sensing*, 44(10):2861–2870, 2006.

[10] M. Guizar-Sicairos, S. Thurman, and J. Fienup. Efficient subpixel image registration algorithms. *Optics Letters*, 33:156–158, 2008.

[11] B. Liu, D. Feng, P. Shu, and N. Wu. Analytic search method for interferometric SAR image registration. *IEEE Geoscience and Remote Sensing Letters*, 5(2):294–298, 2008.

[12] D. Nitti, R. Hanssen, A. Refice, F. Bovenga, and R. Nutricato. Impact of DEM-assisted coregistration on high-resolution SAR interferometry. *IEEE Transactions on Geoscience and Remote Sensing*, 49(3):1127–1143, 2011.

[13] P. Imperatore and E. Sansosti. Multithreading based parallel processing for image geometric coregistration in SAR interferometry. *Remote Sensing*, 13(10), 2021.

[14] G. Solaro, P. Imperatore, and A. Pepe. Satellite SAR interferometry for Earth's crust deformation monitoring and geological phenomena analysis. In *Geospatial Technology*. IntechOpen, 2016.

[15] G. Franceschetti, P. Imperatore, A. Iodice, and D. Riccio. In *Radio-coverage parallel computation on multi-processor platforms*, pages 1575–1578, 11 2009.

[16] P. Imperatore, A. Pepe, P. Berardino, and R. Lanari. A segmented block processing approach to focus synthetic aperture radar data on multicore processors. 07 2015.

[17] A. Plaza, Q. Du, Y. Chang, and R. King. Foreword to the special issue on high performance computing in Earth observation and remote sensing. *IEEE Journal of Selected Topics in Applied Earth Observations and Remote Sensing - IEEE J SEL TOP APPL EARTH OBS*, 4:503–507, 09 2011.

[18] S. Paul and U. Pati. SAR image registration using an improved SAR-SIFT algorithm and delaunay-triangulation-based local matching. *IEEE Journal of Selected Topics in Applied Earth Observations and Remote Sensing*, 12(8):2958–2966, 2019.

[19] Barbara Z. and Jan F. Image registration methods: a survey. *Image and Vision Computing*, 21(11):977–1000, 2003.

[20] D. Li and Y. Zhang. A fast offset estimation approach for InSAR image subpixel registration. *IEEE Geoscience and Remote Sensing Letters*, 9(2):267–271, 2012.

[21] A.A. Cole-Rhodes, K.L. Johnson, J. LeMoigne, and I. Zavorin. Multiresolution registration of remote sensing imagery by optimization of mutual information using a stochastic gradient. *IEEE Transactions on Image Processing*, 12(12):1495–1511, 2003.

[22] A. Sedaghat, M. Mokhtarzade, and H. Ebadi. Uniform robust scale-invariant feature matching for optical remote sensing images. *IEEE Transactions on Geoscience and Remote Sensing*, 49(11):4516–4527, 2011.

[23] Government of Canada. Radar image properties. https://www.nrcan.gc.ca/maps-tools-and-publications/satellite-imagery-and-air-photos/tutorial-fundamentals-remote-sensing/microwave-remote-sensing/radar-image-properties/9299. Accessed: 2022-01-30.

[24] D. Lowe. Distinctive image features from scale-invariant keypoints. *International Journal of Computer Vision*, 60:91–110, 2004.

[25] H. Bay, T. Tuytelaars, and L. Van Gool. SURF: Speeded up robust features. In A. Leonardis, H. Bischof, and A. Pinz, editors, *Computer Vision – ECCV 2006*, pages 404–417, Berlin, Heidelberg, 2006. Springer Berlin Heidelberg.

[26] W. Zou, Y. Li, Z. Li, and X. Ding. Improvement of the accuracy of InSAR image co-registration based on tie points – a review. *Sensors (Basel, Switzerland)*, 9:1259–81, 02 2009.

[27] C. Harris and M. Stephens. A combined corner and edge detector. In *In Proceedings of Fourth Alvey Vision Conference*, pages 147–151, 1988.

[28] Y. Bentoutou, N. Taleb, K. Kpalma, and J. Ronsin. An automatic image registration for applications in remote sensing. *IEEE Transactions on Geoscience and Remote Sensing*, 43(9):2127–2137, 2005.

[29] T. Lindeberg. Feature detection with automatic scale selection. *International Journal of Computer Vision*, 30:79–116, 1998.

[30] L. Zhao, B. Lü, X. Li, and S. Chen. Multi-source remote sensing image registration based on scale-invariant feature transform and optimization of regional mutual information. *Wuli Xuebao/Acta Physica Sinica*, 64, 06 2015.

[31] D. Mistry and A. Banerjee. Comparison of feature detection and matching approaches: SIFT and SURF. *GRD Journal for Engineering*, 2(4):7–13, 2017.

[32] M. Pourfard, T. Hosseinian, R. Saeidi, S. Motamedi, M.d Abdollahifard, R. Mansoori, and R. Safabakhsh. KAZE-SAR: SAR image registration using KAZE detector and modified SURF descriptor for tackling speckle noise. *IEEE Transactions on Geoscience and Remote Sensing*, 60:1–12, 2022.

[33] Y. Liu, R.and Wang. SAR image matching based on speeded up robust feature. In *2009 WRI Global Congress on Intelligent Systems*, volume 4, pages 518–522, 2009.

[34] W. Lee and A. Kim. An efficient automatic geo-registration technique for high resolution spaceborne SAR image fusion. pages 3566–3569, 07 2011.

[35] H. Zhang, W. Ni, W. Yan, J. Wu, and S. Li. Robust SAR image registration based on edge matching and refined coherent point drift. *IEEE Geoscience and Remote Sensing Letters*, 12(10):2115–2119, 2015.

[36] D. Li and Y. Zhang. On the appropriate feature for general sar image registration. In *Proceedings of SPIE 8536, SAR Image Analysis, Modeling, and Techniques XII*, 2012.

[37] M. Teke and A. Temizel. Multi-spectral satellite image registration using scale-restricted SURF. In *2010 20th International Conference on Pattern Recognition*, pages 2310–2313, 2010.

[38] H. Nguyen, R. Fablet, A. Ehrhold, and J. Boucher. Keypoint-based analysis of sonar images: Application to seabed recognition. *IEEE Transactions on Geoscience and Remote Sensing*, 50(4):1171–1184, 2012.

[39] B. Reddy and B. Chatterji. An FFT-based technique for translation, rotation, and scale-invariant image registration. *IEEE Transactions on Image Processing*, 5(8):1266–1271, 1996.

[40] A. Oppenheim, A. Willsky, and I. Young. *Signals and Systems*. Prentice-Hall, 6th edition, 1983.

[41] D. Casasent and D. Psaltis. Position, rotation, and scale invariant optical correlation. *Applied Optics*, 15(7):1795–1799, 1976.

[42] D. Li, Y. Zhang, and X. Shi. On feature-based SAR image registration: Appropriate feature and retrieval algorithm. In *Advanced Remote Sensing Technology for Synthetic Aperture Radar Applications, Tsunami Disasters, and Infrastructure*. IntechOpen, 2018.

[43] H. Moravec. *Obstacle Avoidance and Navigation in the Real World by a Seeing Robot Rover*. PhD thesis, Stanford University, March 1980.

[44] Wikipedia Contributors. Corner detection. https://en.wikipedia.org/wiki/Corner_detection. Accessed: 12-12-2021.

[45] P. Corke. Introduction to corner features (harris). https://robotacademy.net.au/lesson/introduction-to-corner-features-harris/, 2015. Accessed: 2021-12-21.

[46] B. Romeny. *Front-End Vision and Multi-Scale Image Analysis*. Springer Dordrecht, 2003.

[47] Wolfram Mathworld. Hermitian matrix. https://mathworld.wolfram.com/HermitianMatrix.html, 2022. Accessed: 2022-02-05.

[48] Math.NET Numerics. Distance metrics. https://numerics.mathdotnet.com/Distance.html, 2022. Accessed: 2022-01-18.

[49] I. Rey-Otero and M. Delbracio. Anatomy of the SIFT method. *Image Processing On Line*, 4:370–396, 2014.

[50] A. Singh. A detailed guide to the powerful SIFT technique for image matching. https://www.analyticsvidhya.com/blog/2019/10/detailed-guide-powerful-sift-technique-image-matching-python/, 2019. Accessed: 2021-11-12.

[51] T. Lindeberg. Scale-space theory: A basic tool for analysing structures at different scales. *Journal of Applied Statistics*, 21(2):224–270, 1994.

[52] J. Koenderink. The structure of images. *Biological Cybernetics*, 50:363–396, 1984.

[53] P. Viola and M. Jones. Rapid object detection using a boosted cascade of simple features. In *Proceedings of the 2001 IEEE Computer Society Conference on Computer Vision and Pattern Recognition. CVPR 2001*, volume 1, pages I–I, 2001.

[54] Wikipedia Contributors. Summed-area tables. https://en.wikipedia.org/wiki/Summed-area_table. Accessed: 2022-01-03.

[55] F. Crow. Summed-area tables for texture mapping. In *Proceedings of the 11th Annual Conference on Computer Graphics and Interactive Techniques*, SIGGRAPH '84, pages 207—212, New York, NY, USA, 1984. Association for Computing Machinery.

[56] J. Walker. *A Primer on Wavelets and their Scientific Applications*. Chapman and Hall / CRC, 1999.

[57] U. Durgam, S. Paul, and U. Pati. SURF based matching for SAR image registration. In *2016 IEEE Students' Conference on Electrical, Electronics and Computer Science (SCEECS)*, pages 1–5, 2016.

[58] E. Rublee, V. Rabaud, K. Konolige, and G. Bradski. ORB: an efficient alternative to SIFT or SURF. pages 2564–2571, 2011.

[59] D. Lowe. Method and apparatus for identifying scale invariant features in an image and use of same for locating an object in an image, 2000. United States Patent No. 6711293B1.

[60] R. Funayama, H. Yanagihara, L. Van Gool, T. Tuytelaars, and H. Bay. Robust interest point detector and descriptor, 2009. United States Patent No. 20090238460A1.

[61] E. Rosten and T. Drummond. Machine learning for high-speed corner detection. In A. Leonardis, H. Bischof, and A. Pinz, editors, *Computer Vision – ECCV 2006*, pages 430–443, Berlin, Heidelberg, 2006. Springer Berlin Heidelberg.

[62] M. Calonder, V. Lepetit, Christoph Strecha, and P. Fua. BRIEF: Binary robust independent elementary features. In K. Daniilidis, P. Maragos, and N. Paragios, editors, *Computer Vision – ECCV 2010*, pages 778–792, Berlin, Heidelberg, 2010. Springer Berlin Heidelberg.

[63] E. Karami, S. Prasad, and M. Shehata. Image matching using SIFT, SURF, BRIEF and ORB: performance comparison for distorted images. *CoRR*, abs/1710.02726, 2017.

[64] Wolfram MathWorld. Least squares fitting. https://mathworld.wolfram.com/LeastSquaresFitting.html. Accessed: 01-31-2022.

[65] M. Fischler and R. Bolles. Random sample consensus: A paradigm for model fitting with applications to image analysis and automated cartography. *Communications of the ACM*, 24(6):381—395, jun 1981.

[66] S. Dawn, V. Saxena, and B. Sharma. Remote sensing image registration techniques: A survey. In *Internation Conference on Image and Signal Processing*, pages 103–112, 2010.

[67] D. Li and Y. Zhang. A rigorous sar epipolar geometry modeling and application to 3d target reconstruction. In *IEEE Selected Topics in Applied Earth Observations and Remote Sensing*, volume 6, pages 2316–2323, 2013.

[68] Wiki Contributors. Random sample consensus. https://en.wikipedia.org/wiki/Random_sample_consensus. Accessed: 1-18-2022.

[69] P. Rousseeuw and A. Leroy. *Robust Regression and Outlier Detection*. John Wiley and Sons, 1987.

[70] L. Li. An algorithm for computing exact least-trimmed squares estimate of simple linear regression with constraints. *Computational Statistics and Data Analysis*, 48(4):717–734, 2005.

[71] P. Rousseeuw and K. Van Driessen. Computing lts regression for large data sets. In Springer Science, editor, *Data Mining and Knowledge Discovery*, volume 12, pages 29–45, 2006.

[72] P. Rousseeuw and K. Van Driessen. A fast algorithm for the minimum covariance determinant estimator. *Technometrics*, 41:212–223, 1999.

[73] G. Pison, S. Van Aelst, and G. Willems. Small sample corrections for lts and mcd. *Metrika*, 55:111–123, 2002.

Chapter 7

Performance Considerations

The overall performance of a SAR system may be assessed in various ways and by different metrics. For example, the radar general image quality equation (RGIQE) is a figure of merit concerned with the maximum information content of a resolution cell in SAR images, whereas the Radar National Imagery Interpretability Rating Scale (RNIIRS) is a subjective assessment of SAR image quality. Ultimately, the metrics should be in terms of features that define or differentiate the performance of one SAR system from another, and include spatial resolution, SNR, signal and data processing capabilities, image storage and downlink capacity, and revisit times. The calculation of performance metrics depends on several parameters that are interdependent and often nonlinear. Trade-offs must be made between the different parameters that lead to performance bounds and may be generalized to aid in understanding. This chapter begins with an analysis of spatial resolution and associated system parameters. SNR, image contrast, sensitivity, and dynamic range are investigated in terms of RF hardware, environment, and signal processing elements. Image geometry and transmitted pulse characteristics are then presented. Next, an overview of data processing, storage, and downlink limitations is given. The chapter concludes with several Python and MATLAB examples to reinforce key parameter trade-offs and performance bounds associated with those trade-offs.

7.1 SPATIAL RESOLUTION

Spatial resolution is directly related to the ability to resolve objects in the imaged scene, which greatly affects image analysis and interpretation. Figures 7.1–7.4 illustrate the results of imaging a scene at resolutions from 15 cm to 1 m [1]. The loss of detail and image quality between 15-cm resolution and 1-m resolution is significant. Also note the difference in image speckle between Figures 7.1 and 7.4. This is due to there being more scatterers in the larger resolution cells. The scattered wavefronts from the scatterers coherently combine and cause interference patterns.

Sections 2.2.1 and 2.3.1 covered spatial resolution in the ground-range and cross-range planes for stripmap and spotlight SAR modes. Those results are summarized here for ease of reference

$$\delta_g = \frac{c}{2B\cos(\theta_i)} \qquad \text{(both modes)} \qquad \text{(m)}, \tag{7.1}$$

$$\delta_a = \frac{D_a}{2} \qquad \text{(stripmap)} \qquad \text{(m)}, \tag{7.2}$$

$$\delta_a = \frac{\lambda}{2\,\theta_{sa}} \qquad \text{(spotlight)} \qquad \text{(m)}. \tag{7.3}$$

The expression in (7.1) indicates the ground-range resolution depends on the SAR operating bandwidth. This would seem to indicate that a SAR designer is free to devise a system with any arbitrary resolution by adjusting the system bandwidth. However, the International Telecommunications Union (ITU), which coordinates the shared global use of the radio spectrum and satellite orbits, limits the bandwidth of spaceborne radars to 1,200 MHz [2]. In addition, individual countries may have further restrictions. For example, the United States only allocated 600 MHz for spaceborne SAR systems [3]. The Federal Communications Commission (FCC) granted Umbra Lab an experimental license to operate high-bandwidth SAR using the 1,200 MHz band centered at 9.8 GHz and low-bandwidth SAR with the 600 MHz band centered on 9.6 GHz [1, 3]. There are also hardware and signal processing limitations associated with ground-plane range resolution given in later sections of this chapter.

Figure 7.1 Umbra SAR imagery captured over Santa Barbara, CA at 15-cm resolution. Source: Umbra Lab, Inc. (Licensed under Creative Commons 4.0.)

Figure 7.2 Umbra SAR imagery captured over Santa Barbara, CA at 25-cm resolution. Source: Umbra Lab, Inc. (Licensed under Creative Commons 4.0.)

Figure 7.3 Umbra SAR imagery captured over Santa Barbara, CA at 50-cm resolution. Source: Umbra Lab, Inc. (Licensed under Creative Commons 4.0.)

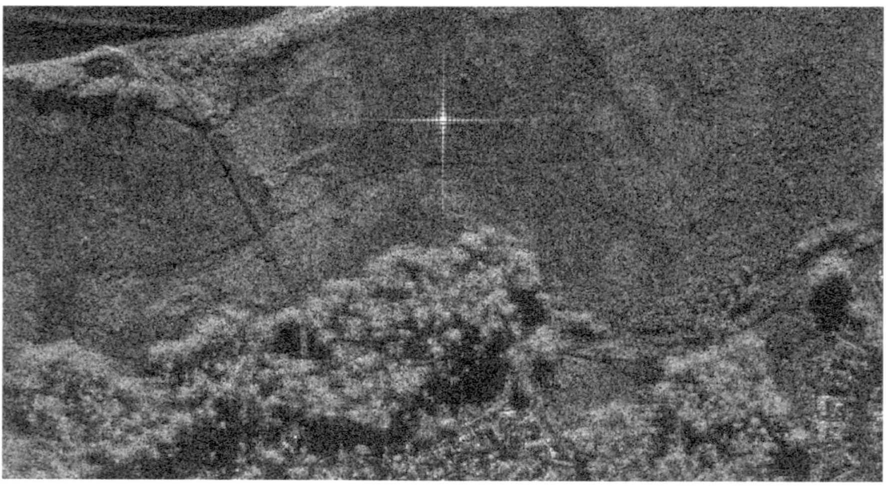

Figure 7.4 Umbra SAR imagery captured over Santa Barbara, CA at 1-m resolution. Source: Umbra Lab, Inc. (Licensed under Creative Commons 4.0.)

Using (7.1) with $B = 1200$ MHz and $\theta_i = 55°$ results in a ground-range resolution of $\delta_g \approx 0.22$ m, including a broadening factor of approximately 1.4 for a Hamming window, results in $\delta_g \approx 0.3$ m. Reducing the bandwidth to 600 MHz results in a ground-range resolution of approximately 0.6 m. In addition to image resolution, the SAR operating bandwidth affects the design and performance of hardware components including the transmitter, receiver, antenna, signal processor, and downlink. Cross-range resolution for stripmap mode of operation is given in (7.2) and is simply one half the dimension of the SAR antenna in the azimuth direction. This indicates that a smaller antenna results in a finer cross-range resolution. A smaller physical antenna has a larger beamwidth and therefore illuminates a target for a longer time. This is effectively a larger virtual antenna and results in finer cross-range resolution [4]. However, smaller physical antennas have lower gain and efficiency, which lowers the SNR and worsens image contrast.

Cross-range resolution for spotlight mode of operation is given in (7.3) and is related to the total integration angle θ_{sa} and operating frequency. The higher the frequency, the smaller the wavelength, and the finer the cross-range resolution. For a fixed aperture size, higher frequencies result in higher antenna gain, thus improving the SNR and image contrast.

7.2 SNR

Another key factor for studying SAR performance is the SNR, which directly affects dynamic range and image contrast. From Chapter 2, the SNR of an image depends on many parameters that are interrelated and arise from different phenomena and sources.

One method for quantifying noise in SAR sensors is the NESZ, which is often written as σ_n and is given by

$$\sigma_n = \frac{8\pi\, \lambda R^3\, v\, k\, T_0\, F\, L}{P_{avg}\, A_e^2\, \delta_r}. \tag{7.4}$$

This parameter gives an indication of the noise floor of an image. All received signals must be stronger than the NESZ value to rise above the noise level. Ideally, the NESZ for a SAR system will be a small value and images with high NESZ values will appear grainy [5]. Typical values of NESZ for spaceborne SAR systems are in the -15 to -25 dB range [5, 6]. Figures 7.5 and 7.6 illustrate the change in image contrast as the NESZ is varied from -20 dB to -5 dB. As shown, changes in the noise floor do not affect the spatial resolution of the system but tend to fill in the shadows or low-intensity regions [7].

Watch this animation illustrating image contrast as the NESZ is varied.

From (2.24), (2.26), and (2.47), some general trends associated with the SNR are observed and are summarized in Table 7.1. When studying radar systems, the parameters in the radar equation are often taken as independent values to be freely varied. While this approach is valuable for first draft designs and sizing studies, these parameters are interrelated (e.g., δ_a is a function of λ and R for spotlight mode), and in some cases the relationship is nonlinear. The subsequent sections cover the sources and limits of the parameters given in Table 7.1, which influence overall SAR performance limitations.

7.2.1 Antenna System

In the design of SAR systems, trade-offs must be made among factors such as size, weight, power, and cost. Accordingly, the antenna system will have space constraints, which influence aperture size and operating frequency. Depending on the application, the constraints may be severe. For example, OneSAR is an extremely small and lightweight system to be used on small unmanned aerial systems and has dimensions of $18.3 \times 22.9 \times 25.6$ cm [8]. In contrast, spaceborne systems typically have much larger antenna apertures. The Umbra SAR antenna system has an area greater than 10 m^2 [1]. Partly due to the aperture size, very small systems operate at a few kilometers, whereas spaceborne systems may operate at ranges over 1,000 km.

The exact antenna design is specific to the SAR mission and may be a reflector type configuration, such as those used by Capella and Umbra [1, 9], or phased array designs employed in ICEYE and COSMO-SkyMed *2nd* Generation systems [6, 10]. Phased array type of antenna designs, as illustrated in Figure 7.7, allows for greater flexibility in steering the radar beam and accommodates a number of imaging modes [11, 12]. The first spaceborne SAR system to use an active antenna system

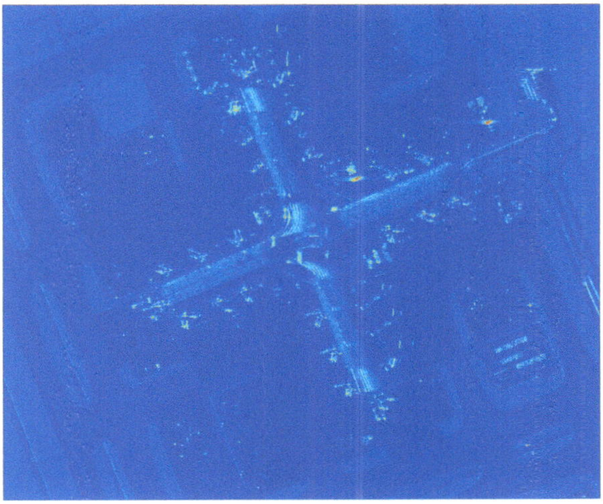

(a) $\sigma_n \approx -18$ dB.

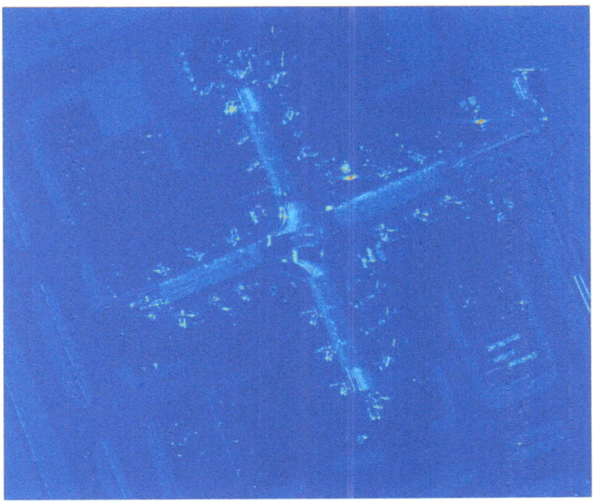

(b) $\sigma_n \approx -15$ dB (simulated).

Figure 7.5 Comparison of SAR imagery of Muscat, Oman with varying values of NESZ (simulated). Courtesy of ICEYE.

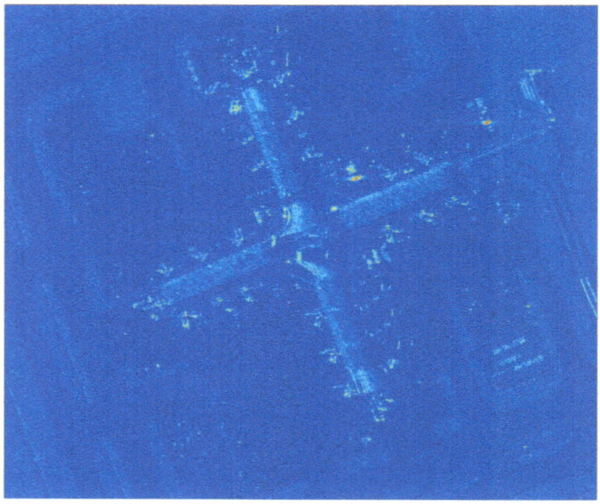

(a) $\sigma_n \approx -10$ dB (simulated).

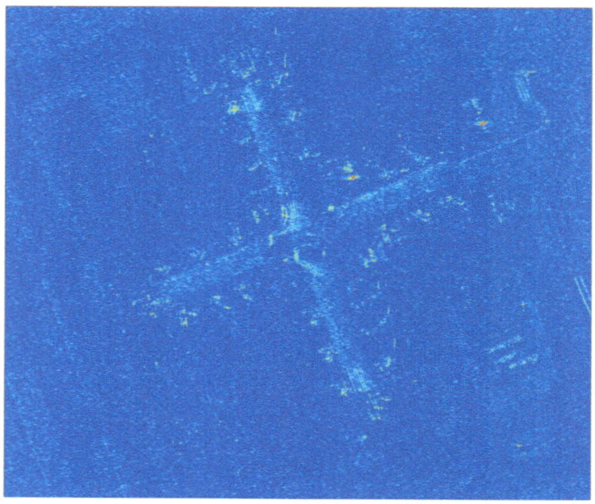

(b) $\sigma_n \approx -5$ dB (simulated).

Figure 7.6 Comparison of SAR imagery of Muscat, Oman with varying values of NESZ (simulated). Courtesy of ICEYE.

Table 7.1
General Trends in SNR

Parameter	*Point-like Targets*	*Distributed Targets*
Average Power (P_{avg})	Increase	Increase
Effective Aperture (A_e)	Increase	Increase
Platform Velocity (v) (for a fixed PRF, fewer pulses are transmitted/received over the same aperture length)	Decrease	Decrease
Wavelength (λ) (for a fixed aperture size, antenna gain is reduced)	Decrease	Decrease
Noise Figure (F)	Decrease	Decrease
Range (R)	Decrease	Decrease
Cross-range Resolution (δ_a) (finer resolution requires longer synthetic apertures, resulting in more transmitted/received pulses)	Increase	Not affected
Range Resolution (δ_r) (increased bandwidth results in a larger time-bandwidth product)	Not affected	Increase

with transmit/receive modules (TRM) was the ENVISAT/ASAR, which was a C-band system capable of operating in a wide variety of imaging modes [13]. For an antenna element to be an effective radiator, its size must be on the order of a tenth of a wavelength or greater [14]. This pushes very small SAR platforms toward higher operating frequencies, such as Ku-band. Referring again to (2.24), (2.26), and (2.47), the SNR is improved by increasing the antenna aperture's electrical size, A_e/λ^2, which again leads to higher frequencies. For spaceborne systems, this is less of a concern and a number of systems operate at C- and L-band [13, 15].

The effective aperture may be written in terms of the physical aperture and the aperture efficiency as [14]

$$A_e = \eta A_p \qquad (\text{m}^2), \tag{7.5}$$

where

η = antenna aperture efficiency,
A_p = antenna physical aperture (m^2),
A_e = antenna effective aperture (m^2).

The aperture efficiency includes factors such as conductor losses, feed losses, and radiation efficiency. Phased array type antenna systems have efficiencies in the 0.6–0.8 range, while reflector type antennas are closer to 0.5 [16, 17]. Figure 7.8 illustrates the increase in SNR versus effective aperture size normalized by wavelength.

7.2.2 Transmitter

The radar transmitter is a device or set of devices used to produce high-power radar pulses, which are then radiated by the antenna system. Transmitters may be characterized by peak output power, maximum duty factor, efficiency, frequency range, and stability. For SAR systems, peak transmitted power may range from 0.5 milliwatts for small, short-range systems to a few kilowatts for airborne and spaceborne systems. Referring again to (2.24), (2.26), and (2.47), the SNR depends on the average transmitted power, which is related to the peak power by [14]

$$P_{avg} = P_t \, \frac{\tau}{T} \qquad (\text{W}), \tag{7.6}$$

where

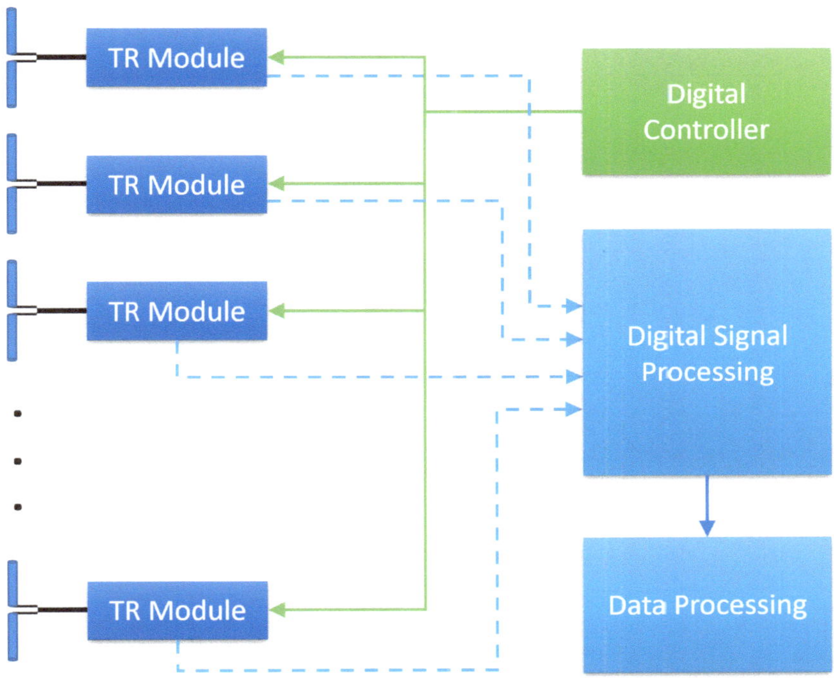

Figure 7.7 Active phased array with digital beamforming. Signal generation, transmission, reception, and sampling performed by each TRM.

P_t	=	peak transmitted power (W),
τ	=	pulse width (s),
T	=	is the pulse repetition interval, PRI (s),
P_{avg}	=	average transmitted power (W).

The term τ/T is the duty cycle and Figure 7.9 illustrates the relationship between peak and average transmitted power. For spaceborne SAR systems, duty cycles of 10% to 25% are typical, while specialty systems may have higher duty cycles. The capability and suitability of a transmitter for SAR applications is highly dependent on the type and design of transmitter under consideration.

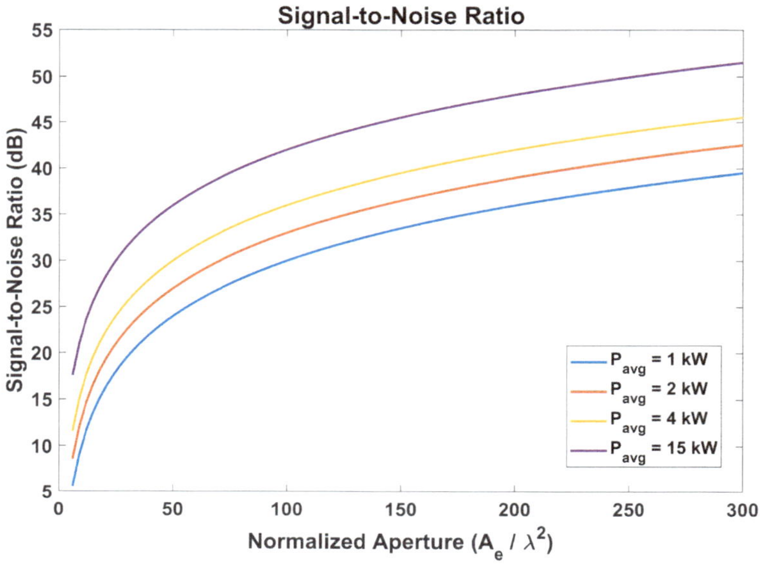

Figure 7.8 SNR vs effective aperture size for varying average transmitted power.

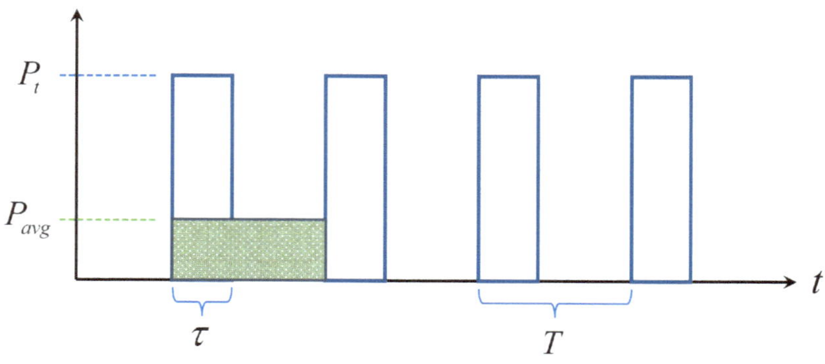

Figure 7.9 Illustration of peak power, average power, and duty cycle.

7.2.2.1 Microwave Tubes

The development of many practical radar systems capable of detecting targets at long range relied on microwave tubes to produce very strong transmitted signals. These devices are typically used with dish antennas and passive arrays, as illustrated in Figure 7.10. Microwave tubes come in a wide variety of designs and configurations but may be separated into two broad categories, linear-beam and crossed-field tubes. Linear-beam devices include traveling-wave tubes, klystrons, and gyrotrons. Crossed-field devices include magnetrons and injected-beam tubes [7, 18]. Table 7.2 gives some examples of microwave tubes and general performance bounds [7, 19]. While improvements in performance reliability have been made to microwave tubes, there are still drawbacks including cooling requirements, high voltages, phase instability, warm-up time, size, weight, and limited lifespan [19].

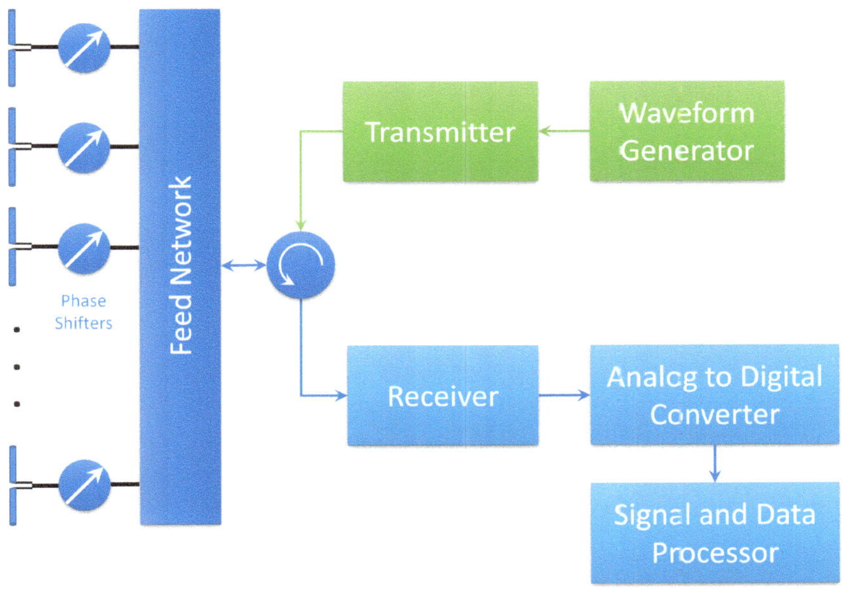

Figure 7.10 Passive array design using a microwave tube transmitter.

Table 7.2
Typical Performance of Microwave Tube Devices

Device	*Maximum Frequency (GHz)*	*Bandwidth* (%)	*Peak Power* (W)	*Average Power* (W)
Magnetron	95	10	10 M	500
Klystron	400	10	10 M	10 k
Gyrotron	500	10	5 M	5 M
Traveling Wave Tube	500	20	250 k	25 k
Crossed-Field Amplifier	30	15	5 M	1 k

7.2.2.2 Solid-State Amplifiers

To overcome some of the limitations of microwave tubes, radar designers began investigating solid-state transmitter modules as an alternative. While individual solid-state transistors cannot reach the kilowatt range needed for many SAR systems, these may be arranged into groups, referred to as transmitter modules, capable of delivering the required radiated power levels [19]. Depending on the specific mission, the outputs of tens to hundreds of transmitter modules are accumulated with microwave power combiners, as illustrated in Figure 7.11. A main advantage of this approach is redundancy. SAR systems may remain operational, with some loss of performance, even with failure of 10% of transmitter modules.

Gallium Nitride (GaN) and Silicon Carbide (SiC) have become ideal candidates for high-power, high-temperature microwave applications and are being used more and more in radar systems [20]. Solid-state amplifiers are also used in TRMs for active-array radar systems, shown in Figure 7.7. These systems are comprised of a few hundred to tens of thousands of TRMs. For example, the CSG antenna is comprised of 2,560 TRMs [6]. Each TRM outputs a few watts and is connected to an antenna element, or a small group of elements, such as printed dipoles or patches. This approach results in lower loss, improved beam agility, better management of

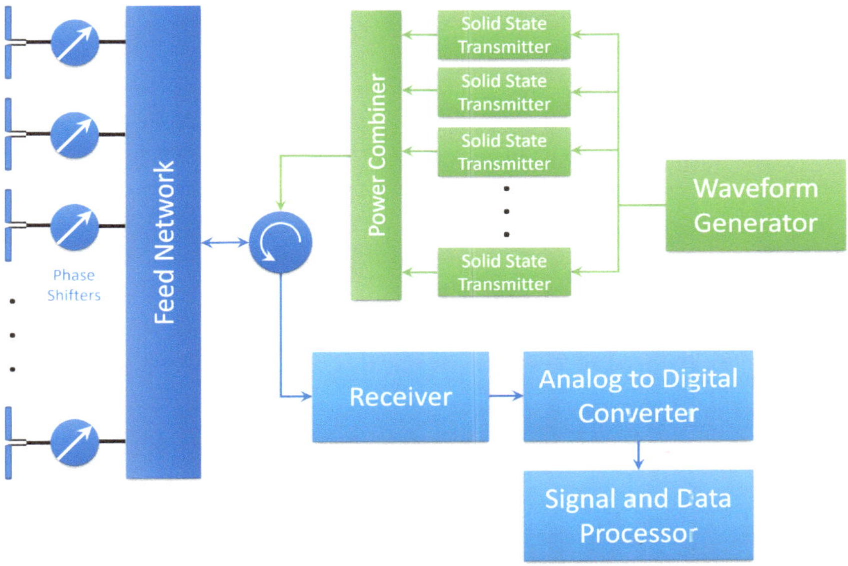

Figure 7.11 Passive array design using solid state transmitters.

radar resources, and the ability to create the radar waveform at each transmitting element [18, 19].

7.2.3 Receiver

The function of the receiver is to take the signals received by the antenna, amplify, filter, downconvert, and digitize these signals, then deliver the signals to the radar signal and data processors [14]. With modern antenna and receiver hardware, the reference boundaries of the radar receiver can become ambiguous. For example, conventional radar systems make use of a duplexer to provide the input to the receiver from the antenna, illustrated in Figures 7.10 and 7.11. However, active array antennas include low-noise amplifiers prior to forming receive beams, shown in Figure 7.7. While these amplifiers are part of the antenna system, they are sometimes included as part of the receiver for analysis. While there are numerous receiver configurations, these may be characterized by performance parameters including dynamic range, bandwidth, gain control, filtering, noise factor, sampling rates, and quantization.

As illustrated in Chapter 2, noise is one of the fundamental limiting factors in the radar range equation. The received signal is corrupted by noise generated internally to the receiver. The major source of noise in receivers is thermal noise that can obscure weak signals [14, 21]. The noise generated in the receiver is random, and statistical techniques are used to characterize the effects. The noise level at the input to the receiver is primarily determined by the antenna noise temperature and its associated loss. The noise level is usually specified by a power spectral density, or alternatively as an average power level over a specified bandwidth [14]. The system noise is then the combined antenna noise and receiver noise and is expressed as

$$T_s = T_a + T_r = T_a + L_r\, T_0\,(F - 1) \qquad \text{(K)}, \tag{7.7}$$

where

T_s = total system noise temperature (K),
T_a = antenna noise temperature (K),
T_r = receiver noise temperature (K),
L_r = receiver loss,
T_0 = receiver operating temperature (K),
F = receiver noise factor.

The noise due to the receiver is typically small compared to the noise input to the receiver. Therefore, receiver noise only has a small effect on the total system noise temperature. This is important when calculating SNR due to the fact the SNR referenced to the receiver noise can be very different from the SNR referenced to the total system noise [22]. Noise figure varies by input power, frequency band, design, and amplifier design, and Table 7.3 provides some typical noise figure values by frequency band.

7.2.3.1 Digital Receivers

With increasing sample rates available in analog-to-digital converters (ADC), direct sampling of the signals in the TRMs is becoming common [14]. In digital receivers, an ADC is used to sample the signal, and downconversion to baseband in-phase

Table 7.3
Typical Receiver Noise Figures by Frequency Band

Frequency Band	*Noise Figure (dB)*
L-band	1–2
S-band	1.2–2.5
C-band	2.2–3
X-band	2.8–3.5
Ku-band	4–6

Table 7.4
Brief Survey of Analog-to-Digital Converters

Device	*Channels*	*Bits*	*Sampling Rate*
Texas Instruments — AD9224R	2	16	3M
Texas Instruments — ADS554J20	2	12	1G
Texas Instruments — ADS1675	1	24	4M
Analog Devices — AD9697	1	14	1.3G
Analog Devices — AD9213	1	12	10G
Linear Technology — LTC2208	1	16	130M
Maxim Integrated — MAX19777	2	12	3M
Maxim Integrated — MAX11284	2	24	4K

and quadrature signals is performed in digital signal processing. One of the main advantages of direct digital sampling is the virtual elimination of the amplitude and phase imbalance between the in-phase and quadrature channels. Other advantages include reduction of DC offset errors, improved linearity, flexibility of bandwidth, improved filtering, and reduction in size, weight, and power [22]. Table 7.4 provides a brief survey of currently available analog-to-digital converters.

Referring again to Figure 7.7, the digitized signals from the TRMs are sent to the digital signal processor where the actual received beams are formed in software. This allows for greater flexibility as a variety of SAR operating modes is possible on a single platform. There is also the possibility of simultaneous modes if the digital signal processor and data recorder meet the necessary requirements. These are discussed in the subsequent sections. Additionally, new modes and functions may be added by upgrading the system software [19]. There are two main methods for using digital signal processing to generate baseband in-phase and quadrature signals from the sampled data. The first is through a process known as direct digital downconversion, and the second method is through the use of the Hilbert transform [23].

Direct Digital Downconversion

The process of direct digital downconversion is illustrated in Figure 7.12. The first step is to directly sample the band-limited signal. Next, the sampled signal is frequency shifted to zero frequency by complex multiplication with $e^{-j\omega_0 n}$, as shown in Figure 7.12. This step is typically accomplished through the use of a numerically controlled oscillator [22]. After this translation to baseband, the signals are passed through two low-pass digital filters to remove the unwanted image frequency.

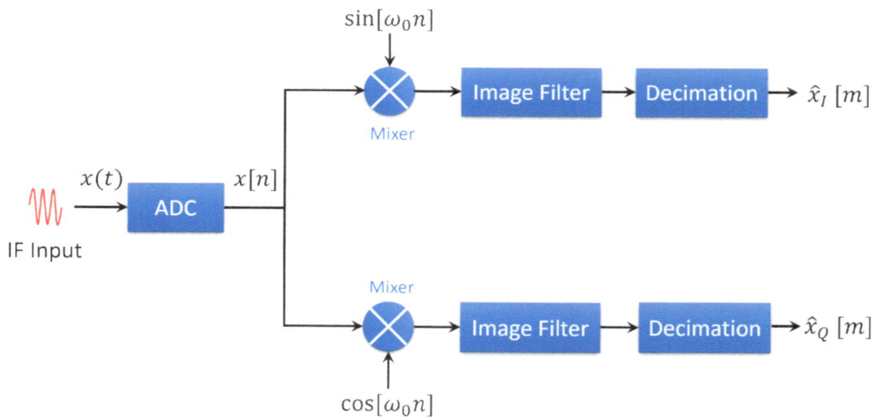

Figure 7.12 Generation of IQ data with direct digital downconversion.

Hilbert Transform

The second method for producing baseband I and Q signals employs the Hilbert transform [24–26], and is shown in Figure 7.13. The Hilbert transform is implemented as two filters $h[n]$ and $g[n]$, as shown in Figure 7.13, with frequency responses given as [22]

$$\left|H(\omega)\right| \approx \left|G(\omega)\right| \approx 1, \tag{7.8}$$

$$\frac{H(\omega)}{G(\omega)} = \begin{cases} -j & \text{for} \quad |\omega - \omega_0| \leq B \\ +j & \text{for} \quad |\omega + \omega_0| \leq B. \end{cases} \tag{7.9}$$

Referring to (7.8) and (7.9), this method of digital filtering produces outputs that form the complex I and Q representation of the signal centered at ω_0 and rejects the signal at $-\omega_0$. The final step in this type of receiver is to shift the spectrum of the signal located at ω_0 to zero frequency by decimation.

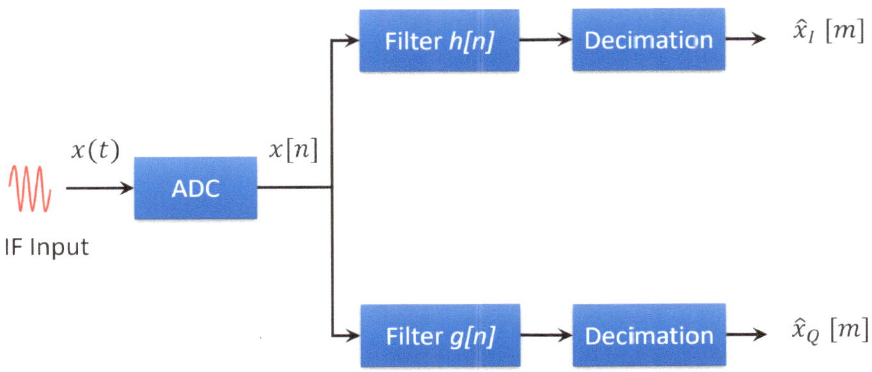

Figure 7.13 Generation of IQ data with the Hilbert transform.

7.3 LOSSES

Referring to (2.47), the SNR is written in terms of a total loss L_{total}, which is comprised of many sources of loss in the system. These include losses in the transmitter and receiver, wave propagation, signal processing, and beam steering.

7.3.1 Hardware Loss

For purposes of this book, hardware losses are considered to be from thermal loss due to conductive materials, dielectric losses, and impedance mismatches. For example, impedance mismatches often occur between the antenna and the feeding network, resulting in a reflection of some of the energy on the transmission line that is not available to the antenna. System losses are frequency dependent and generally increase with frequency. Depending on the type of feed network being used, losses in the 0.5–2.0 dB range are common for the transmitter and receiver path [17, 22].

7.3.2 Propagation Loss

There are several mechanisms contributing to the attenuation of propagating electromagnetic energy. These can include antenna radomes, rain, clouds, fog, the atmosphere, and vegetation. Radome loss is heavily dependent on the specific design, where values in the 0.1–1.0 dB range are common [27]. Electromagnetic waves are attenuated in the atmosphere primarily due to oxygen (O_2) and water vapor (H_2O) [14]. The first peak in attenuation occurs at approximately 22 GHz and is due to water vapor. The second peak is located at 63 GHz due to oxygen. The atmospheric absorption in the frequency range from 0 to 1000 GHz is shown in Figure 7.14. This data is for a temperature of 290° K, a dry air pressure of 1013.25 hPa, and a water vapor density of 7.5 g/m^3 [14].

Another propagation loss mechanism that should typically be considered is rain attenuation. Following the recommendation from the International Telecommunication Union Radiocommunication Sector (ITU-R) [28], the specific attenuation due to rain is given as

$$\gamma = kR^\alpha \qquad \text{(dB/km)}, \tag{7.10}$$

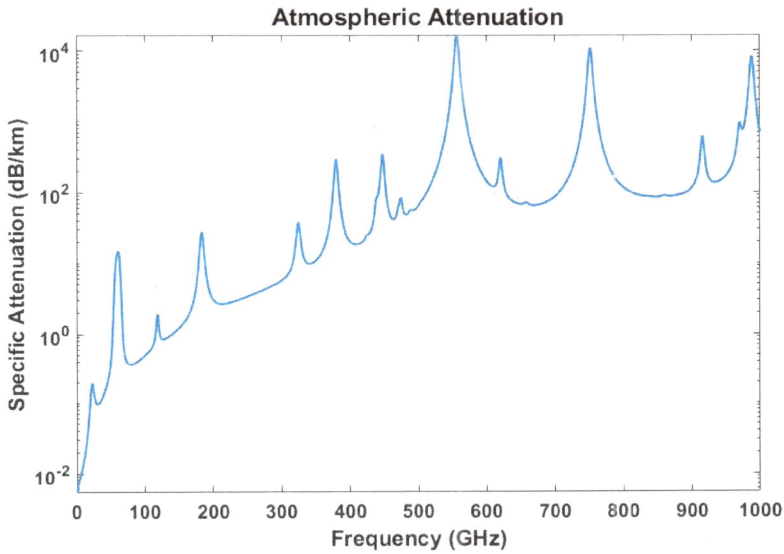

Figure 7.14 Atmospheric absorption as a function of frequency.

where R is the rain rate in millimeters per hour. The factors k and α are functions of frequency and their expressions are derived by curve fitting scattering calculations to the form in (7.10). From [28], the factors k and α are given by

$$k = 0.5 \times \left[k_H + k_V + (k_H - k_V)\cos^2\theta\cos(2\tau)\right], \tag{7.11}$$

$$\alpha = 0.5 \times \left[k_H\alpha_H + k_V\alpha_V + (k_H\alpha_H - k_V\alpha_V)\cos^2\theta\cos(2\tau)\right]/k, \tag{7.12}$$

where θ is the elevation angle and τ is the polarization tilt angle. The expressions in (7.11) and (7.12) are seen to be a combination of horizontal and vertical polarization components and are valid for both linear and circular polarization and for all path geometries. The terms k_H and k_V are given in [28] as

$$\log_{10} k = \sum_{i=1}^{4} a_i\, e^{-\beta^2} + d_k \log_{10} f + e_k, \tag{7.13}$$

$$\alpha = \sum_{i=1}^{5} a_i \, e^{-\beta^2} + d_\alpha \log_{10} f + e_\alpha, \tag{7.14}$$

where

$$\beta = \frac{\log_{10} f - b_i}{c_i}, \tag{7.15}$$

and f is the frequency in GHz, k is either k_H or k_V, and α is either α_H or α_V. The values of the coefficients, a, b, c, d, and e are given in [28], and are listed here in Tables 7.5–7.8 for reference. Figure 7.15 shows the attenuation due to varying rain rates in the frequency range from 1 to 100 GHz, circular polarization, and an elevation angle of 10°. The relative SNR as a function of rain is then calculated by frequency band, as shown in Figure 7.16. As illustrated, higher frequencies are more well suited for short range systems, while lower frequencies tend to perform better at far ranges.

There are several other loss mechanisms for propagation of electromagnetic energy that are beyond the scope of this book and the reader is referred to the following sources for a detailed coverage [14, 29].

7.3.3 Signal Processing Loss

Signal processing losses include imperfect filtering, pulse integration, and windowing. Ideal range-processing gain, due to the pulse compression process, is not

Table 7.5

Coefficients for Calculating k_H

a_i	b_i	c_i	d_k	e_k
−5.33980	−0.10008	1.13098	−0.18961	0.71147
−0.35351	1.26970	0.45400		
−0.23789	0.86036	0.15354		
−0.94158	0.64552	0.16817		

Table 7.6
Coefficients for Calculating k_V

a_i	b_i	c_i	d_k	e_k
−3.80595	0.56934	0.81061	−0.16398	0.63297
−3.44965	−0.22911	0.51059		
−0.39902	0.73042	0.11899		
0.50167	1.07319	0.27195		

Table 7.7
Coefficients for Calculating α_H

a_i	b_i	c_i	d_α	e_α
−0.14318	1.82442	−0.55187	0.67849	−1.95537
0.29591	0.77564	0.19822		
0.32177	0.63773	0.13164		
−5.37610	−0.96230	1.47828		
16.1721	−3.29980	3.43990		

Table 7.8
Coefficients for Calculating α_V

a_i	b_i	c_i	d_α	e_α
−0.07771	2.33840	−0.76284	−0.053739	0.83433
0.56727	0.95545	0.54039		
−0.20238	1.14520	0.26809		
−48.2991	0.791669	0.116226		
48.5833	0.791459	0.116479		

achievable and a loss term must be included, as given in (2.44). Regardless of the pulse-compression technique being employed (i.e., matched filtering or stretch processing), timing errors, phase instability, and nonideal filtering results in a lower than ideal range-processing gain. Often, a windowing function is used to reduce sidelobes in the range and azimuth dimensions, which result in a broadening of the main lobe and lowering of the peak value. Table 7.9 gives typical processing loss

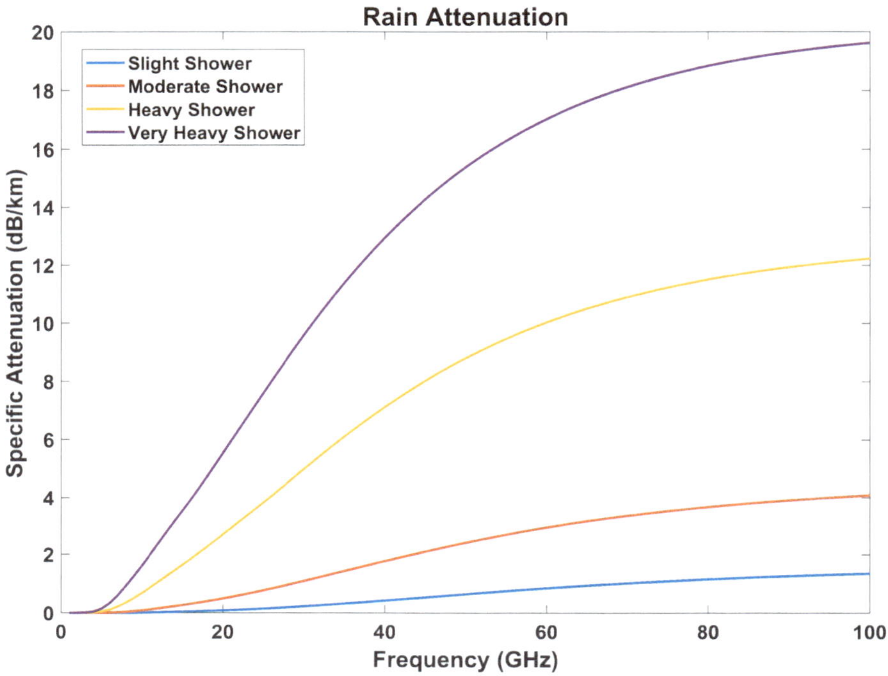

Figure 7.15 Rain attenuation as a function of frequency for circular polarization.

for well-known windowing functions [30]. For point-like scatterers in an imaging scene, there could also be *straddling* loss. This loss occurs when the target is not centered in a range bin. If the radar collects samples at a rate of once per range-resolution cell, the loss may be as large as 3 dB, as shown in Figure 7.17. On the other hand, distributed target areas are not affected by straddling loss.

7.3.4 Antenna Scan Loss

Referring to Section 2.2.5, squint angle operation involves pointing the antenna beam either ahead of or behind the platform. This mode of operation was illustrated in Figure 2.8. Steering the antenna beam in this manner results in a loss in the effective aperture and thus SNR. Consider a uniformly spaced N-element linear

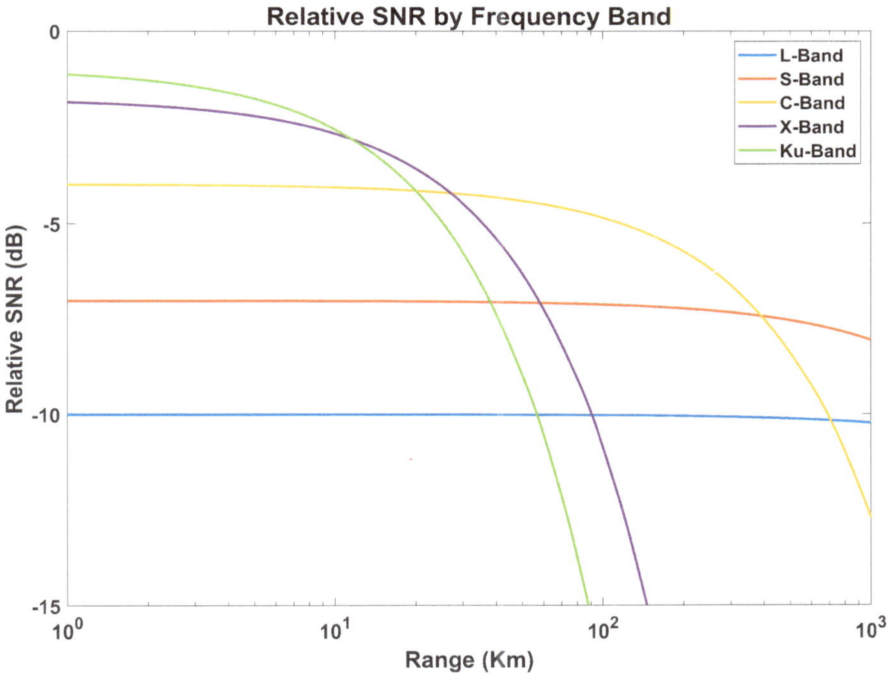

Figure 7.16 Relative SNR by frequency band for circular polarization (rain rate = 5 mm/hr).

Table 7.9

Typical Processing Loss for Windowing Functions

Window	*Processing Loss (dB)*	Peak Sidelobe Level (dB)
Rectangular	0	−13
Bartlett-Hann	1.6	−36
Blackman	2.4	−58
Cosine	0.9	−23
Hamming	1.4	−42
Hanning	1.8	−31
Triangular	1.3	−26

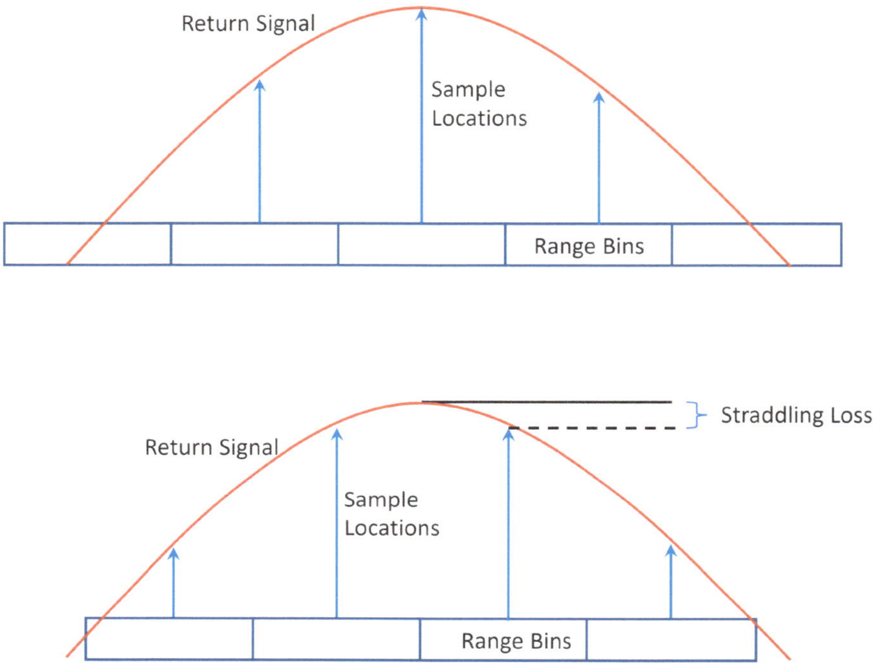

Figure 7.17 Range gate straddling loss for point-like targets.

array shown in Figure 7.18. Steering the beam off boresight (90°) results in the antenna patterns given in Figure 7.19. Note the loss in the peak of the main beam as well as the broadening of the beam. Figure 7.20 gives the value of the main beam loss as a function of steering angle.

7.4 UNAMBIGUOUS RANGE

So far, the assumption has been made that the target return reaches the radar before the next pulse is transmitted. If this is the case, then the range to the target is unambiguous. This is reasonable for land-based SAR systems or in the case

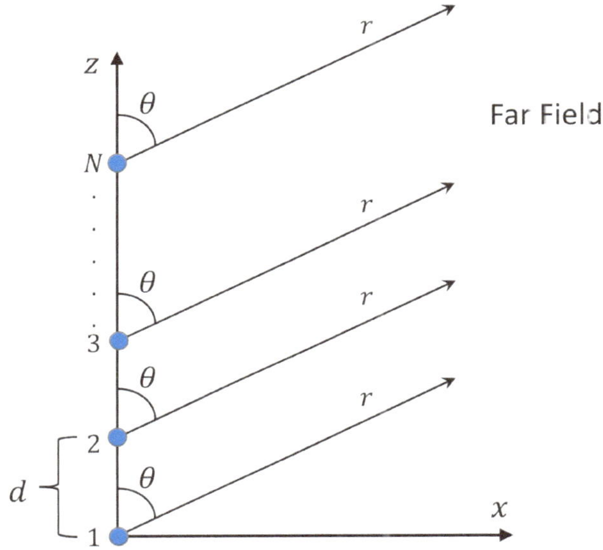

Figure 7.18 N-element linear array antenna.

of short-range, low-speed airborne platforms such as drones. However, for high-velocity, long-range spaceborne platforms, this mode of operation is not feasible. For example, a satellite at an altitude of 300 km requires a velocity of 7.8 km/s to stay in orbit. From (2.22), the required PRF for a SAR system with an azimuth antenna dimension of 1.5 m would be approximately 10 kHz. Using (2.21), this would result in an unambiguous range of approximately 10 km, far short of the range from spaceborne platforms to a target scene. To overcome this limitation, several pulses must be transmitted before receiving the return from the first pulse. This mode of operation is common in spaceborne SAR systems and in long-range ground-based radar systems.

Consider a train of transmitted pulses and target returns as shown in Figure 7.21. For a constant PRF, the return from the target scene falls between two transmitted pulses when

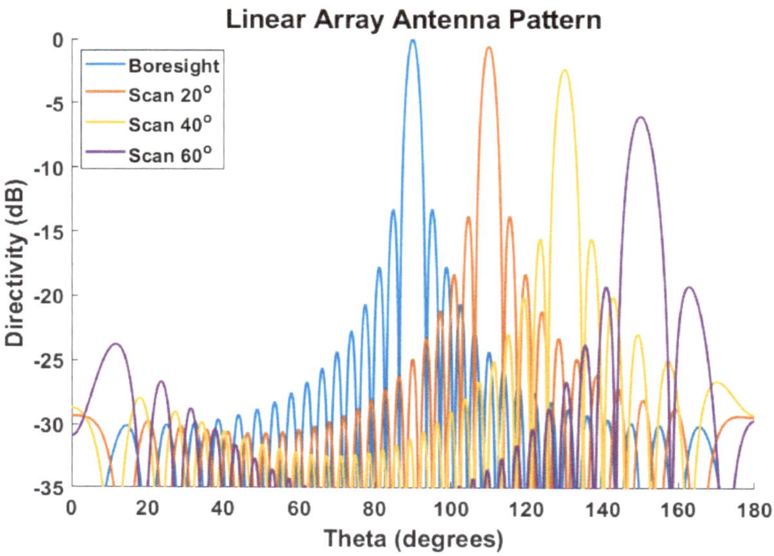

Figure 7.19 Antenna pattern by scan angle for an N-element linear array antenna.

$$\left(N + \frac{1}{2}\right)T = t_s = \frac{2R}{c}, \tag{7.16}$$

where

t_s = time delay to target scene (s),
T = PRI (s),
R = range to target scene (m),
N = number of pulses transmitted prior to the return from the target scene.

This places a constraint on the choice of PRFs for the specific mission. Another concern is the return from nadir, which is typically very strong compared to other returns. Also, the choice of PRF must ensure the nadir return from later pulses does not interfere with returns from the target scene. This is accomplished by choosing

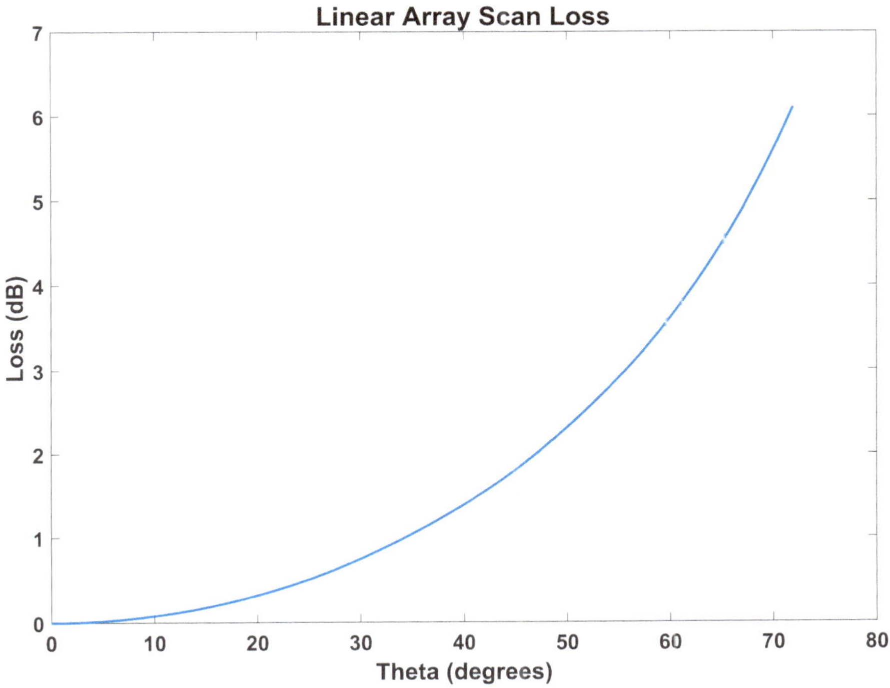

Figure 7.20 Scan loss for an N-element linear array antenna.

PRFs that cause the nadir return time to coincide with the transmitting time of the radar. This is accomplished by requiring

$$MT = \frac{2h}{c} \qquad (s), \tag{7.17}$$

where

M = number of pulses transmitted prior the the return from nadir,
h = platform altitude (m),
T = PRI (s).

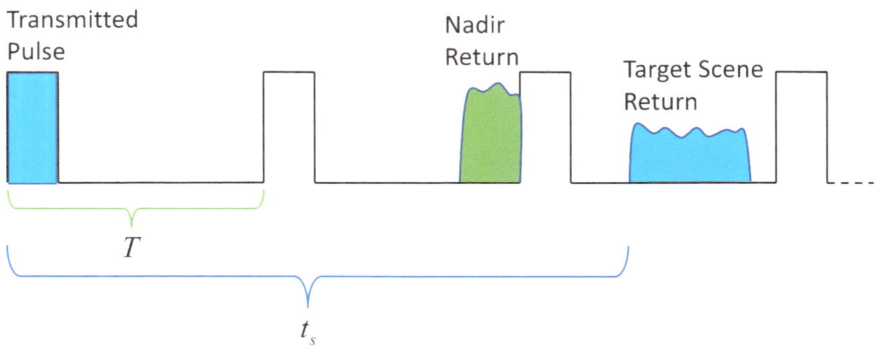

Figure 7.21 Transmitted radar pulses and return pulses from the nadir and the target scene.

Using (7.16) and (7.17) gives the requirement for nadir to be eclipsed by the radar transmitting time and the target scene return to fall between transmitted pulses. This requirement is expressed as

$$\frac{N + 1/2}{M} = \frac{R}{h}. \tag{7.18}$$

For spaceborne systems, the imaging geometry and missions are specified and the system designer must then select PRFs to be radiated by the radar, with the constraints given above.

7.5 DATA HANDLING

SAR systems must handle the acquisition, storage, processing, and downlink of data generated by the sensor. For example, COSMO SkyMed Second Generation (CSG) requires acquisition data rates up to 2 × 1.2 Gbit/s [6]. Processed sensor data and associated auxiliary data must be stored, formatted, encoded, and often encrypted before being sent to the downlink element. The downlink instrumentation must modulate, amplify, and filter the baseband data stream to be transmitted to the ground station. The downlink element must be accounted for in the overall system size weight and power constraints.

Table 7.10
Storage Capacities and Downlink Rates

System	*Storage* (Gbits)	*Downlink* (Mbits/s)
COSMO-SkyMed *2nd* Generation	1530	560
TerraSAR-X	256	300
ICESat-2	704	220
RADARSAT-2	150	210
Sentinel-5	600	781

Ever increasing data sizes along with limitations in link capacity have led to more onboard processing and storage. This requires the use of data compression techniques. Available digital signal processors are fast enough to handle data in the range of gigabytes per second. The onboard memory capacity must be sized to ensure the storage of very large SAR imagery datasets.

Spotlight mode operation does not typically have a limited acquisition time, as the data collection is determined by the synthetic aperture length needed to produce the required azimuth resolution, and typically results in datasets of 0.5–1.0 GB. In contrast, stripmap and scan modes of operation have limited collection times based on the onboard memory, which therefore limits the maximum scene length. Stripmap and scan data sizes may be in the tens of GB [5, 6, 31–33]. Table 7.10 gives some storage capacities and downlink rates for various spaceborne SAR systems. The downlinks are typically X-band, however some of the downlink make use of S- and Ka-band [34].

With advances in the development of graphical processing units (GPUs) and parallel image formation algorithms, real-time onboard processing is now feasible [35]. Table 7.11 provides some examples of GPU processing of SAR data [36–39]. More capable onboard signal and data processing and compression techniques show promise in alleviating some of the limitations in dealing with very large SAR datasets. Chapter 8 discusses some future directions for SAR data downlink systems to further extend the capabilities of airborne and spaceborne systems.

Table 7.11
GPU Performance for SAR Imaging

Source	*Number of Samples*	*GPU*	*Execution Time (s)*
ERS2	26,880 × 912	Tesla C1060	4.4
COSMO-SkyMed	16,384 × 8,192	Tesla C1060	6.7
Sentinel-1	22,018 × 18,903	Tesla K40	6.5
ENVISAT ASAR IM	30,000 × 6,000	Tesla K20	8.5
Sentinel-1 IW	52,500 × 20,000	Tesla K20	65

7.6 EXAMPLES

The sections that follow illustrate the concepts of SAR performance limits with a few Python/MATLAB examples. The Python examples for this chapter are in the directory *software/python/Chapter7* and the matching MATLAB examples are in the directory *software/matlab/Chapter7*. The reader should consult Chapter 1 for information on how to execute the Python and MATLAB code associated with this book.

7.6.1 Spatial Resolution

As a first example, system parameters reported in the ICEYE product guide [5] are used along with (7.1) to calculate the resolution in both the slant plane and ground plane. SAR systems achieve the finest slant-plane resolution with the highest bandwidth waveform. From [5], 300 MHz is the widest bandwidth used. This results in a slant-plane resolution of

$$\delta_r = \frac{c}{2B} \approx \frac{3\times 10^8}{2 \times 300\times 10^6} \approx 0.5 \qquad \text{(m)}. \tag{7.19}$$

The incident angle is mode dependent and may vary from 55°–85° For this example, assume the 300-MHz bandwidth is employed over the possible incident angle coverage. The expression in (7.1) is used to calculate the ground-plane resolution and the results are shown in Figure 7.22.

For spotlight mode operation, the ICEYE product guide gives an azimuth resolution of 0.25 m. For an altitude of 570 km and an incident angle of 55° the slant range is found by

$$R = \frac{570\times10^3}{\cos(90^\circ - 55^\circ)} \approx 695\times10^3 \qquad \text{(m)}. \tag{7.20}$$

Solving (7.3) for the angle subtended by the synthetic aperture gives

$$\theta_{sa} = \frac{\lambda}{2\delta_a} \approx \frac{0.03}{0.5} \approx 0.06 \qquad \text{(rad)}. \tag{7.21}$$

This leads to a synthetic aperture length of

$$L_{sa} = 2 \times 695 \times 10^3 \times \tan(\theta_{sa}/2) \approx 42\times10^3 \qquad \text{(m)}. \tag{7.22}$$

With a velocity of 7.8 km/s, this would require approximately 6 seconds of data collect. This example is given in the Python notebook *spatial_resolution_example.ipynb* and the MATLAB live script *spatial_resolution_example.mlx.*

7.6.2 Windowing Functions

Windowing functions are often used with the range and cross-range data in order to suppress sidelobes and improve overall image quality. The trade-off for lower sidelobes is a widening of the main beam response of the point spread function. This broadening of the main beam degrades the spatial resolution according to the values given in Table 7.9. Figure 7.23 illustrates a few commonly used windowing functions for SAR range and cross-range data, and Figure 7.24 gives the frequency responses of the windowing functions. Note the greater the sidelobe suppression, the greater the broadening of the main beam and lowering of the peak value.

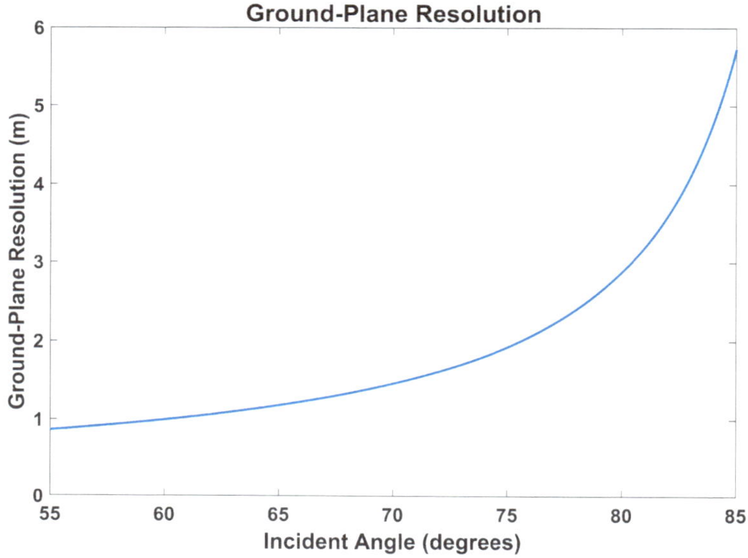

Figure 7.22 Calculated ground-plane resolution vs incident angle using typical spaceborne system parameters.

This example is given in the Python notebook *windowing_example.ipynb* and the MATLAB live script *windowing_example.mlx.*

7.6.3 NESZ

In this example, the effect of normalized effective aperture size and range to the imaging scene are studied. The SAR system has a noise figure of 3 dB along with losses of 3 dB. The platform is in low Earth orbit with a velocity of 7.8 km/s. The operating frequency is 9.65 GHz, the nominal bandwidth is 300 MHz, and the average transmitted power is 320 W. Figure 7.25 shows the change in NESZ as the normalized effective aperture, A_e/λ^2, is varied across nominal values from 1–500. Figure 7.26 is given to illustrate the relationship between the normalized effective aperture and scene range required to maintain a constant NESZ level. This example

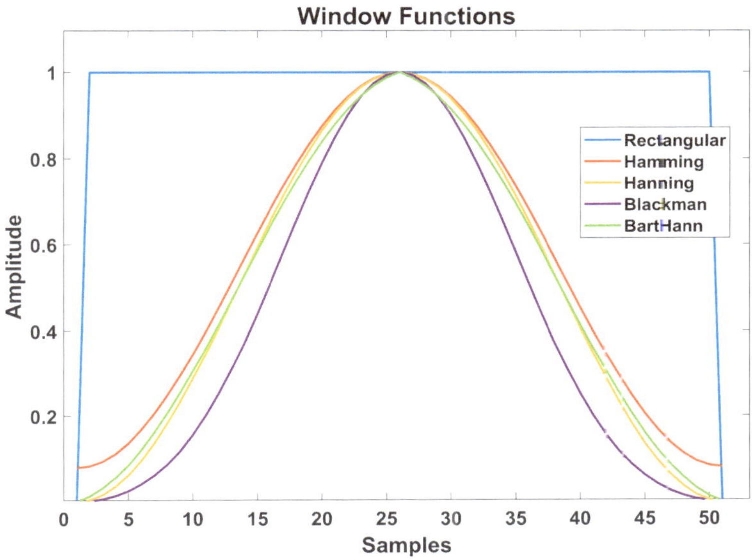

Figure 7.23 Commonly used windowing functions for SAR range and cross-range data.

is given in the Python notebook *nesz_example.ipynb* and the MATLAB live script *nesz_example.mlx*.

7.6.4 Rain Attenuation

For this example, the attenuation due to rain is examined. Table 7.12 gives the United States Geological Survey (USGS) categories for rain rates [40]. There are additional definitions concerning drizzle, but the focus here is on the specific attenuation due to varying rain rates. Figure 7.27 shows the specific attenuation for slight, moderate, heavy, and very heavy rain rates as defined in Table 7.12. The lower frequencies are less affected by rain as the wavelengths become much larger than the size of the rain drops [28]. This example is given in the Python notebook *rain_attenuation_example.ipynb* and the MATLAB live script *rain_attenuation_example.mlx*.

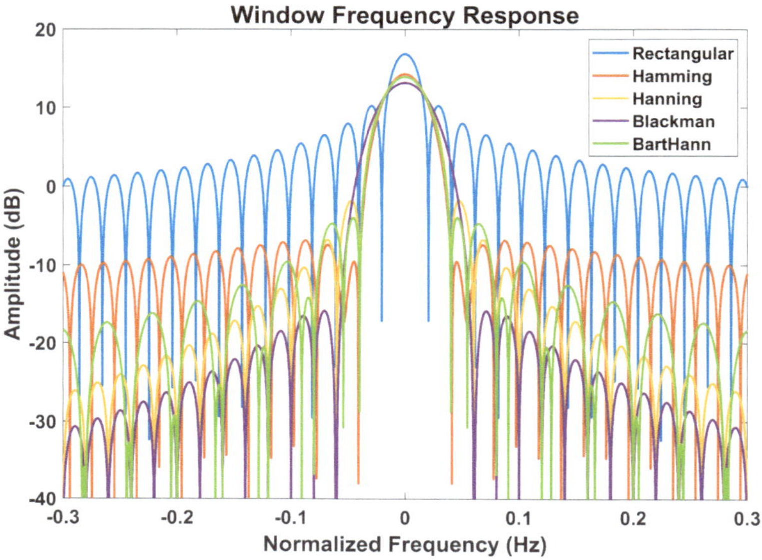

Figure 7.24 Frequency response of commonly used windowing functions.

7.6.5 Unambiguous Range

In terms of the PRI, the number of pulses to extend the unambiguous range to the target scene is given in (7.16) and the number of pulses to eclipse the nadir return is given in (7.17). For a platform operating at an altitude of 500 km, a PRF of 1 kHz, and a scene range of 750 km, the number of pulses to extend the unambiguous range to the target is calculated as

$$N = \frac{2R}{c} PRF - 1/2 = \frac{2 \times (750{\times}10^3) \times (1{\times}10^3)}{3{\times}10^8} = 4, \tag{7.23}$$

and the number of pulses to eclipse the nadir return is

$$M = \frac{(N + 1/2)h}{R} = \frac{(5 + 1/2) \times 500{\times}10^3}{750{\times}10^3} = 5. \tag{7.24}$$

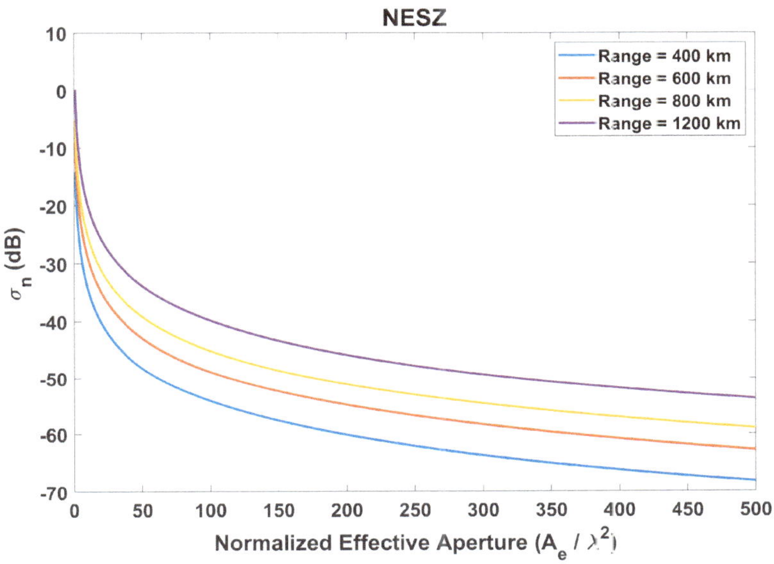

Figure 7.25 NESZ as a function of normalized effective aperture for various scene ranges.

This example is given in the Python notebook *extended_range_example.ipynb* and the MATLAB live script *extended_range_example.mlx*.

7.6.6 RGIQE

The main idea behind RGIQE is to use maximum channel capacity as a figure of merit for SAR imagery. The maximum information that can be carried on a channel is given by the Shannon-Hartley Theorem and is written as [41]

$$C = B\,\log_2\left(1 + SNR\right) \qquad \text{(bits/m}^2\text{)}, \tag{7.25}$$

where B is the channel bandwidth. Adapting this for SAR systems allows the information content in a resolution cell to be expressed as

$$I = B_a\,B_r\,\log_2\left(1 + SNR\right) \qquad \text{(bits/m}^2\text{)}, \tag{7.26}$$

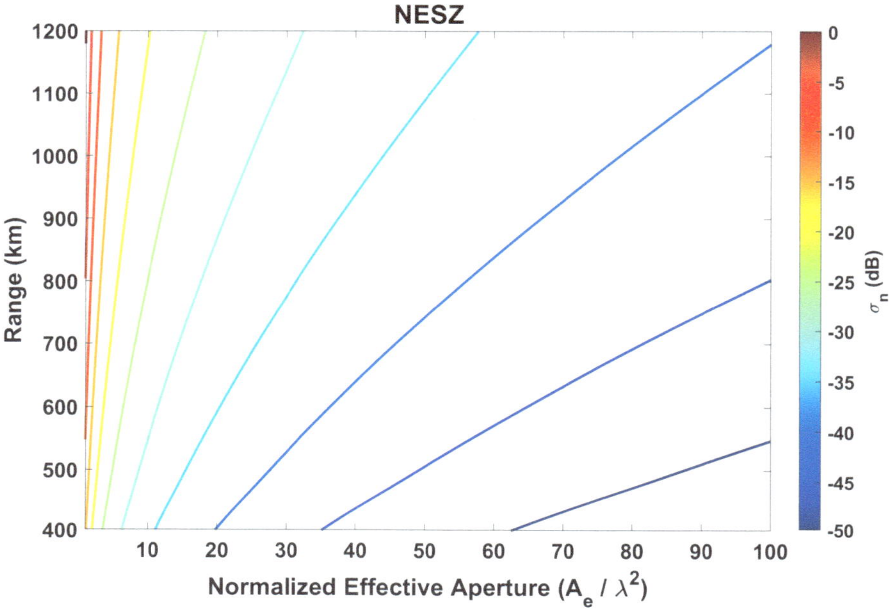

Figure 7.26 NESZ constant contours as a function of range and effective aperture.

where B_a and B_r are the Doppler and ground range bandwidths of a resolution cell, respectively.

As a comparison between imaging modes, RGIQE is calculated using typical spaceborne SAR system values for stripmap, spotlight, and scan SAR modes. The results are summarized in Table 7.13. Note the significant difference between the modes, largely due to the spatial resolution. Even though spotlight mode has the lowest SNR and therefore the poorest image contrast, the significantly better spatial resolution results in a higher performance metric. This example is given in the Python notebook *rgiqe_example.ipynb* and the MATLAB live script *rgiqe_example.mlx*.

Table 7.12
USGS Rain Rate Categories

Category		*Rain Rate Description*
Slight rain	=	Less than 0.5 (mm/hr),
Moderate rain	=	Greater than 0.5 (mm/hr), but less than 4.0 (mm/hr),
Heavy rain	=	Greater than 4 (mm/hr), but less than 8 (mm/hr),
Very heavy rain	=	Greater than 8 (mm/hr),
Slight shower	=	Less than 2 (mm/hr),
Moderate shower	=	Greater than 2 (mm/hr), but less than 10 (mm/hr),
Heavy shower	=	Greater than 10 (mm/hr), but less than 50 (mm/hr),
Violent shower	=	Greater than 50 (mm/hr).

Table 7.13
Comparison of RGIQE for Different Imaging Modes

Mode	δ_a (m)	δ_r (m)	$\psi_g(°)$	*NESZ* (dB)	*RGIQE* (bits/m^2)
Stripmap	3	0.5	60	−20	2.2
Spotlight	0.25	0.5	55	−15	23
Scan	5	5	70	−22	0.1

7.6.7 RNIIRS

RNIIRS provides a definition of the quality of SAR imagery and to some degree a measure of system performance, and has been widely used in the intelligence community [42, 43]. RNIIRS also captures the effects of image processing algorithms used in the creation of the imagery. RNIIRS provides a standard measure, in a ten-step rating scale (0–9), of the interpretability of SAR imagery for its intended use.

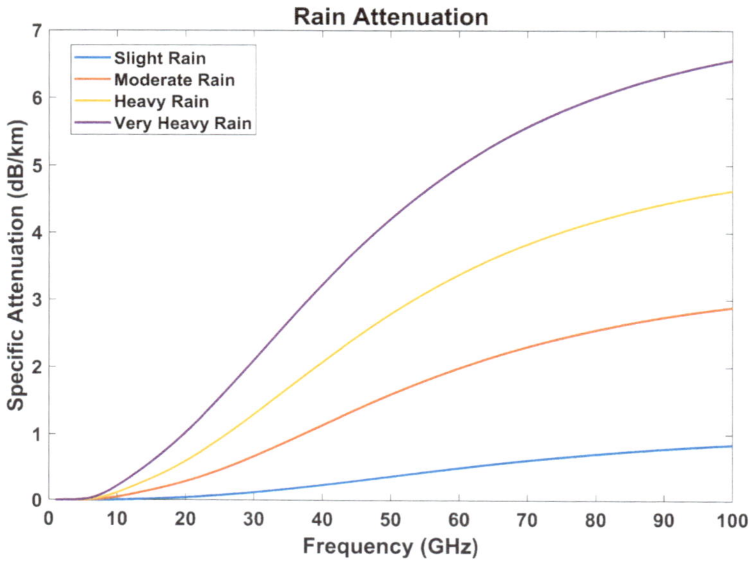

Figure 7.27 Specific attenuation as a function of frequency for different categories of rainfall.

Each level is based on the assessment of subject matter experts and experienced image analysts [42, 43]. Table 7.14 gives some example image exploitability of each rating.

Table 7.14
Example Interpretability of Each RNIIRS Step

Level	*Ground Plane Resolution (m)*	*Interpretability*
0	–	Imagery is not useful due to obstruction, degradation, or very poor resolution.
1	>9.0	Distinguish between various types of very large land areas including urban, agricultural, forest, water, and barren.
2	4.5–9.0	Determine road patterns and detect large buildings such as factories and hospitals.
3	2.5–4.5	Identify individual houses, tornado paths, and harvesting operations.
4	1.2–2.5	Detect large farm equipment and identify commercial greenhouses.
5	0.75–1.2	Detect missile support equipment such as transporter erector launchers (TEL).
6	0.4–0.75	Discriminate small support vehicles and tanks.
7	0.2–0.4	Identify individual railroad ties and steel fence posts.
8	0.1–0.2	Determine automobile features such as grill patterns.
9	<0.1	Identify modifications to aircraft such as fairings and winglets.

PROBLEMS

7.1 Describe the differences in the cross-range resolution between spotlight and stripmap mode SAR.

7.2 What are the major limitations in cross-range resolution associated with both spotlight and stripmap mode SAR?

7.3 Describe the mechanism by which a smaller physical antenna results in finer cross-range resolution for stripmap mode SAR.

7.4 What is the finest ground-range resolution of a spaceborne SAR system operating at an altitude of 500 km, a grazing angle of 55°, and operating at the maximum ITU frequency spectrum allocation for spaceborne SAR systems?

7.5 What are some of the visible feature differences for images at differing resolutions, such as those in Figures 7.1–7.4?

7.6 Describe the visible features that degrade the most in SAR imagery as the NESZ is increased.

7.7 Give specific examples of how the items in Table 7.1 are interrelated.

7.8 What are some of the more commonly used antenna designs in modern SAR systems?

7.9 What are the major advantages and disadvantages of using phased array antenna configurations vs reflector type antennas?

7.10 What are some of the major differences between microwave tube and solid-state amplifiers?

7.11 Give examples of imaging systems that would benefit from using microwave tube amplifiers and systems where solid-state amplifiers would be advantageous.

7.12 What is the major source of noise in a radar receiver?

7.13 Characterize the noise in a radar system with a reflector antenna and a system with an active phased-array antenna.

7.14 Describe the major difference between direct digital downconversion and downconversion via the Hilbert transform.

7.15 What is the general trend of propagation loss as a function of frequency? Include both atmospheric losses and rain losses.

7.16 Why is straddling loss less of an issue for distributed targets?

7.17 For spaceborne SAR, what is the major issue when dealing with the nadir return?

7.18 Under what circumstances would it be advantageous to not eclipse the nadir return with the transmitting time of the system?

7.19 What are some of the major limiting factors for stripmap and scan mode image lengths and what techniques are available to try to reduce these limitations?

7.20 Describe a situation in which a sensor's RGIQE would be significantly inflated.

7.21 Characterize the use of SAR imagery where RNIIRS is a more useful measure of sensor performance than RGIQE.

References

[1] Umbra Lab. Space-based Synthetic Aperture Radar (SAR) Microsatellites - Umbra. http://umbra.space/. Accessed: 2022-03-19.

[2] ITU-R. Spectrum requirements for spaceborne synthetic aperture radar applications planned in an extended allocation to the Earth exploration-satellite service around 9600 MHz. Recommendation RS.2274, International Telecommunication Union, Geneva, 2013.

[3] Federal Communications Commision. FCC reports. https://www.fcc.gov/reports-research/reports. Accessed: 2022-4-5.

[4] A. Moreira, P. Prats-Iraola, M. Younis, G. Krieger, I. Hajnsek, and K. Papathanassiou. "A tutorial on synthetic aperture radar." *IEEE Geoscience and Remote Sensing Magazine*, 1(1):6–43, 2013.

[5] ICEYE. SAR Product Guide. Technical Report Version 4.2, ICEYE, Espoo, 2021.

[6] The European Space Agency. COSMO-SKymed. https://earth.esa.int/eogateway/missions/cosmo-skymed. Accessed: 2022-4-5.

[7] W. Melvin and J. Scheer. *Principles of Modern Radar: Advanced Techniques*. Scitech Publishing, Edison, 2013.

[8] IMSAR. OneSAR data and specifications sheet. Technical report, IMSAR, Springville, UT, 2020.

[9] Capella Space. SAR Made Easy. https://www.capellaspace.com/. Accessed: 2022-4-5.

[10] ICEYE. Example SAR data from ICEYE. https://www.iceye.com/downloads/datasets. Accessed: 2021-09-21.

[11] S. Talisa, K. O'Haver, T. Comberiate, M. Sharp, and O. Somerlock. Benefits of digital phased array radars. *Proceedings of the IEEE*, 104(3):530–543, 2016.

[12] M. Ludwig, S. D'addio, M. Suess, G. Adamiuk, and A. Suriani. Phased array technology developments for next generation European spaceborne SARs with digital beamforming. In *IEEE International Symposium on Phased Array Systems and Technology*. Waltham, 2019.

[13] The European Space Agency. ASAR Overview. https://earth.esa.int/eogateway/instruments/asar/description. Accessed: 03-2022.

[14] A. Harrison. *Introduction to Radar Using Python and MATLAB*. Artech House, Norwood, 2020.

[15] Y. Aoki, M. Furuya, F. De Zan, M. Doin, M. Eineder, M. Ohki, and T. Wright. L-band synthetic aperture radar: Current and future applications to earth sciences. *Earth, Planets and Space*, 73(56), 2021.

[16] B. Mahafaza. *Radar Systems Analysis and Design Using MATLAB,* 3rd ed. Chapman and Hall/CRC, New York, 2015.

[17] G. Stimson. *Introduction to airborne radar*. Hughes Aircraft Company, El Segundo, 1983.

[18] N. Kolias and M. Borkowski. The development of T/R modules for radar applications. In *2012 IEEE/MTT-S International Microwave Symposium Digest*, pages 1–3, 2012.

[19] I. Kassotakis. Moder radar techniques for air surveillance and defense. *Journal of Computations and Modelling*, 4(1):189–205, 2014.

[20] Y. Dora, A. Chakraborty, L. Mccarthy, S. Keller, S. Denbaars, and U. Mishra. High breakdown voltage achieved on AlGaN/GaN HEMTs with integrated slant field plates. *IEEE Electron Device Letters*, 27(9):713–715, 2006.

[21] D. Pozar. *Microwave Engineering, 4th Edition*. John Wiley and Sons, 2012.

[22] M. Skolnik. *Radar Handbook,* 3rd ed. McGraw-Hill, New York, 2008.

[23] S. Hahn. *Hilbert Transforms in Signal Processing*. Artech House, Norwood, MA, 1996.

[24] R. Lyons. *Understanding Digital Signal Processing,* 2nd ed. Prentice Hall, Upper Saddle River, NJ, 2004.

[25] R. Bracewell. *The Fourier Transform and Its Applications,* 3rd ed. McGraw-Hill, New York, 1999.

[26] H. Urkowitz. *Signal Theory and Random Processes*. Artech House, Dedham, MA, 1983.

[27] Antennas for Communications. Radome Transmission Loss and Antenna Pattern Degradation. http://www.radome.net/tl.html, 2020. Accessed: 2021-12-13.

[28] ITU-R. Specific attenuation model for rain for use in prediction methods. Recommendation P.838-3, International Telecommunication Union, Geneva, 2005.

[29] C. Balanis. *Advanced Engineering Electromagnetics,* 2nd ed. John Wiley and Sons, New York, 2012.

[30] M. Viswanathan. Window function figure of merits. https://www.gaussianwaves.com/2020/09/window-function-figure-of-merits/, September 2020. Accessed: 2022-4-1.

[31] Airbus Defence and Space. *TerraSAR-X Image Product Guide*. Airbus, Ottobrunn, 2015.

[32] The European Space Agency. Radarsat-2 - eoPortal Directory - Satellite Missions. https://earth.esa.int/web/eoportal/satellite-missions/r/radarsat-2. Accessed: 2021-01-18.

[33] The European Space Agency. Sentinel-1 SAR user guide. https://sentinel.esa.int/web/sentinel/technical-guides/sentinel-1-sar. Accessed: 2020-10-03.

[34] The European Space Agency. Sharing Earth observation resources. https://directory.eoportal.org/web/eoportal/satellite-missions/. Accessed: 2022-4-10.

[35] V. Krzhizhanovskaya, G. Závodszky, M. Lees, J. Dongarra, P. Sloot, S. Brissos, and J. Teixeira. The challenge of onboard SAR processing: A GPU opportunity. In *Computational Science - 20th International Conference*, pages 46–59, Amsterdam, 2020.

[36] C. Passerone, C. Sansoè, and R. Maggiora. High performance SAR focusing algorithm and implementation. In *2014 IEEE Aerospace Conference*, pages 1–10, 2014.

[37] A. Tiriticco, B. Fratarcangeli, C. Ferrara, and D. Marra. Near-real-time multi-GPU ωk algorithm for SAR processing. *Agency-Esrin, ES (ed.) Big Data from Space (BiDS)*, pages 277–280, 2014.

[38] A. Peternier, J. Peter, M. Boncori, and P. Pasquali. Near-real-time focusing of ENVISAT ASAR Stripmap and Sentinel-1 TOPS imagery exploiting OpenCL GPGPU technology. *Remote Sensing of Environment*, 202:45–53, 2017. Big Remotely Sensed Data: tools, applications and experiences.

[39] M. Di Bisceglie, M. Di Santo, C. Galdi, R. Lanari, and N. Ranaldo. Synthetic aperture radar processing with GPGPU. *IEEE Signal Processing Magazine*, 27(2):69–78, 2010.

[40] The United States Geological Survey. Rainfall calculator, metric units how much water falls during a storm? https://water.usgs.gov/edu/activity-howmuchrain-metric.html. Accessed: 2022-4-9.

[41] C. Shannon. Communication in the presence of noise. *Proceedings of the IRE*, 37(1):10–21, 1949.

[42] FAS Intelligence Resource Program. National image interpretability rating scales. https://irp.fas.org/imint/niirs.htm, 2022. Accessed: 2022-5-20.

[43] J. Irvine and E. Nelson. Image quality and performance modeling for automated target detection. *Proceedings of SPIE - The International Society for Optical Engineering*, 7335, 05 2009.

Chapter 8

Future Directions

Since the origination of SAR concepts in the 1950s, SAR systems, technology, and data have been almost solely under the purview of different government organizations around the world [1–4]. Even though the capability, utility, and versatility of SAR techniques had been shown repeatedly, these systems were only feasible for establishments with extremely large resources of time, money, and manpower, with access to these systems and data tightly controlled. In addition, SAR image processing and analysis has traditionally required vast experience in remote sensing coupled with the use of complicated tools. However, there has recently been a huge surge of interest in commercial SAR applications and systems. As SAR data becomes more affordable and accessible, advancements in data analysis, image processing, and ease of use will follow. This will invariably lead to new and different use cases and will make for exciting times in the world of SAR. This is in part evidenced by the very generous access to data used in this book by ICEYE and Umbra Lab [5, 6]. Academia is often a good indicator for future trends and technologies, and an IEEE search revealed 2,591 conference articles, 2,043 journal articles, 41 magazine articles, and 18 book chapters since 2020. SAR is now, and will continue to be, a growing field with strong interest from commercial, governmental, and academic sectors.

8.1 APPLICATIONS

The utilization of SAR data and imagery in the commercial arena is still quite small compared to the span of applications and high demand for optical data. However, with very high-quality SAR imagery more accessible than ever, many new applications and missions are becoming increasingly important [7, 8]. These new applications span the gambit of spaceborne, airborne, and ground-based systems. Image analysts and professionals in remote sensing are, for the most part, trained in optical imagery and require training to analyze SAR data and imagery and to use the associated processing and display tools. It is expected that advancements in research and development based around SAR data and analysis will expand with the ever increasing access to data and imagery [9]. The following sections highlight some of the latest SAR applications with insight into future missions.

8.1.1 Spaceborne SAR

Recently, there has been rapid growth in the commercial spaceborne SAR industry, with a clear trend towards satellite constellations intended to provide more reliable and more frequent imagery [10–13]. The cost associated with the development and deployment of large satellite systems is typically too prohibitive for startup companies. However, technological advances are making smaller and lighter SAR systems achievable; see Section 7.2.1.

From 2018 to date, ICEYE deployed 16 satellites encompassing both commercial and dedicated customer missions including persistent flood monitoring and oil storage reporting [10]. As part of the Whitney constellation, which serves both the U.S. government and commercial customers, Capella Space has deployed five satellites since 2020 with two more planned in the near future [11]. In 2021, Umbra Lab deployed its first system, capable of 25-cm resolution over a 16 km^2 region. In 2022, Umbra Lab was selected by the U.S. National Reconnaissance Office (NRO) for strategic commercial enhancement efforts [12]. As an example of government organizations also moving towards constellations, the RADARSAT Constellation Mission consists of three satellites and was deployed in 2019. The constellation has daily access to approximately 90% of the world's surface and provides daily images of Canada's vast territory [13].

The exponential growth in the commercial spaceborne SAR industry is in part due to its increasingly important role in international emergency events. In 2019, SAR imagery provided details of the drone attack against oil installations in Saudi Arabia [7]. That same year, SAR imagery identified the extent of flooding in the Bahamas after Hurricane Dorian struck the islands [14]. In 2021, very high-resolution SAR imagery captured the Ever Given container ship blocking the Suez Canal. The ship's containers and support vessels were clearly visible in the imagery [11]. The MV Wakashio, a bulk carrier ship, went off course and grounded on a coral reef off the coast of Mauritius in 2020. Approximately 1,000 tons of oil spilled into the Indian Ocean. Rescue operations, firefighting vessels, and the spread of the leak were all recorded by SAR systems [15]. SAR change detection imagery monitored the devastation and reconstruction in the Port of Beirut after a large explosion destroyed buildings and warehouses [16]. Traditionally, ground deformation measurements have only been taken in response to a landslide or other disaster. With new SAR technology, a deformation study was conducted in Guatemala City from 2018 until 2021. This study allowed for the localization of subsidence phenomena that can identify landslide risks [17].

One area with a very high growth potential for SAR imagery and missions is nighttime and degraded weather imaging. Most of the current optical imaging satellites have sun synchronous orbits and image during peak sunlight hours. This leaves areas of interest without coverage for many hours and even worse if weather conditions are not suitable. With the paradigm shift to SAR satellite constellations rather than large single platforms, revisit times are decreasing. This fact, along with nighttime and all-weather capability leads to greatly improved monitoring, research, and data collection, which is projected to grow and expand into other areas over the next five to ten years [18].

Another growth area for SAR is in the needs of the insurance sector to accurately size losses after floods, hurricanes, and other catastrophes. With SAR constellations providing timely observation data and accurate loss estimates, insurance companies and government organizations can shorten disaster response timelines to better serve customers and more effectively provide aid to the hardest hit areas [19]. With current trends in climate change, losses due to natural disasters are predicted to be even more prevalent over the next 50 years [20].

With population growth and climate change, effective use and management of ecosystems is critical [20]. Agriculture, wetlands, forestry, and coastal changes will need to be carefully monitored and tracked in the decades to come [13, 15, 21].

There are major crises centered on illegal activities in the oceans, seas, bays, and coastal areas. Illegal and unreported fishing is worth several billions of dollars per year [21]. In addition, there are other issues, such as illegal ship transfers, drug trafficking, piracy, and migration [19]. While passenger ships and voyaging ships are required to have Automatic Identification System (AIS) transceivers for tracking purposes, the vessel operator can choose to turn off the AIS device. Along with other issues, the current tracking and monitoring systems are lacking in capability to respond to these crises. Spaceborne SAR systems can be employed to detect and track maritime vessels. This SAR data may then be correlated with AIS tracks and identify so called *dark vessels* to inform authorities and lead to informed actions [5, 13, 22]

With the availability of high-quality SAR imagery, there will be numerous ad hoc applications within the public and private sectors. For example, SAR has been used to help determine the impact of COVID-19 on the global oil market by estimating the amount of world-wide oil stores [5, 9]. Another specialized application centers on the automobile industry. With the shortage of semiconductor chips, the industry suffered in the production of finished vehicles. To get a worldwide measure of this decline, automobile manufacturing performance is determined by monitoring facilities for observable, physical activity including storage of finished cars in lots adjacent to the manufacturing facility [23].

Spaceborne SAR systems have had a huge role in geological investigation, monitoring, and discoveries. Some of these include the characterization and tracking of sea ice, measuring and documenting deforestation, measuring earth deformation due to seismic and volcanic forces, and studying ocean surfaces [24]. Not only does SAR play an important role in the geoscientific work done on Earth, but has also been used on missions to Venus, Europa, and Titan. These geological missions will continue well into the future, and NASA currently has several future missions planned [25].

8.1.2 Airborne SAR

Airborne SAR is experiencing a similar paradigm shift as in the spaceborne arena; smaller, lighter, lower cost systems, with easier access to data and processing tools. These smaller systems are relatively simple to integrate on existing manned and unmanned aircraft, and even small drones [26–28]. There is also a push toward systems that incorporate data from other onboard sensors (e.g., EO, IR, electronic support measures, and full motion video). Multimode imaging systems capable of interferometry, polarimetry, foliage penetration, and ground penetration are beginning to see high interest [26, 28].

Another area of interest in recent years is in the detection and classification of IEDs. Much of the work in this area has focused on material identification and the detection of command wires [29, 30]. These systems often employ UWB waveforms, which require advanced antenna, transmitter, and receiver technologies. UWB systems also perform specialized signal, data, and image processing [28]. These systems will continue to be in high demand and will push the envelope of radar technology.

With recent interest in hypersonic vehicles, research for using SAR in this area has increased [31, 32]. A major hurdle for these systems is the plasma sheath around the hypersonic platform. At very high speeds there is an extreme amount of fiction between the platform and the air, which causes ionization of the air, which forms the plasma sheath. This plasma severely attenuates the transmission and reception of the SAR signal [33]. This leads to degraded imagery and may even lead to the target being masked by noise. This is especially true at low frequencies and high electron density [34]. Various experiments have been conducted, as well as measurements taken from missions, to study the RF blackout experienced by reentry vehicles [35, 36]. Using these data, several RF blackout mitigation techniques have been proposed [37–39]. More research and experiments will be conducted to develop practical imaging systems for hypersonic vehicles.

Another application for airborne SAR sensors that is seeing continued interest is human detection, tracking, and identification. Traditionally, IR sensors have been used for detecting humans as the body emits radiation in a narrow region of the IR spectrum. However, several limiting factors include sunlight, solar gain on fabrics, humidity, wind, and heating on surrounding objects [40]. The dimensions of the human body make microwave SAR imaging a promising solution. Systems operating at lower frequencies (e.g., L-band) are able to penetrate clouds, fog, rain, vegetation, and buildings, while also being capable of imaging human targets [41]. Accurate coordinate location and tracking of human targets is only feasible with low-altitude platforms such as small unmanned aerial vehicles [40]. Research continues to be conducted into identifying individuals in a group. A promising approach uses incremental SAR measurements along with artificial intelligence techniques to differentiate individuals based on movement [40].

A final thought on future trends deals with navigation in GPS-denied environments. As civilian and government applications and operations tend to be over-reliant on GPS for position location and navigation, this can lead to very serious vulnerabilities. One possible solution is a SAR-based navigation aid. Since SAR systems actively transmit and receive signals, there is less sensitivity to various attacks [42, 43]. One approach that is gaining a lot of interest is based on the correlation of the SAR imagery with digital elevation models and often existing maps (e.g., Google Earth) created from satellite and aerial imagery, topography, ocean bathymetry, and other geographic data [44]. The SAR estimates are then used in conjunction with inertial navigation systems to provide accurate navigation when GPS is unavailable [42, 43].

8.1.3 Ground Based SAR

Ground based radar systems, such as NASA's Goldstone Very Large Array (VLA), have been used to create radar reflectivity maps of Mercury, Venus, Mars, and Titan [45]. The now defunct Arecibo Observatory was used to create imagery of various space objects including the JM 8 asteroid from a distance of approximately 9×10^6 km from Earth with a spatial resolution of about 15 m [46]. In 1999, the Arecibo Observatory created three-dimensional imagery of the tumbling asteroid 216 Kleopatra [47]. The National Science Foundation hosted a workshop in 2021 to review concepts for viable replacement options for the Arecibo system [48]. Interest and research into comets, asteroids, moons, and planets in the solar system is not expected to wane in the coming years.

With the current rate of satellite deployments, space situational awareness is increasingly important. To aid in the detection, tracking, and classification of orbiting objects, Lincoln Laboratory's Haystack Ultrawideband Satellite Imaging Radar is capable of achieving a resolution of 3 cm, making it highly effective at characterizing small space debris [49]. Improvements will continue to be made to handle ever increasing amounts of space objects.

With the widespread use of unmanned aerial vehicles and drones, ISAR imaging to characterize and classify these systems is becoming more and more important. Current research includes the use of portable, wide-angle, UWB imaging systems to create ISAR images to be used for characterization and classification of these types of aerial vehicles [50, 51].

The detection and monitoring of individuals in buildings and behind walls has garnered a lot of attention over the years. This type of through-wall technology could potentially be used for a host of applications including search and rescue, antinarcotics operations, intelligence gathering, and explosive entry [52]. Early proposed methods for through-wall sensing, including thermal imagining with IR cameras, were not sufficient as many structures have very poor thermal conductivity [53]. Since radio waves can penetrate forest canopies, buildings, and walls, this lead to much research in the area of through-wall sensing [41, 53]. To create imagery with fine resolution requires large bandwidths, which drives the system design toward higher operating frequencies. However, the penetration of building materials requires systems operating at lower frequencies. The breakthrough came by way of UWB waveforms and systems, which allow for operation at relatively low frequencies while still achieving very wide bandwidth [54, 55]. This allows for penetration capabilities without sacrificing resolution. To help overcome some of the current limitations, future systems are expected to integrate artificial intelligence layers using techniques like machine learning [40, 52, 53]. Another expected direction for the next generation of through-wall sensing is the dispersing of sensors to various locations around the target area to create a multistatic system that provides increased flexibility and more comprehensive imaging.

Other areas of ground-based SAR imaging receiving interest include the imaging of cooperative and noncooperative maritime vessels from ground-based SAR sensors and for ground deformation measurements [56, 57].

8.2 TECHNOLOGY TRENDS

SAR technology is being used more and more in the commercial sectors rather than government only. Thanks to commercial companies including ICEYE, Umbra

Lab, IMSAR, and Capella Space, which are increasingly deploying SAR systems on drones and microsatellites, the technologies associated with these systems are growing rapidly and expanding into new and exciting areas [5, 6, 11, 26]. In fact, several government agencies are now looking toward the commercial sector to meet imaging requirements [12, 58–60].

There are several key elements that make up the current commercial spaceborne SAR framework. These elements include launch vehicles, satellites, downlinks, and signal and image processing. Recent research and development activities and shifts in operational concepts have pushed the long-standing technologies in each of the areas towards more cost effective and far more capable imaging systems. [9]. As these technological advancements come together to form a true cohesive ecosystem, numerous novel missions and applications will follow.

8.2.1 Satellites

Many of the innovations associated with the new wave of SAR satellites are being driven by startup companies backed by venture capitalists [9]. Commercial SAR satellites are pushing the envelope of very low size, weight, power, and cost (SWAP-C), and are possible due to miniaturization of electronics, improved battery performance, and solar panel technologies [61, 62]. Many of the current commercial SAR satellites have a mass less than 100 kg and cost a few million dollars [5, 6, 11, 17]. This is a significant paradigm shift from traditional systems paid for by governments, with typical masses well over 1,000 kg and costs exceeding $500 million dollars [9]. The significant reduction in SWAP-C allows for more satellites to be deployed in constellations, which leads to improved performance, tasking, and revisit rates. Commercial SAR companies are striving to have the most advanced SAR satellite constellation, resulting in innovative designs and cutting-edge technology. Many of these companies have planned several more satellite deployments into constellations [5, 6, 11].

8.2.2 Rockets

SAR satellites do no good without a means of being deployed in orbit. Improvements in rocket technology are making launches more accessible in terms of both cost and frequency [9, 63, 64]. SpaceX's Falcon 9 paved the way for reusable rockets, which drives down the cost of space access, and has been used by commercial SAR systems [5, 11, 63]. Multisatellite ride share options are available and have made it easier for small SAR platforms to deploy into space [63, 64]. The rocket industry has also seen new companies come into the market as more SAR satellite companies increase the demand for launches. As competition grows between rocket companies, the cost for launches goes down. With several SAR companies already planning future launches, the demand for improved technologies and services will increase. A shift to smaller rockets carrying fewer satellites and offering more customized deployment is expected in the future [9]. As the launch cost dropped from tens of millions of dollars to about $100k—the satellite startup companies can refocus the resources to building satellites.

8.2.3 Downlinks

With a host of new satellites being launched, new and expanded types of networks, ground stations, and relays are currently in research and development. These new capabilities are necessary for service providers to deliver high-quality SAR imagery with shorter delays and tasking. With an on-demand type of tasking and near real-time monitoring of events such as natural disasters, search and rescue, and high-value target imaging, high-bandwidth, low-latency downlinks are necessary [9]. SAR systems are now starting to see more onboard processing to lessen the downlink requirements.

The DARPA 100G RF Backbone program is exploring techniques for achieving 100 Gb/s data rates. These rates are desired at air-to-air ranges of 200 km and air-to-ground ranges of 100 km. The goal is to use high-order modulation and spatial multiplexing with millimeter wavelength frequencies [65]. Signal processing algorithms and hardware are also being developed to meet SWAP-C limitations for the airborne platforms.

Traditionally, as part of the downlink of SAR data, ground stations had to be be built, operated, and managed by the individual SAR companies. Future trends are moving toward ground stations as a service. Companies currently providing these services including Amazon, Microsoft, Kongsbert Satellite Services, and Swedish Space Corporation [66–69]. As an example, Amazon Web Services (AWS) has a global network of ground stations and low-latency fiber networks [66]. These types of services relieve SAR satellite companies of the burdens associated with deploying and operating ground station infrastructure.

8.2.4 Signal and Image Processing

SAR data is very complex and can be difficult and time consuming for image analysts to examine and draw conclusions. In addition, the volume of SAR data is growing at a fast pace as the number of satellites is continually increasing. Therefore, technologies such as artificial intelligence and machine learning are starting to play larger roles in image processing and analysis [70–73].

As SAR systems are starting to see more onboard processing, GPUs are starting to be employed for signal and image processing. GPU computational capabilities and algorithms specifically tailored for GPU calculations make real-time onboard processing feasible and will continue to advance for the foreseeable future [74].

With increased use of cloud computing, SAR companies are also taking advantage of these services. For example, Capella Space and Ursa Space are providing on-demand SAR data, all through AWS, which allows for storing, managing, and processing SAR data, including custom machine learning applications [66, 75].

8.3 WORKFLOW AND PROCESSES

Traditionally, working with SAR data required in-depth knowledge of the sensors, electromagnetic phenomena, signal processing, image formation, and data processing. For example, imagery analysts working at government organizations routinely design and implement workflows and processes to take raw SAR data and produce effective analysis products [76]. In the commercial world, these processes are relatively immature. However, much effort is being invested now to make SAR imagery and data much easier to be used by the general public and require less expertise. To make SAR data as useful as GPS information is the ultimate goal. GPS data is commonly used every day by people from all walks of life and on various devices. For example, GPS data used by a smart phone to navigate to a desired location on a map is an extremely useful application that does not require expertise by the end user.

Another consideration in the workflow chain is the actual ordering and receiving of SAR data. This used to be a lengthy process carried out through price quotes, purchase orders, and physical media. Current and future trends are moving toward placing orders online and receiving data electronically. Instead of dedicated ground stations and processing equipment performing image and data processing, more storage, processing, and dissemination of SAR data is being performed on cloud computing, reducing the overall cost and streamlining the workflow [66]. Cloud services are especially beneficial with the number of new satellites being launched, the increased revisit rates, and the enormous amount of associated data.

8.4 MARKET DEMAND

There were many restrictions put in place in 2020 meant to slow the spread of the COVID-19 virus. These restrictions had a negative impact on small firms and contractors in the strategic satellite, rocket, and electronics supply chains. This resulted in launch delays and affected the schedule and funding of many startup commercial SAR companies [77].

In spite of these negative effects, the market demand for SAR imagery is now greater than ever. Both commercial and government sectors are using geospatial information to gain a better understanding of the pandemic's effects on global construction, manufacturing, and transportation [77]. Monitoring these types of changes is well suited for SAR satellite constellations, which have day or night all-weather capability and high revisit rates [5, 6, 11, 23].

According to market analysts, the global SAR market was a $3.3 billion in 2020 and is expected to reach $6.5 billion by 2026 [78]. SAR data and derived products are expected to grow from $700 million in 2018 to more than $1.7 billion by 2028, with the fastest growing market being imagery of less than 50-cm spatial resolution [79]. According to the 11th Satellite-Based Earth Observation report, imagery in the 1- to 5-m resolution has seen a decrease after 2017, while resolutions greater than 5 m have very limited commercial presence [79].

As many new commercial SAR companies join the market, bringing new technological advancements, missions, modes, and operations, it is difficult to predict how far reaching and impactful SAR will be. However, there is great potential, the future is promising, and the next several years will prove to be an exciting time to be in the field of SAR.

References

[1] NASA. NASA's Alaska Satellite Facility Distributed Active Archive Center. http://earthdata.nasa.gov/eosdis/daacs/asf. Accessed: 2020-12-20.

[2] Government of Canada. Radar image properties. https://www.nrcan.gc.ca/maps-tools-and-publications/satellite-imagery-and-air-photos/tutorial-fundamentals-remote-sensing/microwave-remote-sensing/radar-image-properties/9299. Accessed: 2022-01-30.

[3] ESA. The European Space Agency. https://www.esa.int/, 2022. Accessed: 2022-4-24.

[4] CSIRO. Austraila's National Space Agency. https://www.csiro.au/en/, 2022. Accessed: 2022-4-24.

[5] ICEYE. Example SAR data from ICEYE. https://www.iceye.com/downloads/datasets. Accessed: 2021-09-21.

[6] Umbra Lab. Space-based Synthetic Aperture Radar (SAR) Microsatellites - Umbra. http://umbra.space/. Accessed: 2022-03-19.

[7] British Broadcasting Company. Saudi oil attacks: Images show detail of damage. https://www.bbc.com/news/world-middle-east-49718975, 2019.

[8] Capella Space. Ever Given Container Ship Blocks the Suez Canal. https://www.capellaspace.com/gallery/ever-given-container-ship-blocks-the-suez-canal/. Accessed: 2022-4-24.

[9] Ursa Space Systems. SAR Satellite Systems: Perspective On The Year Ahead. https://ursaspace.com/blog/sar-satellite-systems-perspective-on-the-year-ahead/, 2021. Accessed: 2022-4-24.

[10] ICEYE. Press release: 13 January 2022 ICEYE Expands World's Largest SAR Satellite Constellation; Launches First U.S. Built Spacecraft. https://www.iceye.com/press/press-releases/iceye-expands-worlds-largest-sar-satellite-constellation-launches-first/-us-built-spacecraft, 2022. Accessed: 2022-4-23.

[11] Capella Space. SAR Made Easy. https://www.capellaspace.com/. Accessed: 2022-4-5.

[12] UMBRA. NRO selects Umbra for their strategic commercial enhancements effort. https://umbra.space/blog/nro-selects-umbra-for-their-strategic-commercial-enhancements-effort, 2022. Accessed: 2022-4-23.

[13] Canadian Space Agency. RADARSAT Constellation Mission. https://www.asc-csa.gc.ca/eng/satellites/radarsat/default.asphttps://www.asc-csa.gc.ca/eng/satellites/radarsat/default.asp, 2022. Accessed: 2022-4-23.

[14] National Weather Service. Hurricane Dorian. https://www.weather.gov/mhx/Dorian2019, 2019. Accessed: 2022-4-24.

[15] The European Space Agency. Sentinel-1 SAR user guide. https://sentinel.esa.int/web/sentinel/technical-guides/sentinel-1-sar. Accessed: 2020-10-03.

[16] T. ElGharbawi and F. Zarzoura. Damage detection using SAR coherence statistical analysis, application to Beirut, Lebanon. *ISPRS Journal of Photogrammetry and Remote Sensing*, 173:1–9, 03 2021.

[17] Synspective. JICA and Synspective detect new risks in Guatemala using SAR satellite data - The development of a disaster prevention management system makes progress-. https://synspective.com/press-release/2021/jica-and-synspective-new/, 2021. Accessed: 2022-4-24.

[18] J. Rosen. Shifting Ground: Fleets of radar satellites are measuring movements on Earth like neve before. https://www.science.org/content/article/fleets-radar-satellites-are-measuring-movements-earth-never#:~:text=Overall%2C%20nearly%2050%20are%20operational,half%20a%20meter%20or%20less, 2021. Accessed: 2022-4-24.

[19] ICEYE. SAR Product Guide. Technical Report Version 4.2, ICEYE, Espoo, 2021.

[20] D. Herring. Climate Change: Global Temperature Projections. https://www.climate.gov/news-features/understanding-climate/climate-change-global-temperature-projections, 2021. Accessed: 2022-4-24.

[21] D. Agnew, J. Pearce, G. Pramod, T. Peatman, R. Watson, J. Beddington, and T. Pitcher. Estimating the worldwide extent of illegal fishing. *PLOS ONE*, 4(2):1–8, 02 2009.

[22] International Maritime Organization. AIS transponders. https://www.imo.org/en/OurWork/Safety/Pages/AIS.aspx, 2022. Accessed: 2022-4-24.

[23] Ursa Space Systems. Auto Manufacturing Index: Outlook Hurt by Chip Shortage. https://ursaspace.com/blog/auto-manufacturing-index-outlook-hurt-by-chip-shortage/, 2021. Accessed: 2022-4-24.

[24] J. Meyer. New Relevance for Synthetic Aperture Radar (SAR). Technical report, The Alaska Satellite Facility, Fairbanks, 2019.

[25] NASA's Earth Observing System. Missions: Earth System Science Pathfinder Program. https://eospso.nasa.gov/mission-category/11, 2022. Accessed: 2022-4-24.

[26] IMSAR. OneSAR data and specifications sheet. Technical report, IMSAR, Springville, UT, 2020.

[27] NASA: Jet Propulsion Laboratory. Uninhabited aerial behicle synthetic aperture radar. https://uavsar.jpl.nasa.gov/, 2022. Accessed: 2022-4-27.

[28] Mirage Systems. Mirage systems: Platforms. http://www.miragesystems.com/products/platforms/, 2022. Accessed: 2022-4-24.

[29] B. Gonzalez-Valdes, J. Martinez-Lorenzo, and C. Rappaport. Dual band SAR processing for low dielectric contrast buried IED detection. In *2013 IEEE Antennas and Propagation Society International Symposium (APSURSI)*, pages 1080–1081, 2013.

[30] M. Otten, M. Cuenca, and M. Fernandez. IED command wire detection with multi-channel drone radar. In *EUSAR 2021; 13th European Conference on Synthetic Aperture Radar*, pages 1–4, 2021.

[31] J. Li, T. Gong, and L. Guo. SAR imaging of hypersonic platform based on phase screen method. In *2018 IEEE International Symposium on Antennas and Propagation USNC/URSI National Radio Science Meeting*, pages 1555–1556, 2018.

[32] R. Zhou, J. Sun, J. Fu, and J. Wang. Wavenumber domain imaging algorithm for hypersonic platform SAR with curved trajectory. In *2017 IEEE International Geoscience and Remote Sensing Symposium (IGARSS)*, pages 2396–2399, 2017.

[33] L. Song, B. Bai, X. Li, G. Niu, Y. Liu, and L. Zhao. Space-time varying plasma sheath effect on hypersonic vehicle-borne SAR imaging. *IEEE Transactions on Aerospace and Electronic Systems*, page 1, 2022.

[34] L. Song, B. Bai, X. Li, G. Niu, Y. Liu, L. Zhao, and Hu. Zhou. Analysis of hypersonic platform-borne SAR imaging: A physical perspective. *Remote Sensing*, 13(23), 2021.

[35] Japan Aerospace Exploration Agency. Orbital re-entry experiment (OREX). https://global.jaxa.jp/projects/rockets/orex/index.html, 2022. Accessed: 2022-4-27.

[36] R. Savino, D. Paterna, M. De Stefano Fumo, and M. D'Elia. Plasma-radiofrequency interactions around atmospheric re-entry vehicles: Modelling and arc-jet simulation. *The Open Aerospace Engineering Journal*, 3:76–85, 11 2010.

[37] The European Space Agency. Mitigation of RF blackout or re-entry vehicles. https://www.esa.int/Enabling_Support/Space_Engineering_Technology/Shaping_the_Future/Mitigation_of_RF_blackout_for_re-entry_vehicles, May 2017. Accessed: 2022-04-30.

[38] R. Hartunian, G. Stewart, S. Fergason, T. Curtiss, and R. Seibold. Causes and Mitigation of Radio Frequency (RF) Blackout During Reentry of Reusable Launch Vehicles. Technical Report ATR-2007(5309)-1, Aerospace Corporation, El Segundo, 2007.

[39] R. Dutta, R. Biswas, and N. Roy. Reduction of attenuation of e.m wave inside plasma formed during supersonic or hypersonic re-entry of missile like flight vehicles by the application of D.C. magnetic field — a technique for mitigation of RF blackout. In *2011 IEEE Applied Electromagnetics Conference (AEMC)*, pages 1–4, 2011.

[40] Airborne Drones. Detecting humans with SAR remote sensing technologies. https://www.airbornedrones.co/detecting-humans-sar/, April 2020. Accessed: 2022-4-30.

[41] A. Harrison. *Introduction to Radar Using Python and MATLAB*. Artech House, Norwood, 2020.

[42] Z. Reid. Leveraging 3D Models for SAR-based navigation in GPS-denied Environments. Master's thesis, Wright State University, 2018.

[43] T. White. GPS-Denied Navigation Using Synthetic Aperture Radar Images and Neural Networks. Master's thesis, Utah State University, 2021.

[44] Google. Google Earth. https://earth.google.com, 2022. Accessed: 2022-4-30.

[45] D. Muhleman, B. Butler, M. Slade, and A. Grossman. Radar imaging of the planets using the very large array. In J. G. Robertson and W. J. Tango, editors, *Very High Angular Resolution Imaging*, pages 457–468, Dordrecht, 1994. Springer Netherlands.

[46] S. Ostro, S. Hudson, L. Benner, J. Giorgini, C. Magri, J. Margot, and M. Nolan. Asteroid radar astronomy. *Asteroids III*, 01 2002.

[47] S. Ostro, R. Scott, O. Hudson, M. Nolan, J. Margot, D. Scheeres, D. Campbell, C. Magri, J. Giorgini, and D. Yeomans. Radar observations of asteroid 216 kleopatra. *Science*, 288(5467):836–839, 2000.

[48] E. Neus. National Science Foundation Begins Plans for Reviving Arecibo. https://fedtechmagazine.com/article/2021/08/national-science-foundation-begins-plans-reviving-arecibo, August 2021. Accessed: 2022-4-30.

[49] J. Murray, C. Blackwell, J. Gaynor, and T. Kennedy. Haystack Ultra-Wideband Satellite Imaging Radar Measurements of the Orbital Debris Environment: 2014-2017. Technical report, National Aeronautics and Space Administration, 2019.

[50] W. Lee and K. Song. Enhanced ISAR imaging for surveillance of multiple drones in urban areas. In *2018 International Conference on Radar (RADAR)*, pages 1–4, 2018.

[51] E. Giusti, S. Ghio, and M. Martorella. Drone-based 3D interferometric ISAR imaging. In *2021 IEEE Radar Conference (RadarConf21)*, pages 1–6, 2021.

[52] Camero-Tech. Camero Xaver 800. http://www.camero-tech.com/xaver-products/xaver-800/. Accessed: 2021-02-12.

[53] Camero. Through wall imaging. https://camero-tech.com/technology/through-wall-imaging/, 2022. Accessed: 2022-5-1.

[54] Y. Yunqiang and A. Fathy. See-through-wall imaging using ultra wideband short-pulse radar system. In *2005 IEEE Antennas and Propagation Society International Symposium*, volume 3B, pages 334–337, 2005.

[55] D. Kim and S. Shin. Experiments for ultra-wideband imaging radar with one-dimentional synthetic aperture. In *2011 3rd International Asia-Pacific Conference on Synthetic Aperture Radar (APSAR)*, pages 1–4, 2011.

[56] E. Won and K. Ouchi. Ship detection by synthetic aperture radar with ground-based maritime radar with AIS. September 2011.

[57] A. Jungner. Ground-Based Synthetic Aperture Radar Data Processing for Deformation Measurement. Master's thesis, Royal INstitute of Technology, Stockholm, 2009.

[58] U.S. Army. SMDC utilizes no-cost R&D agreements to bring technology to soldiers. https://www.army.mil/article/251985/smdc_utilizes_no_cost_rd_agreements_to_bring_technology_to_soldiers, 2021. Accessed: 2022-4-24.

[59] Breaking Defense. NRO contracts 5 commercial SAR firms for study, considers official acquisition. https://breakingdefense.com/2022/01/nro-contracts-5-commercial-sar-firms-for-study-considers-official-acquisition/, 2022. Accessed: 2022-4-24.

[60] ICEYE. ICEYE and U.S. Army to pursue collaborative research and development. https://iceye.com/press/press-releases/iceye-and-us-army-to-pursue-collaborative-research-and-development, 2021. Accessed: 2022-4-24.

[61] H. Tsuchida, Y. Kawamoto, N. Kato, K. Kaneko, S. Tani, S. Uchida, and H. Aruga. Efficient power control for satellite-borne batteries using Q-learning in Low-Earth-Orbit satellite constellations. *IEEE Wireless Communications Letters*, 9(6):809–812, 2020.

[62] S. Luryi, J. Xu, and A. Zaslavsky. *Future Trends in Microelectronics: Journey into the Unknown*. John Wiley and Sons, 2016.

[63] SpaceX. Falcon 9. First Orbital Class Rockey Capable of Reflight. https://www.spacex.com/vehicles/falcon-9/, 2022. Accessed: 2022-4-28.

[64] Rocket Lab. We Open Access To Space to Improve Life on Earth. https://www.rocketlabusa.com/, 2022. Accessed: 2022-4-29.

[65] P. Zablocky. 100 Gb/s RF Backbone (100G) (Archived). https://www.darpa.mil/program/100-gb-s-rf-backbone, 2022. Accessed: 2022-5-1.

[66] Amazon Web Services. AWS ground station. https://aws.amazon.com/ground-station/, 2022. Accessed: 2022-5-1.

[67] Microsoft. Azure orbital. https://azure.microsoft.com/en-us/services/orbital/, 2022. Accessed: 2022-5-1.

[68] Kongsberg Satellite Services. KSAT ground network services. https://www.ksat.no/ground-network-services/, 2022. Accessed: 2022-5-1.

[69] Sweedish Space Corporation. A word of advanced space services. https://sscspace.com/, 2022. Accessed: 2022-5-1.

[70] C. Messager, Tran V. La, R. Sahl, P. Dupont, E. Prothon, and M. Honnorat. Use of SAR imagery and artificial intelligence for a multi-components ocean monitoring. In *IGARSS 2020 - 2020 IEEE International Geoscience and Remote Sensing Symposium*, pages 3817–3820, 2020.

[71] X. Qian, F. Liu, L. Jiao, X. Zhang, P. Chen, L. Li, J. Gu, and Y. Cui. A hybrid network with structural constraints for SAR image scene classification. *IEEE Transactions on Geoscience and Remote Sensing*, 60:1–17, 2022.

[72] D. Quan, S. Wang, X. Liang, R. Wang, S. Fang, B. Hou, and L. Jiao. Deep generative matching network for optical and SAR image registration. In *IGARSS 2018 - 2018 IEEE International Geoscience and Remote Sensing Symposium*, pages 6215–6218, 2018.

[73] R. Wang, L. Wang, P. Dong, L. Jiao, and J. Chen. Graph-level neural network for SAR image change detection. In *2021 IEEE International Geoscience and Remote Sensing Symposium IGARSS*, pages 3785–3788, 2021.

[74] V. Krzhizhanovskaya, G. Závodszky, M. Lees, J. Dongarra, P. Sloot, S. Brissos, and J. Teixeira. The challenge of onboard SAR processing: A GPU opportunity. In *Computational Science - 20th International Conference*, pages 46–59, Amsterdam, 2020.

[75] Amazon Web Services. Capella uses space to bring you closer to Earth. https://aws.amazon.com/blogs/publicsector/capella-uses-space-bring-you-closer-earth/, 2022. Accessed: 2022-5-1.

[76] National Geospatial-Intelligence Agency. NGA delivers world-class geospatial intelligence that provides a decisive advantage to policymakers, warfighters, intelligence professionals and first responders. https://www.nga.mil/index.html, 2022. Accessed: 2022-5-1.

[77] The Organisation for Economic Co-operation and Development. The impacts of COVID-19 on the space industry. https://www.oecd.org/coronavirus/policy-responses/the-impacts-of-covid-19-on-the-space-industry-e727e36f/, August 2020. Accessed: 2022-5-1.

[78] Mordor Intelligence. Synthetic aperture radar market - growth, trends, COVID-19 impact, and forecasts (2022-2027). https://www.mordorintelligence.com/industry-reports/synthetic-aperture-radar-market, 2022. Accessed: 2022-5-1.

[79] D. Kasaboski. EO SAR: Trick or Treat? https://www.nsr.com/eo-sar-trick-or-treat/, 2022. Accessed: 2022-5-1.

Appendix A

Polarization

Polarization refers to the time-varying orientation of the electric field vector of an electromagnetic wave at a given point in space [1–3]. For plane electromagnetic waves, the electric field vector is situated in a direction perpendicular to the direction of wave propagation. For linearly polarized waves, the direction of the electric field vector is constant. When the direction of the electric field vector rotates with time, the wave is elliptically polarized [1–3]. For SAR applications, linear polarization is decomposed into two important orthogonal polarizations, namely vertical and horizontal, illustrated in Figure A.1. Horizontal and vertical polarizations are, in general, considered to be parallel and perpendicular to the Earth's surface. Circular polarization is a special case of elliptical polarization with a constant electric field intensity and is quite often used in SAR applications. Circular polarization has two orthogonal polarizations, with the electric field rotation being in opposite directions. These are referred to as right-hand circular and left-hand circular polarization and are shown in Figure A.2.

The electromagnetic scattering of a target is dependent on a number of parameters including polarization and requires the full orthogonal polarization information for a complete definition [2–5]. For example, a horizontally polarized wave incident upon a target can result in a scattered wave with both horizontal and vertical components. Since the antenna is the mechanism by which the radar radiates energy, it is the antenna design that determines the polarization of the transmitted electromagnetic wave. Analogous to the antenna pattern, the polarization is also a function of beam-pointing angles and thus varies with direction.

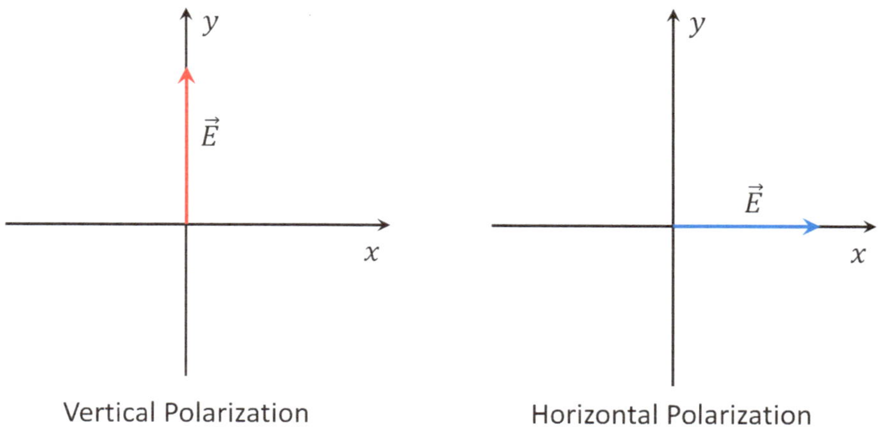

Figure A.1 Linearly polarized electric field.

Watch this animation of linear and circular polarization.

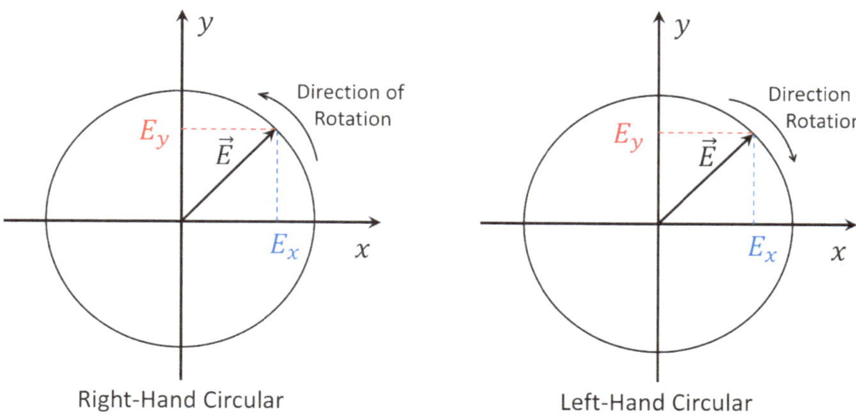

Figure A.2 Circularly polarized electric field.

A.1 COORDINATE CONVENTION

When describing the propagation of electromagnetic energy in a three-dimensional space, the handedness changes upon reflection from a target and must be accounted for. Therefore, the coordinate space has to be defined for the incident and scattered energy, which has led to two conventions. The first is the forward scattering alignment (FSA) in which the positive z axis is in the same direction as the direction of propagation for both incident and scattered waves. The second is the back scattering alignment (BSA), where the positive z axis points toward the target for both the incident and scattered waves. The difference in the conventions is illustrated in Figure A.3. The BSA convention is more commonly used for monostatic radar systems. This difference in convention leads to a slightly different form in the scattering matrix. In the FSA convention, the scattering matrix is the Jones matrix, while in the BSA convention, the scattering matrix is the Sinclair matrix [6–8].

A.2 JONES VECTOR

For many SAR applications, it is convenient to express linear polarization in terms of horizontal and vertical components, this is analogous to the x and y components in Figure A.1. Following the detailed derivations in [9], a fully polarized electric field may be mathematically represented by the Jones vector as

$$\mathbf{J} = \begin{bmatrix} E_H \, e^{j\phi_H} \\ E_V \, e^{j\phi_V} \end{bmatrix} . \tag{A.1}$$

The Jones vector is a concise representation of a monochromatic, uniform plane wave with a constant polarization [8, 10, 11]. Table A.1 gives the normalized Jones vector for common polarizations. In correspondence with the BSA convention, the incident (transmitted) and scattered (received) electric field intensities may now be decomposed into horizontal and vertical components as shown in Figure A.4 and is expressed as

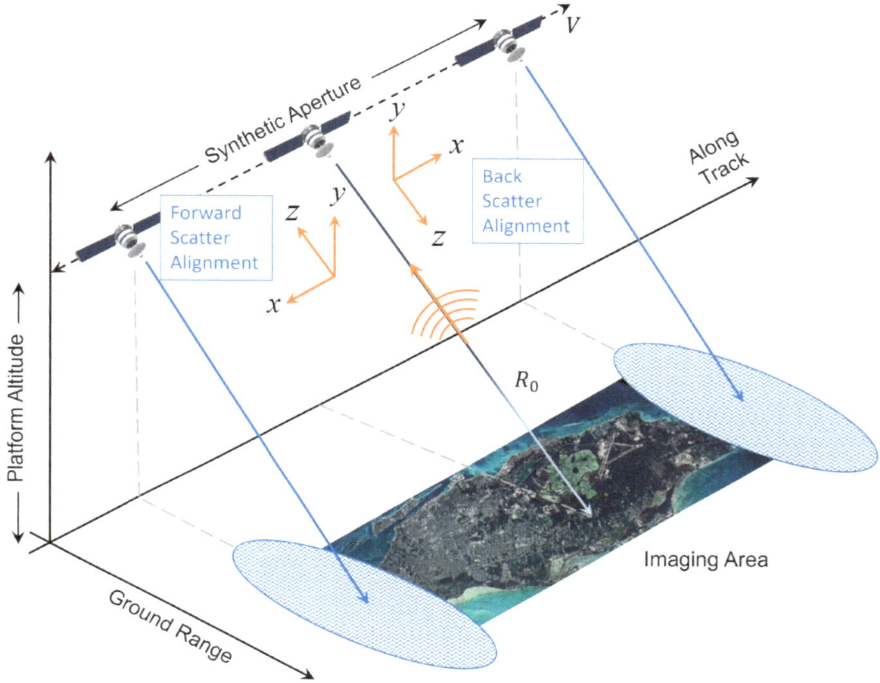

Figure A.3 Comparison of BSA and FSA conventions.

$$\mathbf{E}^i = E_H^i\,\hat{\mathbf{h}} + E_V^i\,\hat{\mathbf{v}}, \tag{A.2}$$

$$\mathbf{E}^s = E_H^s\,\hat{\mathbf{h}} + E_V^s\,\hat{\mathbf{v}}, \tag{A.3}$$

where

$\mathbf{E}^i$ = incident electric field intensity (V/m),
$\mathbf{E}^s$ = scattered electric field intensity (V/m),
$\hat{\mathbf{h}}$ = unit vector in the horizontal direction,
$\hat{\mathbf{v}}$ = unit vector in the vertical direction.

Table A.1

Normalized Jones Vector for Common Polarizations

Polarization	*Jones Vector*	*Typical Notation*
Horizontal	$\begin{bmatrix} 1 \\ 0 \end{bmatrix}$	H
Vertical	$\begin{bmatrix} 0 \\ 1 \end{bmatrix}$	V
Right-hand circular	$\frac{1}{\sqrt{2}}\begin{bmatrix} 1 \\ -j \end{bmatrix}$	R
Left-hand circular	$\frac{1}{\sqrt{2}}\begin{bmatrix} 1 \\ +j \end{bmatrix}$	L

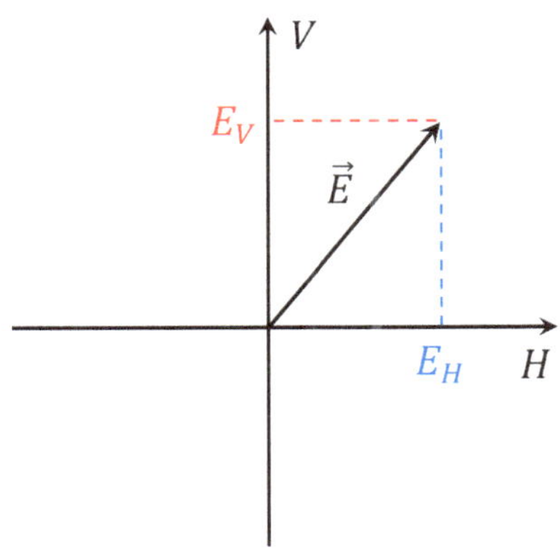

Figure A.4 Electric field decomposed into H and V components.

A.3 SCATTERING MATRIX

As previously stated, target scattering is dependent upon the polarization of the incident wave. Similarly, the energy transmitted and received by the radar is dependent on the antenna polarization. Therefore, it is necessary to represent the target scattering as a tensor that relates the incident and scattered fields [3]. To that end, the incident and scattered fields are related through the normalized scattering matrix as

$$\mathbf{E}^s = \left[\mathbf{S}\right]\mathbf{E}^i. \tag{A.4}$$

Using (A.2) and (A.3) allows (A.4) to be expanded as

$$\begin{bmatrix} E_H^s \\ E_V^s \end{bmatrix} = \begin{bmatrix} S_{HH} & S_{HV} \\ S_{VH} & S_{VV} \end{bmatrix} \begin{bmatrix} E_H^i \\ E_V^i \end{bmatrix}. \tag{A.5}$$

The HH and VH components of the scattering matrix may be measured by transmitting a horizontally-polarized electric field and measuring the scattered fields in both horizontal and vertical polarization. The remaining HV and VV components are similarly measured by transmitting a vertically-polarized field and again measuring the scattered fields in both horizontal and vertical polarization.

The scattering matrix can be thought of as a transformation of the incident electric field to the scattered electric field. A single instance of the scattering matrix is valid for the incident and observation angles and frequency. Once the scattering matrix is determined, the scattered electric field may be computed for any arbitrary polarization of the incident field (e.g., linearly polarized at 45° from the x axis). For monostatic radar systems, the scattering matrix is reciprocal for most targets, meaning $S_{HV} = S_{VH}$. However, this is not true for bistatic systems or targets containing magnetically biased ferrite materials. The scattering matrix is also commonly written in terms of circular polarization, which is expressed as

$$\begin{bmatrix} E_R^s \\ E_L^s \end{bmatrix} = \begin{bmatrix} S_{RR} & S_{RL} \\ S_{LR} & S_{LL} \end{bmatrix} \begin{bmatrix} E_R^i \\ E_L^i \end{bmatrix}. \tag{A.6}$$

The linear and circular field components are related to each other through

$$\begin{bmatrix} E_R \\ E_L \end{bmatrix} = \frac{1}{\sqrt{2}} \begin{bmatrix} 1 & -j \\ 1 & +j \end{bmatrix} \begin{bmatrix} E_H \\ E_V \end{bmatrix} = [T] \begin{bmatrix} E_H \\ E_V \end{bmatrix}, \tag{A.7}$$

and

$$\begin{bmatrix} E_H \\ E_V \end{bmatrix} = \frac{1}{\sqrt{2}} \begin{bmatrix} 1 & 1 \\ +j & -j \end{bmatrix} \begin{bmatrix} E_R \\ E_L \end{bmatrix} = [T]^{-1} \begin{bmatrix} E_R \\ E_L \end{bmatrix}. \tag{A.8}$$

Using (A.7) and (A.8), the circularly polarized scattering matrix is written in terms of the linear scattering matrix as

$$\begin{bmatrix} S_{RR} & S_{RL} \\ S_{LR} & S_{LL} \end{bmatrix} = [T] \begin{bmatrix} S_{HH} & S_{HV} \\ S_{VH} & S_{VV} \end{bmatrix} \begin{bmatrix} 1 & 0 \\ 0 & -1 \end{bmatrix} [T]^{-1}. \tag{A.9}$$

Writing out each term in (A.9) results in

$$S_{RR} = \frac{1}{2}\Big[(S_{HH} - S_{VV}) - j(S_{VH} + S_{HV})\Big], \tag{A.10}$$

$$S_{RL} = \frac{1}{2}\Big[(S_{HH} + S_{VV}) - j(S_{VH} - S_{HV})\Big], \tag{A.11}$$

$$S_{LR} = \frac{1}{2}\Big[(S_{HH} + S_{VV}) + j(S_{VH} - S_{HV})\Big], \tag{A.12}$$

$$S_{LL} = \frac{1}{2}\Big[(S_{HH} - S_{VV}) + j(S_{VH} + S_{HV})\Big]. \tag{A.13}$$

Similarly, the linearly-polarized scattering matrix is written in terms of the circular scattering matrix as

$$\begin{bmatrix} S_{HH} & S_{HV} \\ S_{VH} & S_{VV} \end{bmatrix} = [T]^{-1} \begin{bmatrix} S_{RR} & S_{RL} \\ S_{LR} & S_{LL} \end{bmatrix} [T] \begin{bmatrix} 1 & 0 \\ 0 & -1 \end{bmatrix}. \tag{A.14}$$

Writing out each term in (A.14) results in

$$S_{HH} = \frac{1}{2}\Big[S_{RR} + S_{LR} + S_{RL} + S_{LL}\Big], \tag{A.15}$$

$$S_{HV} = \frac{1}{2}\Big[-j\big(S_{RR} + S_{LR} - S_{RL} - S_{LL}\big)\Big], \tag{A.16}$$

$$S_{VH} = \frac{1}{2}\Big[j\big(S_{RR} - S_{LR} + S_{RL} - S_{LL}\big)\Big], \tag{A.17}$$

$$S_{VV} = \frac{1}{2}\Big[-S_{RR} + S_{LR} + S_{RL} - S_{LL}\Big]. \tag{A.18}$$

A.4 COHERENT AND NONCOHERENT SCATTERING

Referring to Figure A.5, when a SAR resolution cell contains a single target that is coherent or deterministic, the scattering matrix is sufficient in representing the target returns. However, due to various scattering mechanisms, the return signal from target scenes where a resolution cell contains multiple objects or noncoherent scatterers, the scattering matrix is often insufficient. Since the scattering matrix is difficult to analyze directly and may not adequately represent noncoherent scattering, covariance and coherency matrices are employed to calculate various decompositions [12–14].

A.4.1 Covariance

To calculate the covariance matrix, reciprocity is used to first create a slight variation of the scattering matrix, which is expressed as

$$\mathbf{k} = \begin{bmatrix} S_{HH} \\ \sqrt{2}S_{XX} \\ S_{VV} \end{bmatrix}, \tag{A.19}$$

where S_{XX} is either S_{HV} or S_{VH}. Then, (A.19) is multiplied by its conjugate transpose to give the covariance matrix. This matrix is written as

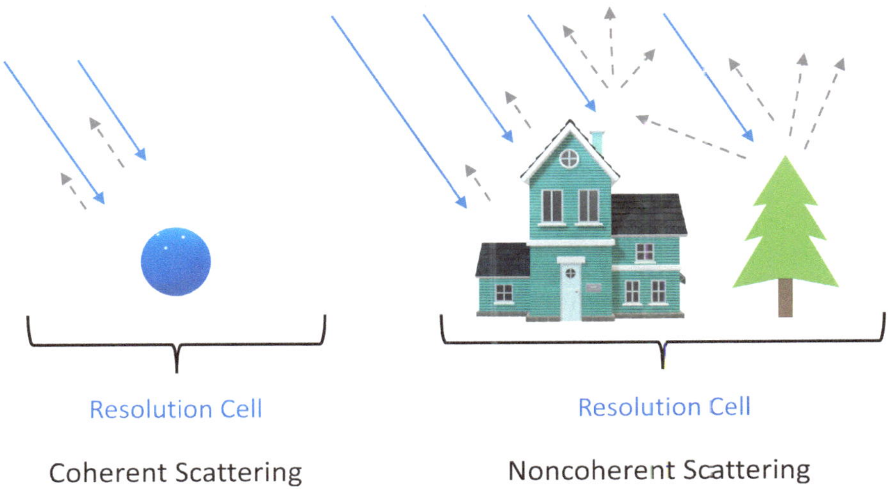

Figure A.5 Coherent and noncoherent scattering.

$$C = \mathbf{k} \times \mathbf{k}^{*T} = \begin{bmatrix} |S_{HH}|^2 & \sqrt{2}S_{HH}S_{XX}^* & S_{HH}S_{VV}^* \\ \sqrt{2}S_{HH}^*S_{XX} & 2|S_{XX}|^2 & \sqrt{2}S_{XX}S_{VV}^* \\ S_{HH}^*S_{VV} & \sqrt{2}S_{XX}^*S_{VV} & |S_{VV}|^2 \end{bmatrix}. \tag{A.20}$$

The diagonal elements are real-valued and represent the intensities of the different polarizations. The off-diagonal elements are complex valued and describe the amplitude and phase difference between different polarizations [9, 14].

A.4.2 Coherency

The coherency matrix is similar to the covariance matrix in that it describes the difference in intensities and phases between the different polarizations. To calculate the coherency matrix, begin with the Pauli vector, which is expressed as [9]

$$\mathbf{p} = \frac{1}{\sqrt{2}} \begin{bmatrix} S_{HH} + S_{VV} \\ S_{HH} - S_{VV} \\ 2S_{XX} \end{bmatrix}. \tag{A.21}$$

Next, the Pauli vector in (A.21) is multiplied by its complex conjugate to give the coherency matrix, which is expressed as [9]

$$T = \mathbf{p} \times \mathbf{p}^{*T} = \frac{1}{2} \begin{bmatrix} |S^+|^2 & (S^+)(S^-)^* & 2(S^+)S_{XX}^* \\ (S^+)^*(S^-) & |S^-|^2 & 2(S^-)S_{XX}^* \\ 2(S^+)^*S_{XX} & 2(S^-)^*S_{XX} & 4|S_{XX}|^2 \end{bmatrix}, \tag{A.22}$$

where $S^+ = (S_{HH} + S_{VV})$ and $S^- = (S_{HH} - S_{VV})$.

A.5 DECOMPOSITIONS

While the details of decomposition methods are beyond the scope of this book, a brief overview of coherent and noncoherent decomposition is given in the subsequent sections. The various decomposition methods are very useful for classification of areas of various sizes and types, including roads, grass, water, developed areas, forests, and pastures. The reader is referred to a few excellent references for more information on the subject [9, 12, 13, 15].

A.5.1 Coherent

Coherent decompositions operate on the scattering matrix **S** to express the scattering as a combination of different types of scatterers. These typically include lakes, roads, urban areas, and other man-made objects. The basic idea is to extract the physical properties of the target scene through the use of elementary responses. This may be written as [14]

$$\mathbf{S} = \sum_{i=1}^{N} \alpha_i \mathbf{S}_i, \tag{A.23}$$

where $\mathbf{S}_i$ are the responses for each of the N scattering types and α_i are the weights assigned to each scattering type.

A.5.2 Noncoherent

Noncoherent decompositions operate on the second-order descriptors, namely the covariance ($\mathbf{C}$) and coherency ($\mathbf{T}$) matrices. These methods are applied to target scenes containing incoherent scattering mechanisms such as forests, agriculture, wetlands, and choppy water [9, 14]. Noncoherent methods are broadly categorized into eigen-based and model-based decompositions with associated advantages and disadvantages.

References

[1] D. Cheng. *Field and Wave Electromagnetics,* 2nd ed. Addison-Wesley, Reading, MA, 1989.

[2] C. Balanis. *Advanced Engineering Electromagnetics,* 2nd ed. John Wiley and Sons, New York, 2012.

[3] G. Ruck, D. Barrick, W. Stuart, and C. Krichbaum. *Radar Cross Section Handbook.* Plenum Press, New York, 1970.

[4] A. Harrison. *Introduction to Radar Using Python and MATLAB.* Artech House, Norwood, 2020.

[5] D. Hess. Introduction to RCS measurements. *Antennas and Propagation Conference*, pages 37–44, 2008.

[6] E. Luneburg, S. Cloude, and W. Boerner. On the proper polarimetric scattering matrix formulation of the forward propagation versus backscattering radar systems description. In *IGARSS'97. 1997 IEEE International Geoscience and Remote Sensing Symposium Proceedings. Remote Sensing - A Scientific Vision for Sustainable Development*, volume 4, pages 1591–1593, 1997.

[7] G. Sinclair. The transmission and reception of elliptically polarized waves. *Proceedings of the IRE*, 38(2):148–151, 1950.

[8] R. Jones. A new calculus for the treatment of optical systems I. Description and discussion of the calculus. *Journal of the Optical Society of America*, 31(7):488–493, July 1941.

[9] C. López-Martínez and E. Pottier. *Basic Principles of SAR Polarimetry*, pages 1–58. Springer International Publishing, Cham, 2021.

[10] H. Hurwitz and R. Jones. A new calculus for the treatment of optical systems II. Proof of three general equivalence theorems. *Journal of the Optical Society of America*, 31(7):493–499, July 1941.

[11] R. Jones. A new calculus for the treatment of optical systems III. The Sohncke theory of optical activity. *Journal of the Optical Society of America*, 31(7):500–503, July 1941.

[12] W. Boerner. Basics of SAR polarimetry I. in radar polarimetry and interferometry. *North Atlantic Treaty Organization Science and Technology Organization Educational Notes*, (3):1–400, 2007.

[13] W. Boerner. Basics of SAR polarimetry II. In radar polarimetry and interferometry. (4):1–30, 2007. Educational Note.

[14] G. Vasile. Independent component analysis data processing framework for polarimetric synthetic aperture radar images. Technical report, 07 2020.

[15] T. Ainsworth, Y. Wang, and J. Lee. Model-based polarimetric SAR decomposition: An L_1 regularization approach. *IEEE Transactions on Geoscience and Remote Sensing*, 60:1–13, 2022.

About the Author

Andy Harrison received a bachelor of science, master of science, and doctor of philosophy in electrical engineering from the University of Mississippi in 1994, 1996, and 1999, respectively. His studies focused on electromagnetic theory, microwave measurements, antenna analysis and design, acoustics, and computational electromagnetics. His graduate research dealt with the use of wavelet-like basis functions in the finite element solution of partial differential equations. He has authored several papers in areas of computational electromagnetics and global optimization techniques. He is the author of *Introduction to Radar Using Python and MATLAB®* and has taught short courses in radar system design, simulation, and analysis.

Dr. Harrison served as an associate editor for the Applied Computational Electromagnetics Society and the Southeastern Symposium on System Theory. He is a senior member of the Institute of Electrical and Electronics Engineers and served as vice chair as well as education chair for the Mississippi section. He also served on the Electrical Engineering Industrial Advisory Board. He is a member of the electrical engineering honor society, Eta Kappa Nu (Epsilon Omega Chapter), the scientific research honor society, Sigma Xi, and the Antenna Measurements Techniques Association.

Dr. Harrison received recognition for his outstanding performance while supporting both the Ground-based Midcourse Defense X-Band Radar System Engineering Division and the Sea-Based X-Band Radar Requirements and Analysis team. He was also recognized for his significant contributions to the Missile Defense Agency's Project Hercules.